PROCESSING AND SYNTHESIS OF HYDROGEOLOGICAL DATA

Abacus House, Speldhurst Road, Tunbridge Wells, Kent, England

Revised, up-dated English edition of the Romanian work PRELUCRAREA ŞI SINTEZA DATELOR HIDROGEOLOGICE first published by Editura Tehnică, Bucharest 1972

Translated by Emil Gelles
Translation edited by Patricia Monahan

British Library Cataloguing in Publication Data

Gheorghe, Alexandru
Processing and synthesis of hydrogeological data.
1. Hydrogeology — Statistical methods
I. Title II. Gelles, Emil
551.4'9'0182 GB1001.72.S7

ISBN 0-85626-107-6

BSDCS7

Printed in Romania

Alexandru Gheorghe

PROCESSING AND SYNTHESIS OF HYDROGEOLOGICAL DATA

ABACUS PRESS

Tunbridge Wells, Keut

PREFACE

The object of hydrogeological research, on site or in the laboratory, is to establish the groundwater resources available for general consumption or for particular areas of the economy. It is also necessary to establish whether the hydrogeological conditions are suitable for the exploitation of various mineral deposits or for building purposes.

The present work applies working methods derived from other sciences and disciplines to the processing and synthesizing of site and laboratory data. The common objective of all hydrogeological research is the achievement of a model which offers an accurate interpretation of the hydrogeological synthesis of the structure or characteristic being studied. This book endeavours to solve some intricate hydrogeological situations using mathematical statistical procedures, the theory of information, underground hydraulics, together with classical geological methods.

Accumulations of underground waters are geological bodies whose properties are determined by their interactions with the reservoir rock and by the geological structure, so particular attention is paid to both the hydraulic and hydrological factors, as well as to the lithological, petrophysical and structural factors.

The quantitative study of the permeable lithofacies is dealt with by means of mathematical statistical procedures and *trend surfaces*. By establishing the distribution laws of the permeable bodies within the framework of a well-defined geological population, it is possible to estimate quantitatively the lithological complexity as well as the degree of certainty of the investigation. The method described in Chapter 1 can also be used in other areas of theoretical and applied geology.

The hydrodynamic study of aquifers and of hydrogeological wells (Chapter 2) is based on the latest achievements in underground hydraulics and emphasizes the limits of error and the high cost of certain classical methods, in contrast to the higher technical and economic efficiency of the new methods. The influence of the petrophysical factors upon the draining capacity of the sands (Chapter 3) is dealt with on the basis of our own laboratory researches with a view to defining new laboratory techniques and methods of processing experimental data.

Particular attention is paid to the correlations between the natural factors and the aquiferous potential of the hydrogeological structures (Chapter 4), which constitute an essential condition for schematizing the hydrodynamics of aquifers, as well as for estimating the water balance.

This book is practical in structure and content, stressing the solution of actual problems by means of numerical methods of processing, analysis and synthesis.

In its treatment of the problems as well as the original solutions it presents, this book breaks new ground in the field of hydrogeology. It is meant for geologists, hydrogeologists and hydrotechnicians who are involved in the research, design and exploitation of groundwater capture and drainage systems.

CONTENTS

1

STUDY OF THE DISTRIBUTION OF PERMEABLE LITHOFACIES BY MEANS OF MATHEMATICAL STATISTICAL PROCEDURES *

One of the main factors determining the complexity of hydrogeological conditions is the distribution of the lithofacies, which in turn determines the distribution of the permeable deposits. Thus, in the case of a stable permeable lithofacies, the thickness of the aquifers are relatively uniform; this means less exploratory work and simpler principles for schematizing the hydrogeological conditions, and a simpler choice of hydrodynamic scheme. Very changeable lithofacies raise great difficulties in structure simulation, exploration methods and hydrodynamic forecasts.

Lithofacies variation along the vertical in hydrogeological structures of the detrital kind has important consequences from the hydrodynamic standpoint. Thus, in the case of lithological sequences where the sandy aquifers are separated by thin argyllaceous-marly layers, the structure as a whole should be considered as a stratified aquiferous complex, with the aquiferous horizons hydraulically related along the vertical, through the intermediate clays and marls. Such sequences refer to the quaternary deposits of the fluviatile-torrential type, as well as the many pre-quaternary formations with great lithofacies variation. For instance, the Dacian and Levantine deposits, where the principal Romanian lignite deposits lie, are characterized by a quite non-uniform space distribution of the permeable deposits. That is why an accurate elaboration of structural-lithological maps of the water bearing formation in the neighbourhood of the lignite strata (so very useful for designing the drainage systems) with significant cartographic images is almost impossible. In situations such as this, when a geological parameter is characterized by random changes, which could not clearly express the established geological laws, it can be regarded as a random variable and so be quantitatively studied by the methods of mathematical statistics. The use of statistical models in the study of geological parameters characterized as random variables is important in the process of geological mathematizing. Usually, at the exploratory stage, the total number of drillings allows for selection, and therefore, an objective estimation of these parameters.

In solving this problem, mathematical statistics can intervene by:

— systematizing factual data, determining the typical selection values and estimating the laws of theoretical distribution;

* On the basis of lignite deposit Şotînga-Margineanca, Romania.

— establishing the certainty degree of the statistical estimations;

— separating the trend from local effects;

— setting up stochastic relations between lithological parameters by means of the correlation theory.

The solutions of these problems permit not only a quantitative characterization of the macroheterogeneity of the permeable deposits, but an elaboration of the graphical material liable to show the variation of the parameters along both the horizontal and the vertical.

When applying the methods of mathematical statistics to geological problems, it is necessary to take into account the limits of the population before establishing which of the geological conditions should be considered first. For instance, lignite deposits can easily be separated along the vertical because the populations of the sandy body is bounded by two lignite layers with a continuous distribution. In complex situations, these limits are only given by lithological, stratigraphical, hydrogeological, etc. reference points with a large extension area. In fact, the application of mathematical statistics to the problems of economic geology involves some peculiarities. This determined G. Matheron to introduce the term geostatistics [45], which differs from variational statistics because the geological parameter is here considered as a space function called a *regionalized variable*.

In applied hydrogeology, a quantitative study of the lithofacies can be made with the help of certain lithological parameters such as: the total thickness of the lithological interval (H), the cumulated thickness of the sandy bodies (Σh_i), the centre of gravity of the sandy bodies (y_0), the lithological coefficient sand/clay (C), the thickness of the elementary sandy bodies (h_i) and the number of the sandy bodies in a drilling (N). The values of these parameters can be computed for each of the geological and hydrogeological drillings.

Parameters Σh_i and C characterize the variability along the vertical of the sandy bodies in the considered interval, but the quality of this information is not sufficient since the computation takes into account only the total thickness of these bodies, while their individual position and identity are lost. W. C. Krumbein and W. G. Sloss [38] show that this deficiency can be partially restored by computing the first *moment* which gives the centre of gravity of the permeable bodies with regard to the lower or higher limit of the respective interval. In the case of worked underground lignite deposits, the drawing of such a parametric map is highly important since the value of the centre of gravity (expressed in percentages of the total thickness of the interval) indicates the zone having the highest frequency of distribution of the sandy bodies. The drainage from the top requires us to consider the reference plane as being at the lower limit of the interval. The determination of the centre of gravity is easy if we use relation (Figure 1)

$$y_0 = \frac{\Sigma h_i y_i}{\Sigma h_i} \tag{1}$$

where

h_i is the thickness of the elementary sandy bodies;

y_i is the distance between the reference plane and the centre of gravity of each sandy body.

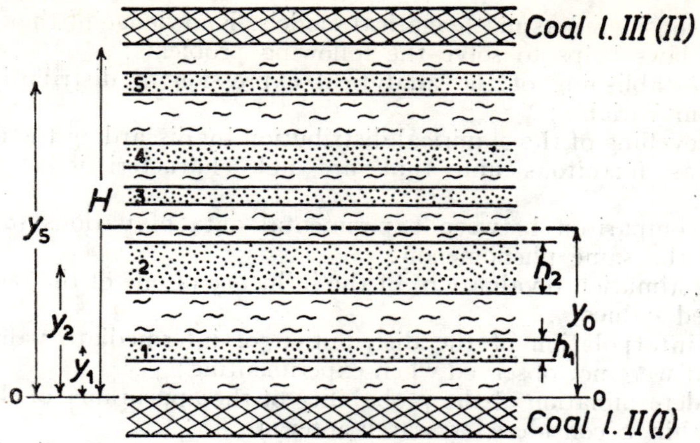

Figure 1. Scheme of the lithological condition between coal layers II—III.

In the problems of water supply, it is recommended that the reference plane should coincide with the higher limit of the geological population.

The above mentioned lithological parameters, as well as the morphology of the aquifer top and bottom, represent the size characteristics of the aquifer. Knowing these space variables provides the best solution of any hydrogeological problem of an economic or a technical nature.

1.1 Determination of the numerical characteristics of the statistical distributions and estimation of the laws of theoretical distribution

An objective characterization of the macroheterogeneity of the lithological structure can be supplied by statistical estimation of the level around which the values of the various geological parameters oscillate, and of their degree of variation. Mathematical statistics defines a series of numerical characteristics (typical values) of the statistical distributions which in the case of large volume sampled populations can be estimated by means of the method of moments. These numerical characteristics determine the essential peculiarities of the parameter characteristics. The random character of these distributions is justified because in practice exploratory drillings are usually located in fixed patterns, which allows one to obtain a systematic sampling.

A more complete characterization of the macroheterogeneity of the lithological structure can be achieved with the help of the theoretical distributions of the lithological parameters. These distributions are law-like in character and can be considered as mathematical models. Knowledge of the theoretical distribution laws helps to solve the following problems:

— the establishing of the analytical expression of distribution of the given random variable;

— the levelling of the empirical distribution by discarding the differences considered as fortuitous and that are not characteristic of the given distribution;

— the comparison between experimental data of various sources but referring to the same phenomenon;

— the estimation (evaluation) of the influence of one of the factors upon the measured values;

— the interpolation of the distribution of intermediate values of the variable that were not observed when experimenting;

— the determination of the probability of the appearance of the random variable within a considered research interval.

Besides these practical purposes, knowledge of the distribution law is also necessary in order to achieve the main objective of scientific research, namely to establish a law of the given phenomenon.

1.1.1 General distribution of the permeable deposits

There are many cases where the frequent lithofacies variations along the lateral results in highly complex hydrogeological conditions. In these cases it is, however, possible to make a thorough analysis in smaller areas so as to emphasize certain distribution trends of the permeable deposits. This is dependent on the existence of a high density of exploration wells and of geological reference points which permit the localization of the permeable deposits.

An example of this is the Șotînga-Mărgineanca lignite deposit of the Dacian age in Romania (Figure 2) where the weight of the aquiferous sands, which have to be drained in order to mine the three lignite layers, is great. On the other hand, these lignite layers constitute conclusive reference points for the separation of the lithologic intervals (populations).

The predominantly clayey-marly complex was intercepted by drillings and shows various thicknesses. In this complex, the sands are of a considerable weight; they form layers and horizons of thicknesses reaching up to 15 m. As a whole, however, the sands are subordinated to the marly-clays. The *basic complex* is separated from the sandy horizon at the bottom of the coal layer I by a continuous interbeded level of clayey-marly of 10—20 m-thick. At one point, the basic complex was crossed by drilling along approximately 180 m; owing to the lithological uniformity of both the late-Pontian and the late-Dacian it is possible that the stratigraphic limit between these two stages could be found within the limits of this complex.

The sandy horizon at the bottom of the coal layer I is, in the great majority of cases, in direct contact with the coal; only in a small number of drillings this is separated from the bottom of the coal layer I, by a layer of marly-clays 2—4 m-thick. As the continuity is practically uniform across all the explored area, the lithologic structure of the horizon shows certain variations. Thus, there are areas where the sands appear as a single level 20—40 m-thick, which forms the whole horizon, while in other areas sandy banks of about one metre thick, are separated by 1—4 marly-clayey intercalations. These characteristics show that in the early-Dacian, up to the appearance of the coal facies, sedimentation had been relatively constant across large areas.

In the late-Dacian, as a result of the warm and relatively humid climate, the sedimentation worked concomitantly with the formation of the whole geological structure. The variable amplitudes of the oscillatory movements resulted in the non-uniform structure of both the sandy deposits and the coal layers (constrictions, frayings, fasciculations, welding of the layers, etc.).

The lithologic interval between coal layers I—II offers large variations of the lithofacies. In the north-eastern sector, the whole interval is sandy, as opposed to the north-western sector (the western plunge of anticline Mărgineanca and of syncline Valea Popii) where only marly-clay appear. The absence of the sands in this interval is also recorded on a transverse strip of the structures in the Cărbunaru and Șotînga mining fields. In the rest of the perimeter (the mining fields of Gheboeni and most of those in Șotînga) sands are concentrated in one or two horizons of 10—20 m-thick. A typical case of a limited (in the plane) development of the sandy deposits is also to be seen along fault F_1, in the axial area of syncline Valea Popii: the fraying lines, obvious here, delimit two sandy areas.

The lithologic interval between coal layers II—III is characterized by a non-uniform distribution of the sandy deposits as compared to the former interval and by their marked tendency to approach the top of coal layer II. In certain areas, the sands are absent, the whole interval consisting only of marly clays. These areas form the western plunge of anticline Mărgineanca and the southern part of the mining field Șotînga, corresponding to anticline Gheboeni-Șotînga. Beside these two areas, the fraying lines of the sandy deposits are quite uncommon; the non-uniform character of the distribution is due to strangulations, ramifications and layer weldings which show some continuity of the sands along both the vertical and the horizontal. It should be noted that the lithofacies, which is purely sandy in the north-eastern sector in interval I—II, changes into a sandy-pelitic facies in coal interval II—III.

The lithologic interval at the top of the coal layer III includes a large number of sandy horizons, many of which have but a limited development in the horizontal plane. This is emphasized in the northern area (Figures 3 and 4) where the complex is fully developed up to the limit of the Levantine.

The structure and distribution of the lignite layers are important elements in the appreciation of the natural hydrogeological conditions. For instance, owing to its simple structure, its great thickness and its continuity, a lignite layer may constitute the regional vertical limit of an aquifer with all its

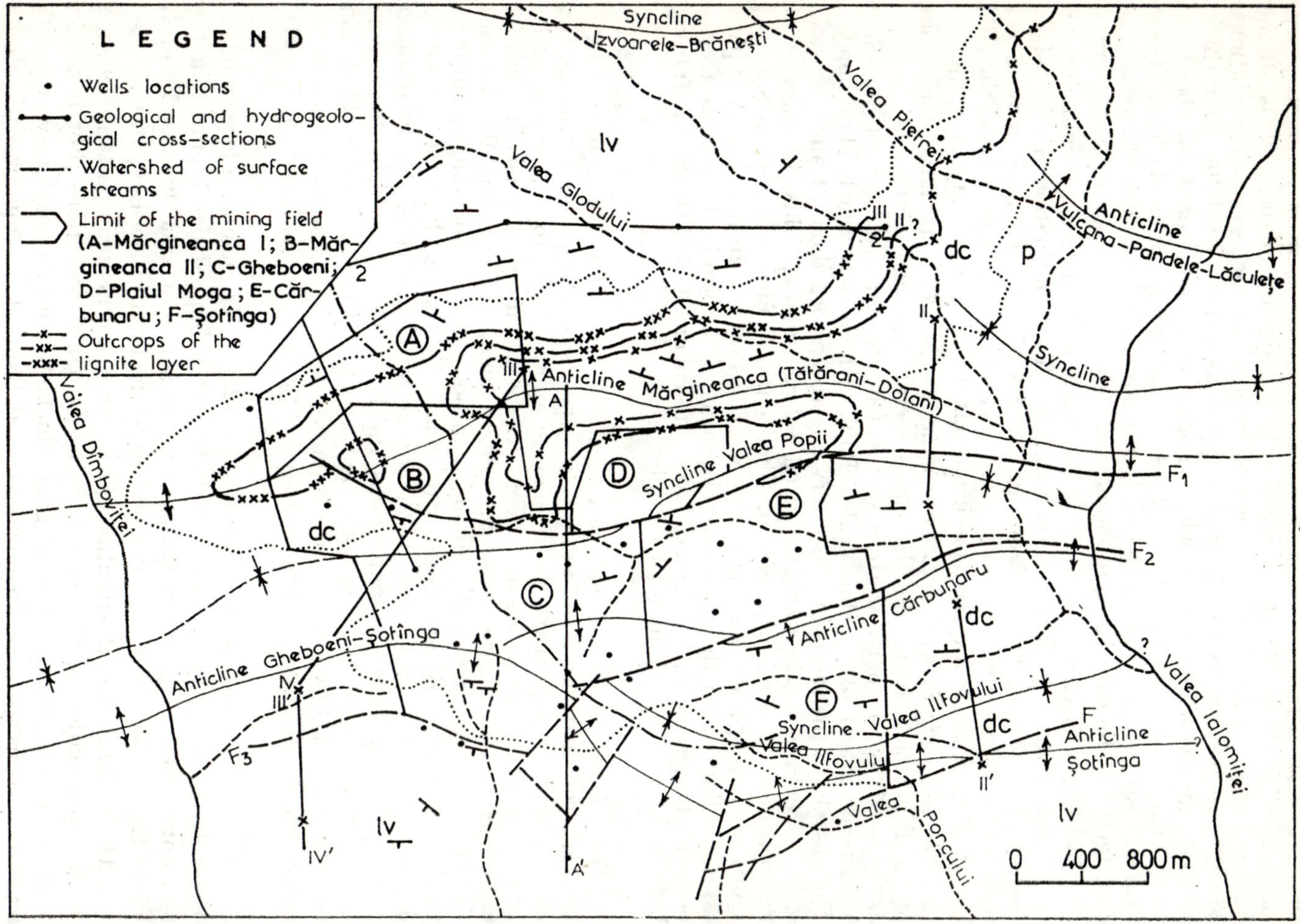

Figure 2. Geological map of Şotînga-Gheboeni-Mărgineanca area.

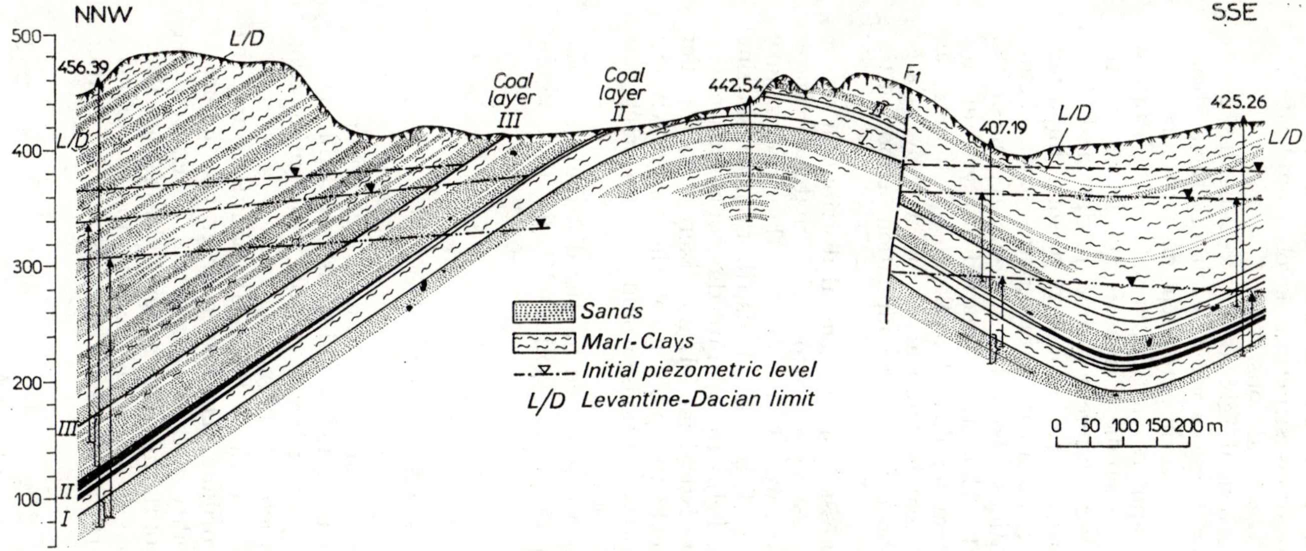

Figure 3. Hydrogeological cross-section *1—1'*.

Figure 4. Hydrogeological cross-section *2—2'*.

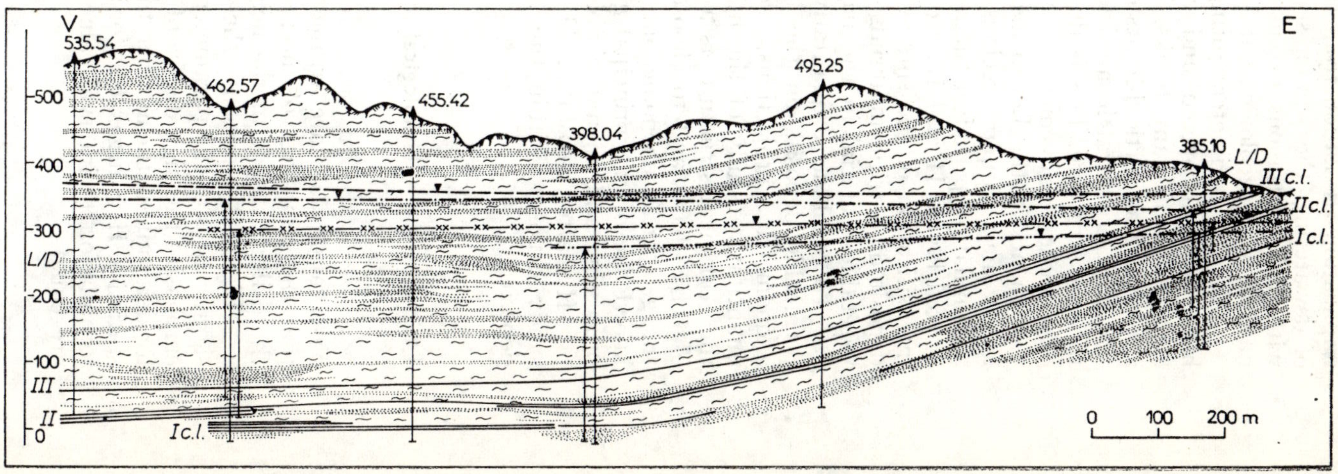

implications (isolation from the rest of the aquiferous system, the necessity of dewatering the aquifer separately, etc.). The presence of certain layers with a discontinuous extension at various structural levels complicate the natural hydrogeological and drainage conditions.

At the perimeter Șotînga-Gheboeni-Mărgineanca three lignite layers have been discovered and are now being worked, each of them being constituted of many banks. In spite of their heterogeneous structure, they are worked as vertical limits of the aquifers with the view to being drained.

The structural localization of the sandless areas is not related to the tectonic structure. This is due to a very different régime, in time and space, of oscillatory movements in the upper-Dacian as well as to some important movements in the Levantine. The geometrical characteristics of the lignite layers may supply precious indications about the tectonic régime in the accumulation period of the sediments. Thus, the formation in a certain area of coal layers II and III from a great number of banks (6—8) signifies an intensification of the sinking movement in those areas followed by a large deposit of detritic materials that shortened the accumulation periods of the vegetable material in the basin.

To conclude, a qualitative impression of the general distribution of the permeable deposits on the basis of suitable cartographic material (maps, axonometric projections, sections) would constitute the preliminary quantitative study of the lithofacies. This general distribution, considering the geological structure as a whole, helps to separate the population on an objective basis and to choose geostatistical models not likely to contradict the geological phenomenon. To tackle the problem of the model without knowledge of the principal geological conditions would lead to irrelevant processing and an unprobable synthesis.

Hereafter, geostatistical models are proposed for the distribution of the sandy deposits within the framework of the Șotînga-Mărgineanca lignite deposit.

1.1.2 Lithological interval between coal layers II and III

In order to compute the statistical characteristics, the observation data are arranged by mining fields and wells (see Table 1). So that the sampling volume should be larger, we also took into consideration the drillings made in the neighbourhood of the mining fields.

Figure 1 represents the diagram of the lithological conditions and the significances of the parameters of a drilling. The dispersion graphs (Figure 5) emphasize the high degree of variability of these parameters. Several criteria for determining the number and size of the intervals are given by mathematical statistics. H. Sturges [62], for instance, recommends the following approximate formula for the interval value when the intervals of the statistical distribution are equal:

$$d = \frac{w}{n} = \frac{x_{max} - x_{min}}{1 + 3.322 \log N} \tag{2}$$

where d is the interval value, x_{max} and x_{min} the highest and lowest values of the sampled population, w the selection amplitude, n the number of intervals and N the population volume.

Obviously, this relation cannot be applied without reservation. Indeed it is known that the information loss for a normal distribution is determined

Table 1

Lithological parameters of interval between coal layers II—III (Şotînga mining field) computed by wells data

No.	Well No. / elevation	Interval thickness, H [m]	Distance y_i [m] from the centre of the sandy body in top of layer II	Thickness of the sandy body, h_i[m]	$h_i y_i$	Centre of gravity $y_0 = \frac{\Sigma h_i y_i}{\Sigma h_i}$ $\left(\frac{y_0}{H} \cdot 100\right)$	Clay thickness $\Sigma a_i = H - \Sigma h_i$	$C = \frac{\Sigma h_i}{\Sigma a_i}$ (%)
1	2	3	4	5	6	7	8	9
(4)	1001 / 428.06		26.3	6.1	159.6			
(3)			13.3	0.7	9.2			
(2)			10.2	5.1	52.3			
(1)			6.2	2.2	13.6			
		30.0		14.1	234.7	16.6(55.3)	15.9	88.7
	1004 / 447.67	14.9	—	—	—	—	14.9	—
(2)	1005 / 418.99		12.5	3.4	41.8			
(1)			5.5	2.0	11.2			
		27		5.4	53.0	9.8(36.3)	21.6	25.0
(2)	1006 / 408.15		25.3	8.7	220.3			
(1)			11.4	7.4	84.2			
		31.6	36.7	16.1	304.5	18.9(59.8)	15.5	103.9
	1007 / 444.48	13.6	8.2	1.6	—	8.2(60.3)	12.0	13.3
(6)	1012 / 434.00		18.7	1.2	22.6			
(5)			16.8	0.7	12.5			
(4)			15.9	0.4	7.0			
(3)			13.9	2.5	33.8			
(2)			7.5	0,9	6.7			
(1)			4.3	1.4	6.0			
		28.3		7.1	88.6	12.5(44.2)	21.2	33.5

by the ratio between the interval value and the standard deviation S. For interval size $d < 1/4\,S$, the information loss is about 1% while for $d = 1/3\,S$, the loss reaches 2.3%. For this reason one should always ensure that the interval size supplied by Sturges formula is smaller than the standard deviation S.

The classification of the typical values is made according to the feature they emphasize:

the central trend of grouping;
the dispersion of the random variable values;
the form of the frequency curves.

The typical values most commonly used to characterize the central trend of grouping are: the mean value, the median and the mode. In order to establish how much the values of the variables deviate from the position of the grouping centre, the following typical values of the dispersion are used: the selection amplitude (w), the dispersion (S^2), the standard deviation (S) and the coefficient of variation (V). The determination of the form of the frequency curves is made with the help of the asymmetry degree (α) and of the excess (E).

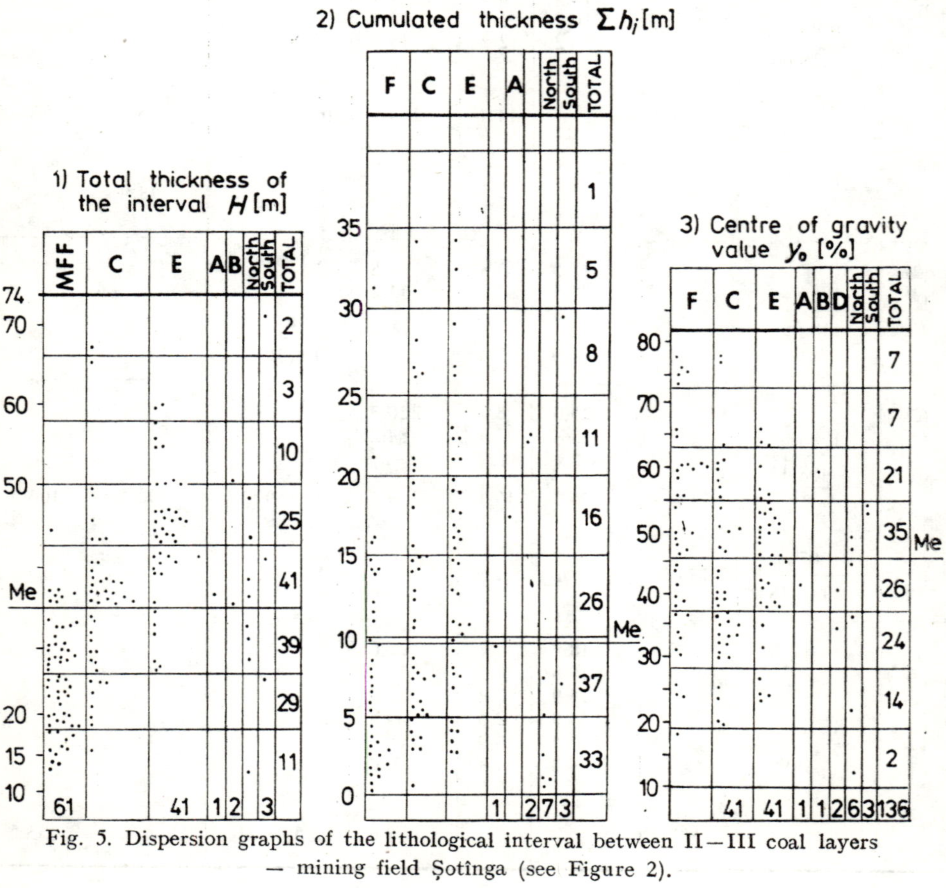

Fig. 5. Dispersion graphs of the lithological interval between II—III coal layers — mining field Șotînga (see Figure 2).

Total thickness of the interval (H). The statistical (empiric) distribution of continuous variable H and the computation of conventional selection moments V_1—V_2—V_3—V_4 are recorded in Table 2.

Table 2

Interval $a-b$	$\frac{a+b}{2}$	n	$f_i(\%)$	α_i	$f_i\alpha_i$	$f_i\alpha_i^2$	$f_i\alpha_i^3$	$f_i\alpha_i^4$
10–18	14	11	6.9	−3	−20.7	62.1	−186.3	558.9
18–26	22	29	18.1	−2	−36.1	72.2	−144.4	288.8
26–34	30	39	24.4	−1	−24.4	24.4	−24.4	24.4
34–42	38	41	25.6	0	0	0	0	0
42–50	46	25	15.6	+1	+15.6	15.6	+15.6	15.6
50–58	54	10	6.2	+2	+12.4	24.8	+49.6	99.2
58–66	62	3	1.9	+3	+5.7	17.1	+51.3	153.9
66–74	70	2	1.3	+4	+5.2	20.8	+83.2	332.8
Total		160	100.0		−81.2 +38.9 −42.3 $V_1=-0.42$	237.0 $V_2=2.37$	−155.4 $V_3=-1.55$	1 473.6 $V_4=14.74$

The computation method of the initial and centred moments is simplified owing to the equality of the intervals by introducing a new quantity α_i (where the name conventional numbers method comes from), which is supplied by relation

$$\alpha_i = \frac{x_i - x_c}{c}$$

where x_i is the middle of the given interval; x_0, the middle of the zero interval; and c, the interval value.

Therefore for interval 66—74, $\alpha = \frac{38-70}{8} = \frac{-32}{8} = -4$

The sample mean is computed by relation

$$\bar{x} = x_0 + V_1c \tag{3}$$

where V_1 is the conventional moment of the first order and

$$\bar{x} = 38—0.42\times 8 = 34.6 \text{ m}$$

The variance estimation S^2 being the centred moment of the second order (m_2) will be supplied by relation

$$m_2' = (V_2 - V_1^2)\,c^2 \tag{4}$$

Since the computation makes use of the middle of the intervals, the values of the moments are affected by certain errors. The passage from the rough centred moments (m'_r) to the adjusted ones is made by Sheppard's corrections. In this case

$$S^2 = m_2 = m'_2 - \frac{c^2}{12}$$

whence

$$S^2 = (V_2 - V_1^2 - 0.08)c^2$$
$$S^2 = [2.37 - (0.42)^2 - 0.08]8^2 = 135 \ \text{m}^2 \qquad (5)$$

The standard deviation of the density is

$$S = \sqrt{135} = 11.6 \text{ m}.$$

Since the frequency polygon (Figure 6) shows a normal distribution, it was possible to compute both the parameters which account for the arching (flatening) of the frequency curve, that is the asymmetry degree (α) and the excess (E), by means of the normalized moments of the third and fourth order [1]:

$$\alpha = r_3 = \frac{m_3}{S^3} \qquad (6)$$

$$E = \beta - 3 \qquad (7)$$

where β is the arching coefficient [2], supplied by relation

$$\beta = r_4 = \frac{m_4}{S^4} \qquad (8)$$

As a function of the sign and values of α, the frequency curve is considered to have: a *positive asymmetry* when $\alpha > 0$; a *negative asymmetry* when $\alpha < 0$; a practically *symmetrical distribution* for $|\alpha| < 0.1$; a *slightly asymmetrical distribution* for $|\alpha| \approx 0.25 - 0.5$ and an *asymmetrical distribution* for $|\alpha| > 0.5$. The computation of these characteristics by means of the conventional moments proceeds as follows:

$$m_3 = (V_3 - 3V_1V_2 + 2V_1^3)c^3$$
$$m_3 = [-1.55 - 3(-0.42)2.37 + 2(-0.42)^3]8^3 = 662 \ \text{m}^3 \qquad (9)$$

$$\alpha = \frac{662}{11.6^3} = 0.42$$
$$m'_4 = (V_4 - 4V_3V_1 + 6V_2V_1^2 - 3V_1^4)\ c^4 \qquad (10)$$

Sheppard's correction for the centred moment of the fourth order is [47]

$$m_4 = m'_4 - \frac{1}{2}\ m'_2c^2 + \frac{7}{240}c^4. \qquad (11)$$

---

[1] The general relation of these moments is $r_k = \dfrac{m_k}{S^k}$

[2] For a normal distribution, $r_3 = 0$, $E = 0$, hence $r_4 = 0$.

Gathering them together in a single relation we get:

$$m_4 = [V_4 - V_1(4V_3 - 0.5V_1) + V_2(6V_1^2 - 0.5) - 3V_1^4 + 0.03]c^4$$

$$m_4 = \{14.74 + 0.42\,[4(-1.55) - 0.5(-0.42)] + 2.37[6(-0.42)^2 - \quad (12)$$

$$- 0.5] - 3(-0.42)^4 + 0.03\}\; 8^4 = 55\,255 \text{ m}^4$$

$$r_4 = \frac{55\,255}{11.6^4} = 3.03$$

$$E = 3.03 - 3 = 0.03$$

To conclude, the frequency curve of total thickness H of the complex has a pronounced positive asymmetry (its ascending branch having a steeper slope than the descending one) while the excess value is practically equal to zero.

So as to exclude the values which have gross errors and which deviate widely from the bulk of the observed values, we can in a normal distribution, apply the following relations:

$$x_{(n)P} = \bar{x} + tS \quad (13)$$

$$x_{(1)P} = \bar{x} - tS \quad (14)$$

where t is a coefficient (according to Student's distribution) the value of which varies as a function of probability P. To $P = 95\%$ corresponds [1] $t = 2$, therefore:

$$x_{(n)P} = 34.6 + 2\times 11.6 = 57.8 \text{ m}$$

$$x_{(1)P} = 34.6 - 2\times 11.6 = 11.4 \text{ m}$$

To $P = 99.7\%$ corresponds $t = 3$, hence

$$x_{max} = 34.6 + 3\times 11.6 = 69.4 \text{ m.}$$

According to the last hypothesis it follows that a single value (Figure 5: $x_n > x_{(n)P}$) should be excluded, because it can no longer be considered as a random error but a gross one.

To emphasize the degree of asymmetry, the sample mean value but also the mode value (Mo) and the median value (Me) have to be calculated. For distributions with a small asymmetry the following approximate estimation relation of the modulus [47] could be applied:

$$Mo = \bar{x} - 3(\bar{x} - Me) \quad (15)$$

The median value is supplied by the dispersion graphic $Me = 34.3$ m. Whence

$$Mo = 34.6 - 3(34.6 - 34.3) = 33.7 \text{ m}$$

---

[1] In paper [54], see table X ($t = 1.96$ for which the number of the degrees of freedom is infinite).

The following arrangement of the three values $Mo < Me < \bar{x}$ corresponds, therefore, to a distribution with a positive asymmetry.

So as to be able to compare and appreciate the dispersion degree of two or more distributions, the variables of which possess various units, the statistical practice uses the *coefficient of variation* which is a dimensionless quantity

$$V = \frac{S}{\bar{x}} \cdot 100(\%) \tag{16}$$

With variable H,

$$V = \frac{11.6}{34.6} \cdot 100 = 33.5\%$$

A random variable is considered statistically determined only if the theoretical function of the given population distribution is estimated too. In order to verify the hypothesis for the empirical distribution and the theoretical one (assumed as a hypothesis) the following tests are usually used: the Kolmogorov's criterion, criterion χ^2 (chi-square) and in certain cases the method of the probabilistic diagrams (*"goodness of fit"* tests).

The law of distribution, denoted by χ^2 has a universal form with a single parameter k, which represents the number of the degrees of freedom. The value χ_e^2 (acquired from experimental data) is given by the square of the sum of the empirical frequencies deviations n_i with regard to the theoretical ones n_i' related to the theoretical frequencies:

$$\chi_e^2 = \Sigma\left[\frac{(n_i - n_i')^2}{n_i'}\right] \tag{17}$$

The empirical distribution and the theoretical one are related if quantity χ_e^2 does not exceed a given value χ_q^2 corresponding to the significant level q. The significant level q has a small enough value[1] and represents the probability of the values of the random variable belonging to the critical interval. For the analysis of the geological data we can consider as the deciding line between acceptance and rejection of the proposed hypothesis, the value $q = 0.05$ which represents a small enough probability.

The probability is therefore

$$P(\chi_e^2 > \chi_q^2) = q \leqslant 0.05$$

When the value χ_e^2 (supplied by relation 17) is smaller than the table value χ_q^2, the concordance between the distribution taken as a hypothesis and the observed data can be accepted; in the contrary case the hypothesis should be rejected.

The computations for the determination of χ_e^2 in the case of variable H (the total thickness of the lithological complex) are shown in Table 3.

---

[1] The probability $P = 1 - q$; for this probability, the variable values are at a certain interval called a "reliability interval".

Table 3

Centre of intervals $x_i = \frac{a+b}{2}$	Number of measurements n_i	$z_i = \frac{x_i - \bar{x}}{S}$	$f(z)$	Theoretical frequency $n_i' = \frac{f(z)}{\frac{\Sigma f(z)}{n}}$	$n_i - n_i'$	$(n_i - n_i')^2$	$\frac{(n_i - n_i')^2}{n_i'}$
14	11	−1.77	0.0833	9	2	4	0.44
22	29	−1.08	0.2227	26	3	9	0.35
30	39	−0.39	0.3697	40	−1	1	0.02
38	41	+0.30	0.3814	41	0	0	0
46	25	+0.99	0.2444	28	−3	9	0.32
52	10	+1.68	0.0973	11	−1	1	0.09
62	0						
70	5	+3.06	0.0037	5	0	0	0
Total	160		1.4025	160			1.22

The computation of the absolute theoretical frequencies (n_i) was made with the help of the standard normal density $f(z) = \frac{1}{\sqrt{2\pi}} e^{-\frac{z^2}{2}}$ the values of which are tabulated [54].

For the successive computations of the standard normal variable z_i we can make use of the recurrence relation

$$z_i = z_{i-1} + \frac{c}{S} \tag{18}$$

where c is the interval value.

For the theoretical frequency of each interval (n_i) the use of the following formula is advised

$$n_i = \frac{nc}{S} f(z) \tag{19}$$

The drawback of this formula is that the sum of the theoretical frequencies is not always equal to the number of the measurements. To avoid this discrepancy we have recourse to another property of the normal distribution as expressed by equality

$$\Sigma f(z) \approx \frac{S}{c} \tag{20}$$

Substituting S in formula (19) results in

$$n_i = \frac{nc}{c\Sigma f(z)} f(z) = \frac{nf(z)}{\Sigma f(z)} = \frac{f(z)}{\frac{\Sigma f(z)}{n}} \tag{21}$$

Since the normal distribution possesses two parameters ($\bar{x}$ and S), the statistics χ_e^2 has $f = k - 1 - 1 = 7 - 2 - 1$ degrees of freedom ($k = 7$ is the number of the computing intervals). Considering that value $\chi_e^2 = 1.22$ is

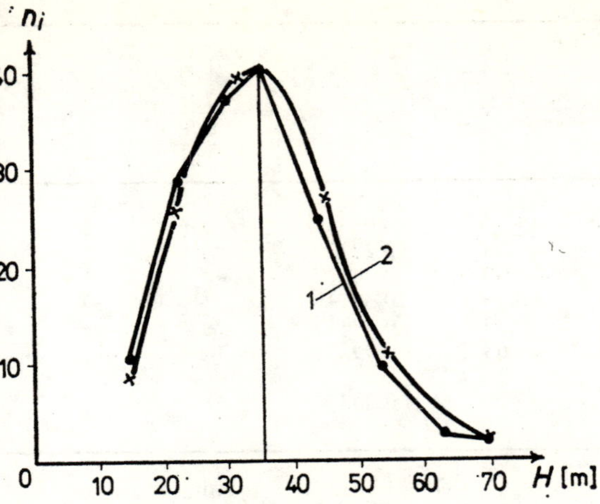

Figure 6. Polygon (*1*) and frequency curve (*2*) for parameter *H*.

lower than the table value (See [62] Appendix 14) $\chi^2_{0.05} = 9.49$, it follows that the theoretical model of the parameter H distribution is the normal law (Figure 6).

Cumulated thickness of the sandy bodies (Σh_i). The polygonal form of the absolute frequencies (Figure 7) suggests a very asymmetrical distribution with its vertex shifted to the left, which is a characteristic element of the normal logarithmic (lognormal) law with density function

$$f(x) = \frac{1}{\sigma_{\ln x}\sqrt{2\pi}} - \left[\exp \frac{(\ln x - \ln \gamma)^2}{2\sigma^2_{\ln}}\right] \tag{22}$$

where $\ln \gamma$ and $\sigma^2_{\ln}$ represent the mean and the variance of variable $\ln x$

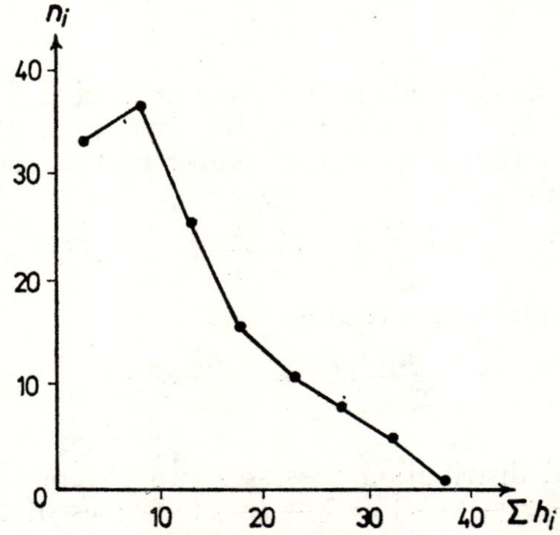

Figure 7. Polygon of absolute frequencies for Σh_i.

The lognormal populations can be treated in the same way as the normal ones provided that, in the computations, the logarithmic values of the parameter are considered, hence a new random variable namely $\log(\Sigma h_i)$. By so doing all the statistical parameters will be dimensionless since they are expressed by logarithmic values. For the normal population of variable $\log(\Sigma h_i)$ the typical sample values can be computed with the help of Table 4.

The mean selection, computed by relation (3) is

$$\log \gamma = 1.15 - 0.74 \times 0.3 = 0.93$$

The logarithmic variance (formula 5) is:

$$\sigma^2_{\log} = [2.52 - (-0.74)^2 - 0.08]0.09 = 0.17$$

$$\sigma_{\log} = \sqrt{0.17} = 0.41$$

Table 4

Interval $a - b$		Interval centre log	Frequency		α_i	$f_i\alpha_i$	$f_i\alpha_i^2$	$f_i\alpha_i^3$	$f_i\alpha_i^4$
Normal values	Log values		absolute n_i	relative f_i (%)					
0.16– 0.32	$\bar{1}.2...\bar{1}.5$	$\bar{1}.35$ (−0.65)	1	0.7	−6	− 4.2	25.2	−151.2	907.2
0.32– 0.63	$\bar{1}.5...\bar{1}.8$	$\bar{1}.65$ (−0.35)	2	1.5	−5	− 7.5	37.5	−187.5	937.5
0.63– 1.25	$\bar{1}.8...0.1$	$\bar{1}.95$ (−0.05)	4	2.9	−4	−11.6	46.4	−185.6	742.4
1.25– 2.50	0.1...0.4	0.25	6	4.4	−3	−13.2	39.6	−118.8	356.4
2.50– 5.00	0.4...0.7	0.55	20	14.6	−2	−29.2	58.4	−116.8	233.6
5.00–10.00	0.7...1.0	0.85	37	27.0	−1	−27.0	27.0	− 27.0	27.0
10.00–20.00	1.0...1.3	1.15	42	30.6	0	0	0	0	0
20.00–40.00	1.3...1.6	1.45	25	18.2	+1	+18.3	18.3	+ 18.3	18.3
			137	100.0		−92.7 +18.3 −74.4 $V_1 = -0.74$	252.4 $V_2 =$ 2.52	−786.9 + 18.3 −768.6 $V_3 = -7.69$	3 222.4 $V_4 =$ 32.22

So as to eliminate the values affected by gross errors, if $P = 99.7\%$ (according to formulae 13—14) is considered, it follows

$$x_{(n)P} = 0.93 + 3 \times 0.41 = +2.16$$

$$x_{(1)P} = 0.93 - 3 \times 0.41 = -0.30$$

Consequently it is necessary to eliminate the first three values of the selection which are insignificant in comparison with its large volume. The centred moment of the third order is (formula 9):

$$m_3 = [-7.69 - 3(-0.74)\,2.52 + 2(-0.74)^3]\,0.3^3 = -0.067$$

The degree of asymmetry (formula 6) is

$$\alpha_{\log} = \frac{-0.067}{0.41^3} = -0.96$$

The centred moment of the fourth order (formula 12) is

$$m_4 = \{32.22 - (-0.74)\,[4(-7.69) - 0.5(-0.74)] + 2.52\,[6(-0.74)^2 - \\ - 0.5] - 3(-0.74)^4 + 0.03\}\,0.3^4 = 0.073$$

The arching coefficient is

$$\beta_{\log} = \frac{-0.073}{0.41^4} = 2.53$$

The excess value is

$$E_{\log} = 2.53 - 3 = -0.47$$

To conclude, the logarithm distribution of the cumulated thickness of the sandy bodies of each drilling ($\log \Sigma h_i$) shows a large negative asymmetry (the curve end is extended to the left, Figure 8) while the respective frequency curve is slightly sloped.

In order to verify the correspondence between the empirical distribution of $\log(\Sigma h_i)$ and the normal law, test χ^2 was used. It follows from Table 5 that this assumption[1] can be admitted since for $q = 0.05$ and $f = 2$ the computed value of λ_e^2 (3.64) is lower than the table value λ_q^2 (5.99). As a consequence, the distribution of parameter h_i follows the lognormal law.

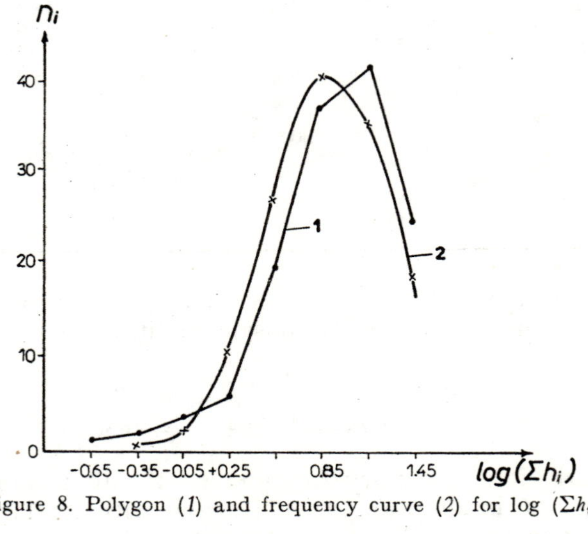

Figure 8. Polygon (*1*) and frequency curve (*2*) for log (Σh_i).

---

[1] The grouping of the first intervals is ascertained by the fact that the low frequencies are not admitted (generally not lower than 5)

Table 5

$\log x_i$	n_i	$z_i = \dfrac{\log x_i - \log\gamma}{\sigma_{\log}}$	$f(z)$	$n'_i = \dfrac{f(z)}{\Sigma f(z)}\, n$	$n'_i - n_i$	$(n'_i - n_i)^2$	$\dfrac{(n'_i - n_i)^2}{n'_i}$
−0.65	1	−3.85	0.0002	0.02			
−0.35	2	−3.12	0.0031	0.32			
−0.05	4	−2.39	0.0229	2.40	0.32	0.10	0.01
+0.25	6	−1.66	0.1006	10.58			
+0.55	20	−0.92	0.2613	27.50	7.50	5.62	0.02
+0.85	37	−0.19	0.3918	41.20	4.20	17.64	0.43
+1.15	42	+0.54	0.3448	36.25	−5.75	33.06	0.91
+1.45	25	+1.27	0.1781	18.73	−6.27	39.31	2.09
	137		1.3028	137.00			3.64

The $\log\gamma$ value is, in fact, the arithmetic mean of the logarithms of the cumulated thickness of the sandy bodies, i.e.

$$\log\gamma = \frac{1}{n}\,\Sigma \log(\Sigma h_i) \tag{23}$$

Passing to antilogarithms we obtain the defining equation of the mean

$$\gamma = [(\Sigma h)_1\,(\Sigma h)_2 \ldots (\Sigma h)_n]^{1/n} \tag{24}$$

This is the geometric mean as well as the median of the geological variable Σh_i. The arithmetic expectation (mean) as well as the other statistical quantities derive from the defining equation of the initial moments in case of a lognormal distribution [32]:

$$\mu_n = \gamma^n \exp\left(\frac{n^2\sigma^2_{\ln}}{2}\right) \tag{25}$$

where n is the moment order.

Since the lognormal mean is the moment of the first order ($n = 1$), its expression (by using the common logarithms) is

$$\bar{x}_l = \exp(2.3\,\log\gamma + 2.65\,\sigma^2_{\log}) \tag{26}$$

The variance of lognormal variable x is

$$\sigma^2_x = \mu_2 - \mu^2_1 = \bar{x}^2_l \cdot (e^{5.3\sigma^2_{\log}} - 1) \tag{27}$$

By introducing the values of parameter Σh_i we obtain

$\gamma = 8.5$ m $\qquad \sigma^2_x = 258.3$ m^2

$\bar{x}_l = 13.3$ m $\qquad \sigma_x = 16.1$ m

The coefficient of variation can be computed in two ways:

— for normal variable $\log(\Sigma h_i)$

$$V_{\log} = \frac{\sigma_{\log}}{\log \gamma} = \frac{0.41}{0.93} = 0.44 = 44\% \tag{28}$$

— for lognormal variable Σh_i

$$V_x = \frac{\sigma_x}{\bar{x}_l} = \sqrt{e^{5.3\sigma_{\log}} - 1} = \frac{16.1}{13.3} = 1.21 = 121\% \tag{29}$$

The modulus of variable Σh_i is

$$Mo = \exp(2.3 \log \gamma - 5.3\ \sigma^2_{\log}) \quad (30$$

$$Mo = 3.46\ \text{m}$$

It is interesting to estimate the errors liable to appear should selection of population Σh_i be considered as a normal population, on the ground that the information volume is large. To this end the typical selection values $\bar{x}$, S^2, S and V are computed (Table 6).

Table 6

$a - b$	x_i	n_i	f_i	$\alpha_i = \frac{x_i - x_0}{c}$	$f_i\alpha_i$	$f_i\alpha_i^2$
0– 5	2.5	33	24.0	−2	−48.0	96.0
5–10	7.5	37	27.0	−1	−27.0	27.0
10–15	12.5	26	19.0	0	0	0
15–20	17.5	16	11.7	+1	+11.7	11.7
20–25	22.5	11	8.0	+2	+16.0	32.0
25–30	27.5	8	5.9	+3	+17.7	53.1
30–35	32.5	5	3.7	+4	+14.8	59.2
35–40	37.5	1	0.7	+5	+ 3.5	17.5
		137	100.0		−75.0 +63.7 −11.3 $V_1 = -0.11$	296.5 $V_2 = 2.96$

$$\bar{x} = 12.5 - 0.11 \times 5 = 11.95\ \text{m}$$

$$S^2 = [2.96 - (-0.11)^2 - 0.08]\ 5^2 = 71.75\ \text{m}^2$$

$$S = \sqrt{71.75} = 8.47\ \text{m}$$

$$V = \frac{8.47}{11.95} \cdot 100 = 71\%$$

Let us first note that the value of the lognormal mean is higher than that of the arithmetic mean ($\bar{x}_e > \bar{x}$). In works on mathematical statistics it is shown, however, that the order relation is $\gamma < \bar{x}_e < \bar{x}$. This anomaly can be explained by two factors:

1. The computation of the arithmetic mean by means of the data grouped in a frequency distribution of positive asymmetry leads to higher values than in the case of non-grouped data ($\bar{x}_n$). To ascertain this we computed

$$\bar{x}_n = \frac{\sum_1^n x_i}{n} = \frac{1\,613}{137} = 11.80 \text{ m.}$$

The difference ($11 \cdot 95 - 11 \cdot 80 = 0.15$ m) being negligible, the influence of this factor can be neglected.

2. If the sampling distribution has a truncated form with one or both ending intervals open, the arithmetic mean has only an approximate character. This is the case of parameter Σh_i the form of which is truncated towards the lower values.

An appreciable difference appears however in the case of the coefficient of variation, its order being $V_{\log} < V < V_x$; the true value is of course $V_x = 121\%$.

Centre of gravity of the sandy bodies (y_0). The absolute position of the centre of gravity of the sandy bodies can be computed by relation (1). It is, however, more useful to use its relative position within the lithological interval expressed as a percentage of its total thickness:

$$y_0(\%) = \frac{y_0}{H} \cdot 100$$

Formula (2) supplies both the number and the size of the intervals:

$$d = \frac{80 - 12.5}{1 + 3.322 \log 137} = \frac{67.5}{8} \approx 9$$

The computation of the conventional moments is shown in Table 7. On this basis the typical sample values were computed.

Table 7

$a - b$	x_i	n_i	f_i	$\alpha_i = \frac{x_i - x_0}{c}$	$f_i\alpha_i$	$f_i\alpha_i^2$	$f_i\alpha_i^3$	$f_i\alpha_i^4$
10–19	14.5	2	1.5	−4	−6.0	24.0	−96.0	384.0
19–28	23.5	14	10.3	−3	−30.9	92.7	−278.1	834.3
28–37	32.5	24	17.7	−2	−35.4	70.8	−141.6	283.2
37–46	41.5	26	19.2	−1	−19.2	19.2	−19.2	19.2
46–55	50.5	35	25.7	0	0	0	0	0
55–64	59.5	21	15.4	+1	+15.4	15.4	15.4	15.4
64–73	68.5	7	5.1	+2	+10.2	20.4	40.8	81.6
73–82	77.5	7	5.1	+3	+15.3	45.9	137.7	413.1
		136	100.0		−50.6 $V_1 = -0.51$	288.4 $V_2 = 2.88$	−341.0 $V_3 = -3.41$	2030.8 $V_4 = 20.31$

The sample mean (formula 3) is:

$$\bar{x} = 50.5 + (-0.51 \times 9) = 45.9\%$$

The sample variance (formula 4) is:

$$S^2 = [2.88 - (0.51)^2 - 0.08]\, 9^2 = 206$$

$S = \sqrt{207} = 14.4\%$ — which is the standard deviation.

The coefficient of variation (formula 16) is

$$V = \frac{14.4}{45.9} 100 = 31.4\%$$

To ascertain the normal feature of the distribution, the probabilistic diagram (Figure 9) was used, the points of which are determined by the normalized values (z_i) and by the centres of the intervals. The collinearity of the points confirm this hypothesis, i.e. the normal distribution (Table 8). This procedure is also known as *Henry's straight-line*. It is to be emphasized that this test is not always conclusive since it may reject the admitted hypothesis (the normal or lognormal law) while according to other more accurate tests (χ^2 or Kolmogorov) the hypothesis is admitted.

The normalized value in the first interval is

$$z_1 = \frac{14.5 - 45.9}{14.4} = -2.18$$

The following values are determined successively by adding value

$$dz = \frac{c}{S} = \frac{9}{14.4} = 0.625$$

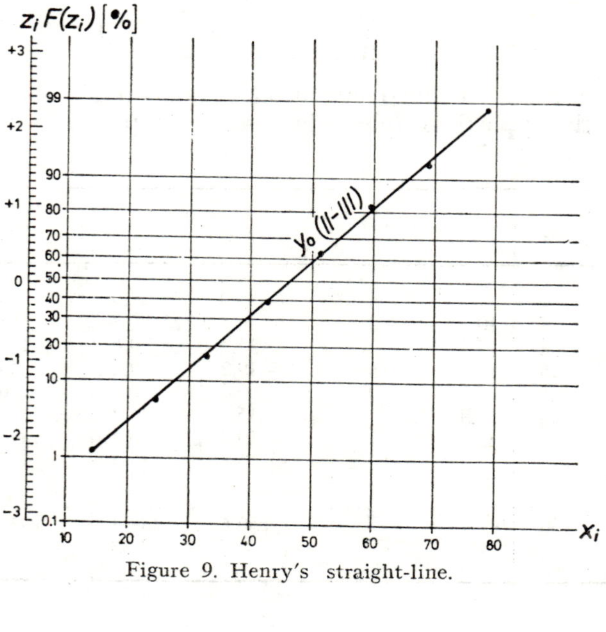

Figure 9. Henry's straight-line.

Table 8

x_i	n_i	z_i	$f(z)$	$n'_i = \frac{f(z)}{\frac{\Sigma f(z)}{n}}$
14.5	2	−2.180	0.0371	3.2
23.5	14	−1.555	0.1200	10.4
32.5	24	−0.930	0.2589	22.0
41.5	26	−0.305	0.3808	32.4
50.5	35	+0.320	0.3790	33.2
59.5	21	+0.945	0.2553	21.7
68.5	7	+1.570	0.1163	9.9
77.5	7	+2.195	0.0358	3.2
	136		1.5832	13.60

To determine the degree of asymmetry α and the excess E, the values of the centred moments m_3 and m_4 (formulae 9 and 13) are necessary:

$$m_3 = [-3.41 - 3(-0.51)\times 2.88 + 2(-0.51)^3]\, 9^3 = 925$$

$$m_4 = \{20.31 - (-0.51)\,[4(-3.41) - 0.5 - (0.51)] + 2.88[6(-0.51)^2 -$$

$$- 0.5] - 3(-0.51)^4 + 0.03\}\, 9^4 = 91\,886.$$

The degree of asymmetry (α) is

$$\alpha = \frac{925}{3\,000} = 0.31$$

The excess value is

$$E = \frac{m_4}{S^4} - 3 = \frac{91\,886}{14.44} - 3 = -0.86$$

The data with gross errors (formulae 13 and 14) can be eliminated. When $P = 99.7\%$ we obtain:

$$x_{max} = 45.9 + 3\times 14.4 = 89.1\%$$

$$x_{min} = 45.9 - 3\times 14.4 = 2.7\%$$

Therefore all selection values are acceptable.

To conclude, the unimodal frequency curve of the relative position of the centre of gravity of the sandy bodies has a small positive asymmetry and a flattened form ($E < 0$) with regard to the ideal normal curve (Figure 10).

Lithological coefficient sand/clay (C). From frequency Table 9 it follows that the values of this parameter are distributed along an amodal curve,

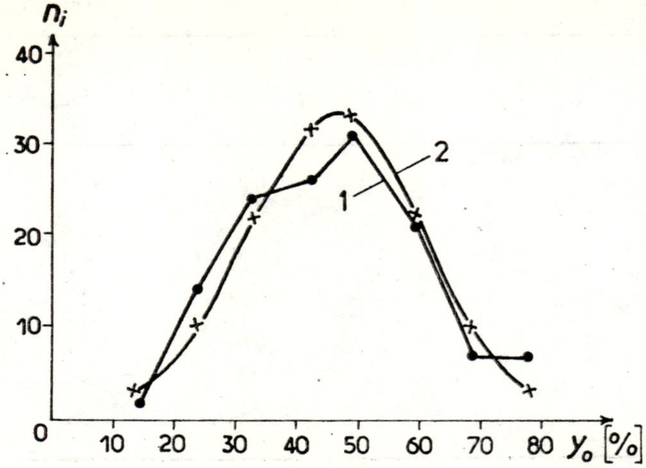

Figure 10. Polygon (*1*) and frequency curve (*2*) for y_0 [%].

Table 9

$a-b$	$x_i = \frac{a+b}{2}$	n_i	f_i	α_i	$f_i\alpha_i$	$f_i\alpha_i^2$	$f_i\alpha_i^3$	$f_i\alpha_i^4$
0. −0.3	0.15	45	31.9	−3	− 95.7	287.1	− 861.3	2 583.9
0.3−0.6	0.45	34	24.3	−2	− 48.6	97.2	− 194.4	388.8
0.6−0.9	0.75	30	21.2	−1	− 21.2	21.2	− 21.2	21.2
0.9−1.2	1.05	11	7.8	0	0	0	0	0.
1.2−1.5	1.35	6	4.3	+1	+ 4.3	4.3	+ 4.3	4.3
1.5−1.8	1.65	5	3.5	+2	+ 7.0	14.0	+ 28.0	56.0
1.8−2.1	1.95	4	2.8	+3	+ 8.4	25.2	+ 75.6	226.8
2.1−2.4	2.25	2	1.4	+4	+ 5.6	22.4	+ 89.6	278.4
2.4−2.7	2.55	2	1.4	+5	+ 7.0	35.0	+ 175.0	875.0
2.7−3.0	2.85	1	0.7	+6	+ 4.2	25.2	+ 151.2	907.2
3.0−3.3	3.15	1	0.7	+7	+ 4.9	34.3	+ 240.1	1 680.7
		141	100.0		− 165.5 + 41.4 − 124.1 $V_1 = -1.24$	565.9 $V_2 = 5.66$	− 1 076.9 + 763.8 − 313.1 $V_3 = -3.13$	7 022.3 $V_4 = 70.22$

namely a unilateral positive one. In a first stage, the data processing was performed by means of the empirical moments, by computing the typical sample values

$$\bar{x} = 0.68;\ S^2 = 0.36;\ S = 0.60$$

Because of the great difference between the highest and the mean values, the possibility of excluding the extreme values, strongly deviated with regard to the bulk of the observed values should be ascertained. Since the only known values are those of the arithmetic mean and of the standard deviation as estimated on the basis of a sample size n, while the distribution is non-normal, the following criterion [54] can be applied:

$$x_{(n)P} \approx \bar{x} + S_f z_{P_1} \left(1 + \frac{3}{f}\right) \sqrt{\frac{f}{n}} \tag{31}$$

where

$x_{(n)P}$ is the highest value of the sample which is only exceeded by a probability q taken as equal to 0.01;

$P = 1 - q = 0.99$;

S_f is the standard deviation of a sampled population of value $f = n - 1$;

$P_1 = 1 - \frac{1 - P}{n}$ for $P > 0.95$

In the case of a large population, $f = n$ and relation (31) is simplified as follows

$$x_{(n)P} \approx \bar{x} + S_n z_{P_1} \tag{32}$$

The application of this criterion supplies

$$P_1 = 1 - \frac{1 - 0.99}{141} = 0.999929$$

The standard normal variable value z_{P_1} is given by tables [54] if the value of Laplace's function $\Phi(z_{P_1})$ is known:

$$\Phi(z_{P_1}) = P_1 - 0.5 = 0.999929 - 0.5 = 0.499934$$

$$z_{P_1} = 3.80$$

$$x_{(n)P} \approx 0.68 + 0.60 \times 3.80 = 2.96$$

It follows that the only value of 3.3 shows a great deviation from the bulk of the observed values and is to be excluded as such. This situation affects only moderately the above computed typical values:

$$\bar{x} = 0.66;\ S^2 = 0.32;\ S = 0.57;\ V = 0.86$$

The lowest value of selection ($x_{(1)P}$) is supplied by the same relations (31 and 32), where the normalized value z_{p_1} enters with the minus sign.

An accurate and complete characterization of the random variables is supplied by the distribution function. The problem of the choice of the statistical model becomes more intricate if the empirical distribution cannot be approximated by the normal or lognormal law. Beside these two statistical models, in the geological practice, other classical distributions can intervene, such as gamma, beta, Weibull, exponential, etc. In order to facilitate the approximation method of the empirical sampling distributions, some of the books of mathematical statistics offer graphs from which the best statistical models results. Such a graph (Figure 11) is based on the square of the normalized moment of the third order (α^2) and the normalized moment of the fourth order which is also called arching coefficient (β). The exact values of these moments cannot be known but can be estimated accurately enough by means of the empirical moments (of selection).

It follows from this graph that certain distributions in plane α^2—β are represented (theoretically) by a point: the normal distribution with co-ordinates $\alpha^2 = 0$ and $\beta = 3$, the exponential distribution with $\alpha^2 = 4$ and $\beta = 9$, and the uniform distribution with $\alpha^2 = 0$ and $\beta = 1.8$. Another peculiarity of this graph is that the gamma and lognormal distributions are represented by neighbouring curves. The largest domain is occupied by distribution β which is represented by a great variety of form.

It should however be pointed out that these graphs only offer information about the choice of statistical model because of the two limitations of the method. The first of these limitations is that the estimations of both the parameters α and β are very sensitive to a small number of extreme values because of the use of the empirical moments. Consequently, it is recommended to use this method, especially when the sample size is $n < 200$. The second limitation is due to the fact that the distribution form is not universally determined by the two normalized moments (of third and fourth order).

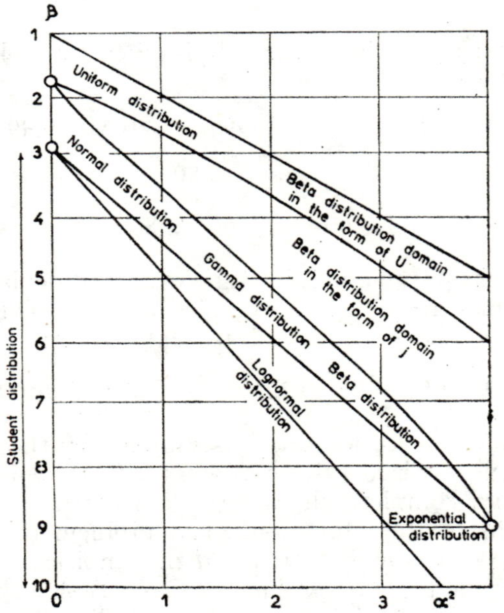

Figure 11. Informative graphs for establishing the population distribution law.

To permit the use of this graph in the case of variable C (sand-clay coefficient) paramerers α^2 and β were computed with the help of Table 9:

$$m_3 = 0.38$$

$$r_3 = \alpha = \frac{m^3}{S^3} = \frac{0.38}{0.57^3} = 2.09$$

$$\alpha^2 = 4.37$$

$$m_4 = 0.79$$

$$r_4 = \beta = \frac{m^4}{S_4} = \frac{0.79}{0.57} = 0.79$$

Since $\alpha^2 > 4$ the only possible conclusion is that the sampled population of the lithological coefficient (C) follows none of the six theoretical distributions marked on the figure. On the other hand, it should be emphasized that although the beta distribution present some forms unilaterally positive and the domain of this distribution is very large, it cannot be applied because the defining limits of the variable are $0 \leqslant x \leqslant 1$ while those of our parameter are $0.15 < C < 3.0$.

Another statistical model, which includes moreover amodal, diminishing frequency curves, is Weibull's distribution frequently used in the safety theory. The density of Weibull's distribution take the form [27]

$$f(x) = \frac{\eta}{\sigma}\left(\frac{\alpha}{\alpha}\right)^{\eta-1} \exp\left[-\left(\frac{x}{\sigma}\right)^{\eta}\right]; \qquad (33)$$

$$x \geqslant 0$$

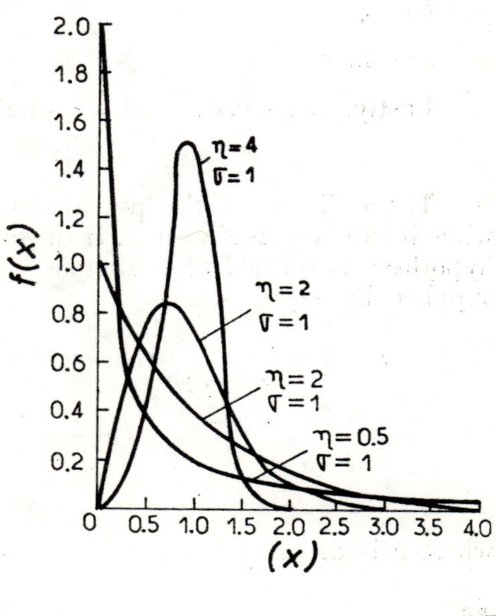

Figure 12. Weibull distribution for $\sigma - 1$ and for various η values.

where σ is the scale parameter and η the form parameter. This distribution assumes various forms as also results from Figure 12, which represents the frequency curves for $\sigma = 1$, as well as various values of η. The verification of the concordance between the empirical distribution and the Weibull as well as the estimation of parameters σ and η can be made by means of the probabilistic diagram drawn up on the basis of the distribution function $F(x)$, the form of which is simpler than that of the density function

$$F(x) = 1 - \exp\left[-\left(\frac{x}{\sigma}\right)^{\eta}\right] \tag{34}$$

$$0 \leqslant x < \infty, \quad \eta > 0, \quad \sigma > 0$$

To linearize the distribution function, the following transformations are required:

$$1 - F(x) = \exp\left[-\left(\frac{x}{\sigma}\right)^{\eta}\right] \tag{35}$$

$$\frac{1}{1 - F(x)} = \exp\left[\left(\frac{x}{\sigma}\right)^{\eta}\right] \tag{36}$$

$$\ln\left[\frac{1}{1 - F(x)}\right] = \left(\frac{x}{\sigma}\right)^{\eta} \tag{37}$$

$$\ln \ln \frac{1}{1 - F(x)} = -\eta \ln \sigma + \eta \ln x \tag{38}$$

$$W = \ln \ln \left[\frac{1}{1 - F(x)}\right] \tag{39}$$

$a = -\eta \ln \sigma$

$b = \eta$

$Z = \ln x$

Lastly, the equation of a straight line is obtained

$$W = a + bZ \tag{40}$$

The ordinate of the probabilistic diagram (Figure 13) is function W while its abscissa is $Z = \ln x$. If the experimental points are collinear the zero hypothesis is verified while sample parameters σ and η of the distribution are supplied by expressions

$$\eta = b = \frac{W_2 - W_1}{Z_2 - Z_1} \text{ (the angular coefficient of the straight line)} \tag{41}$$

$$\ln \sigma = -\frac{a}{\eta} = -\frac{a}{b} \tag{42}$$

$$\sigma = e^{-a/b} \tag{43}$$

where a is the initial ordinate.

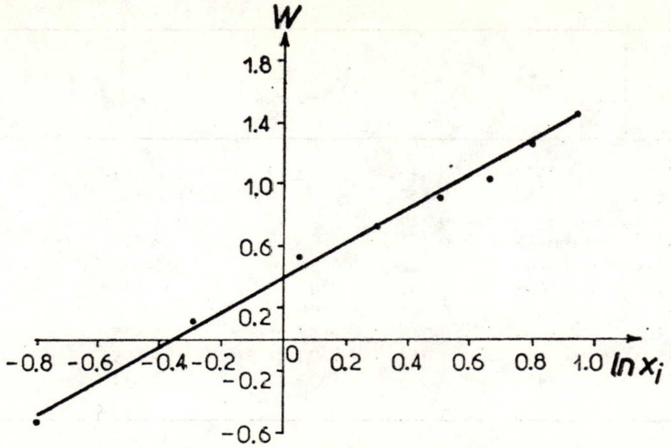

Figure 13. Probabilistic diagram for the Weibull distribution of lithological coefficient C.

The computations are presented in Table 11, that includes moreover the function of empirical distribution $F_n(x)$ from Table 10.

Figure 13 shows that the points are very close to a straight line, which proves the concordance between the empirical distribution and the Weibull theoretical one. Parameters σ and η acquire the following values

$$\eta = b = \frac{1.2740 - 0.7435}{0.9109 - 0.3001} = 1.04$$

$$a = 0.4; \quad \ln \sigma = -\frac{0.4}{1.04}; \quad \sigma = 0.68$$

Table 10

$a - b$	$x_i = \frac{a+b}{2}$	n_i	$f_i = \frac{n_1}{140}$	Cumulated relative frequency	1/2 jump	$F_n(x)$
0 —0.3	0.15	45	0.321	0.321	0.160	0.160
0.3—0.6	0.45	34	0.243	0.564	0.122	0.443
0.6—0.9	0.75	30	0.214	0.778	0.107	0.671
0.9—1.2	1.05	11	0.079	0.857	0.040	0.818
1.2—1.5	1.35	6	0.043	0.900	0.021	0.878
1.5—1.8	1.65	5	0.036	0.936	0.018	0.918
1.8—2.1	1.95	4	0.029	0.965	0.015	0.951
2.1—2.4	2.25	2	0.014	0.979	0.007	0.972
2.4—2.7	2.55	2	0.014	0.993	0.007	0.986
2.7—3.0	2.85	1	0.007	1.000	0.004	0.997
		140	1.000			

Computation examples for $F_n(x)$

$$0.321 + 0.122 = 0.443$$
$$0.564 + 0.107 = 0.671$$

Table 11

x_i	$\ln x_i$	$1 - F_n(x)$	$\frac{1}{1 - F_n(x)}$	$\ln\left[\frac{1}{1 - F_n(x)}\right]$	$W = \ln\ln\left[\frac{1}{1 - F_n(x)}\right]$
0.15	−1.8971	0.840	1.19	0.1740	−1.7487
0.45	−0.7985	0.557	1.79	0.5822	−0.5413
0.75	−0.2877	0.329	3.04	1.1119	+0.1068
1.05	+0.0488	0.182	5.49	1.7029	+0.5326
1.35	+0.3001	0.122	8.19	2.1029	+0.7435
1.65	+0.5008	0.082	12.19	2.5015	+0.9170
1.95	+0.6678	0.049	20.41	3.0155	+1.035
2.25	+0.8109	0.028	35.71	3.5752	+1.2740
2.55	+0.9361	0.014	71.43	4.2683	+1.4516
2.85	+1.0473	0.003	333.33	5.8092	+1.7596

To ascertain this concordance a stronger test was also applied, that is Kolmogorov's criterion the character of which is non-parametric. According to this criterion, if the absolute values of the greatest difference (d_n) between the empirical and the theoretical values of the distribution functions is smaller than quantity $d_a = \frac{\lambda_q}{\sqrt{n}}$, then the respective variable follows Weibull's law of theoretical distribution. In some mathematical statistics works the following function is used

$$\lim_{n\to\infty} P\left[d_n \leqslant \frac{\lambda_q}{\sqrt{n}}\right] = K(\lambda) \tag{44}$$

Since the value of this function is known, and $K(\lambda) = 1 - q$, from tables (see reference 54) we find the value of parameter λ. In this case, for $q = 0.05$ and $K(\lambda) = 1 - 0.05 = 0.95$, we find $\lambda_{0.05} = 1.36$, thence

$$d_a = \frac{1.36}{\sqrt{140}} = 0.115.$$

Table 12

x_i	$\exp\left[-\left(\frac{x}{0.68}\right)\right]^{1.04}$	$F(x)$	$F_n(x)$	$d_n = \lvert F_n(x) - F(x)\rvert$
0.15	0.813	0.187	0.160	0.027
0.45	0.522	0.478	0.443	0.035
0.75	0.333	0.667	0.671	0.004
1.05	0.210	0.790	0.818	0.028
1.35	0.129	0.871	0.878	0.017
1.65	0.081	0.919	0.918	0.001
1.95	0.050	0.950	0.951	0.001
2.25	0.031	0.969	0.972	0.003
2.55	0.019	0.981	0.986	0.005
2.85	0.012	0.988	0.997	0.009

Since $d_n < d_a$ (Table 12) it follows that the hypothesis undergoing the verification can be accepted. Figure 14 presents the polygon of the cumulated relative frequencies and the distribution curve from which the absolute theoretical frequency n_i was graphically deduced.

The knowledge of the law of theoretical distribution offers also the possibility of an accurate determination of the numerical characteristics usually estimated by means of the empirical moments. Thus, in the case of Weibull's distribution, the mathematical expectation ($\bar{x}_w$), the dispersion (σ_w^2) and the normalized moments of the third (α) and fourth (β) order the following relations (27) are given

$$\bar{x}_w = \sigma\,\Gamma\left(\frac{1}{\eta} + 1\right) \tag{45}$$

$$\sigma_w^2 = \sigma^2\left[\Gamma\left(\frac{2}{\eta} + 1\right) - \Gamma^2\left(\frac{1}{n} + 1\right)\right] \tag{46}$$

$$\alpha_w = \frac{\Gamma\left(1 + \frac{3}{\eta}\right) - 3\Gamma\left(1 + \frac{2}{\eta}\right)\Gamma\left(1 + \frac{1}{\eta}\right) + 2\Gamma^3\left(1 + \frac{1}{\eta}\right)}{\left[\Gamma\left(1 + \frac{2}{\eta}\right) - \Gamma^2\left(1 + \frac{1}{\eta}\right)\right]^{3/2}} \tag{47}$$

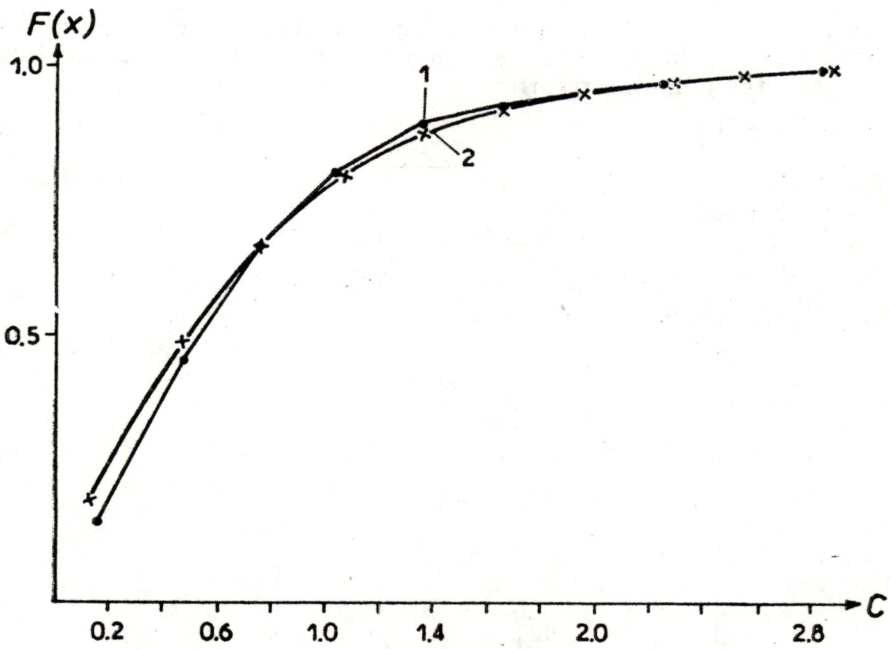

Figure 14. Polygon of the relative frequencies (*1*) and distribution curve (*2*) for parameter *C*.

$$\beta_w = \frac{\Gamma\left(1+\frac{4}{\eta}\right) - 4\Gamma\left(1+\frac{3}{\eta}\right)\Gamma\left(1+\frac{1}{\eta}\right) + 6\Gamma\left(1+\frac{2}{\eta}\right)\Gamma^2\left(1+\frac{1}{\eta}\right) - 3\Gamma^4\left(1+\frac{1}{\eta}\right)}{\left[\Gamma\left(1+\frac{2}{\eta}\right) - \Gamma^2\left(1+\frac{1}{\eta}\right)\right]^2} \tag{48}$$

Table 13

$a - b$	x_i	Cumulated theoretical relative frequency	Theoretical relative frequency f_i	$n_i' = 140 f_i'$
0 —0.3	0.15	0.335	0.335	47
0.3—0.6	0.45	0.575	0.240	34
0.6—0.9	0.75	0.730	0.155	22
0.9—1.2	1.05	0.830	0.100	14
1.2—1.5	1.35	0.895	0.065	9
1.5—1.8	1.65	0.933	0.038	5
1.8—2.1	1.95	0.965	0.032	4
2.1—2.4	2.25	0.980	0.015	2
2.4—2.7	2.55	0.993	0.013	2
2.7—3.0	2.85	1.000	0.007	1
			1.000	140

In these relations, $\Gamma(x)$ represents function gamma, the values of which are usually to be found in tables (reference 75) for $1 < x < 2$. For other values of x, the following recurrence relations may be used:

— for $x < 1$

$$\Gamma(x) = \frac{\Gamma(x+1)}{x} \tag{49}$$

— for $x > 2$

$$\Gamma(x) = (x-1)\ \Gamma(x-1) \tag{50}$$

By applying the above formulae the following exact values of the numerical characteristics were obtained:

$\bar{x}_w = 0.67$ $\alpha_w = 1.90$

$\sigma_w^2 = 0.41$ $\beta_w = 8.30$

$\sigma_w = 0.64$

The coefficient of variation is defined by relation

$$V_w = \frac{\sigma_w}{\bar{x}_w} = \sqrt{\frac{\sigma^2\left[\Gamma\left(\frac{2}{\eta}+1\right) - \Gamma^2\left(\frac{1}{\eta}+1\right)\right]}{\sigma^2\Gamma^2\left(\frac{1}{\eta}+1\right)}} = \sqrt{\frac{\Gamma\left(\frac{2}{\eta}+1\right)}{\Gamma^2\left(\frac{1}{\eta}+1\right)} - 1} \tag{51}$$

$$V_w = 0.96 = 96\%$$

Comparing these estimations of the highest likelihood with the ones obtained on the basis of the empirical moments, it follows that with the exception of the mathematical expectation, all the other characteristics present errors (e):

— for σ_w^2, $e = -22\%$

— for σ_w, $e = -11\%$

— for α_w, $e = +10\%$

— for β_w, $e = -14.5\%$

— for V_w, $e = +11.6\%$

The calculated errors point out the fact that even in the case of great sample size (in our case $n_i = 140$), for an increase of accuracy we must operate with thoroughly specified distributions, which implies knowing the function and all its parameters. The sample size necessary for obtaining $e = 0$, increases at the same time as the asymmetry degree of the frequency distribution.

The frequency curves (see Figure 15) of the unilateral-positive distributions resemble the equilateral hyperbola, a reason for which these distributions are also called hyperbolic distributions. In such a case, it is recommended that the mathematical expectation be approximated by the weighted harmonic mean, the value of which for the lithological coefficient C is given in Table 14:

$$\bar{x}_{arm} = \frac{\Sigma n_i}{\Sigma \frac{n_i}{x_i}} = \frac{140}{437.56} = 0.32 \qquad (52)$$

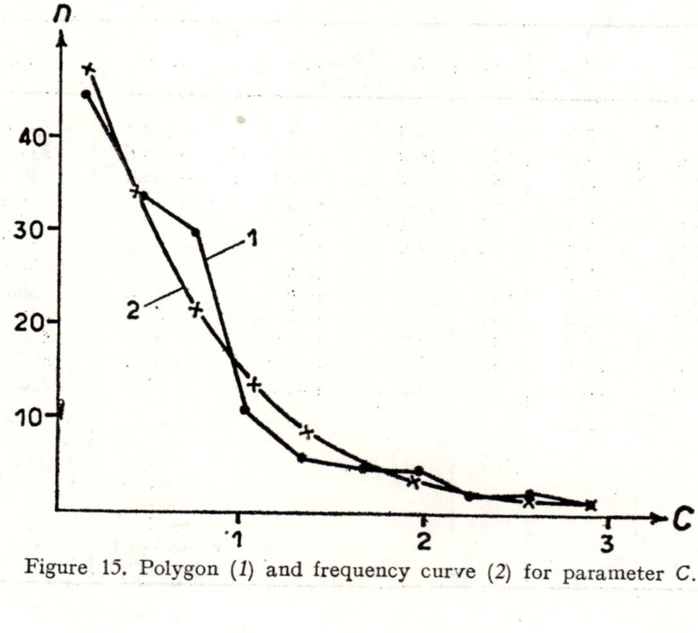

Figure 15. Polygon (*1*) and frequency curve (*2*) for parameter *C*.

The big difference with regard to mean $\bar{x}_w$ does not recommend the generalization of this method. As a matter of fact, in the series of the means, this one has the smallest value.

Table 14

$a-b$	x_i	n_i	$\frac{n_i}{x_i}$	$a-b$	x_i	n_i	$\frac{n_i}{x_i}$
0–0.3	0.15	45	300.00	1.5–1.8	1.65	5	3.03
0.3–0.6	0.45	34	75.55	1.8–2.1	1.95	4	2.05
0.6–0.9	0.75	30	40.00	2.1–2.4	2.25	2	0.88
0.9–1.2	1.05	11	10.48	2.4–2.7	2.55	2	0.78
1.2–1.5	1.35	6	4.44	2.7–3.0	2.85	1	0.35
						$n=140$	437.56

Thickness of the elementary sandy bodies (h_i). The selection amplitude of this parameter is large, between 0.2 m and 32 m; that is why the computations were made with unequal class intervals (Table 15). Under these conditions, the mean sample is supplied by relation

$$\bar{x} = x_0 + V_1 = 5.0 - 0.17 = 4.83 \text{ m} \tag{53}$$

Table 15

$a-b$	x_i	n_i	f_i	$\zeta = x_i - x_0$	$f_i\zeta_i$	$f_i\zeta_i^2$
0.2– 0.8	0.5	121	26.5	− 4.5	−119.3	536.6
0.8– 1.6	1.2	73	16.0	− 3.8	− 60.8	231.0
1.6– 2.6	2.1	49	10.7	− 2.9	− 31.0	89.9
2.6– 4.0	3.3	47	10.3	− 1.7	− 17.5	29.8
4.0– 6.0	5.0	42	9.2	0	0	0
6.0– 8.0	7.0	35	7.6	+ 2.0	+ 15.2	30.4
8.0– 10.0	9.0	18	3.9	+ 4.0	+ 15.6	62.4
0.0– 12.0	11.0	17	3.7	+ 6.0	+ 22.2	133.2
2.0– 14.0	13.0	17	3.7	+ 8.0	+ 29.6	236.8
4.0– 18.0	16.0	16	3.6	+11.0	+ 39.6	435.6
8.0– 25.0	21.5	15	3.3	+16.5	+ 54.4	897.6
5.0– 32.0	28.5	7	1.5	+23.5	+ 35.3	830.0
		457	100 0		−228.6 +211.9	3 513.3 $V_2 = 35.1$
					− 16.7 $V_1 = -0.17$	

and the sample variance by relation

$$S^2 = V_2 - V_1^2 = 35.1 - (-0.17)^2 = 35.0 \text{ m}^2 \tag{54}$$

$$S = \sqrt{35.0} = 5.9 \text{ m} \text{ — sample standard deviation}$$

To exclude the gross errors, the criterion supplied by relation (32) was used which led to $x_{(n)P} = 29.0$ m that only eliminates from the population a single measurement ($x = 30.9$ m). By applying this reduction to the size of the last interval ($n = 6$) the following typical values of the sample were found:

$$\bar{x} = 4.72 \text{ m}; \quad S^2 = 31.4 \text{ m}^2; \quad S = 5.6 \text{ m}; \quad V = 118\%.$$

Since the frequency distribution emphasizes a positive unilateral curve (as in the case of lithological coefficient C), an attempt was made to approximate the empirical distribution also with the help of Weibull's model. The concordance verification and the estimation of the distribution parameters σ and η were made by means of the probabilistic diagram (Figure 16) by using the computed data in Tables 16 and 17. The distribution parameters were computed by relations (41—42):

$$\eta = b = \frac{1.0106 - (-0.8557)}{2.7726 - 0.1823} = 0.72$$

$$a = -0.97$$

$$\ln \sigma = -\frac{-0.97}{0.72}$$

$$\sigma = 3.85$$

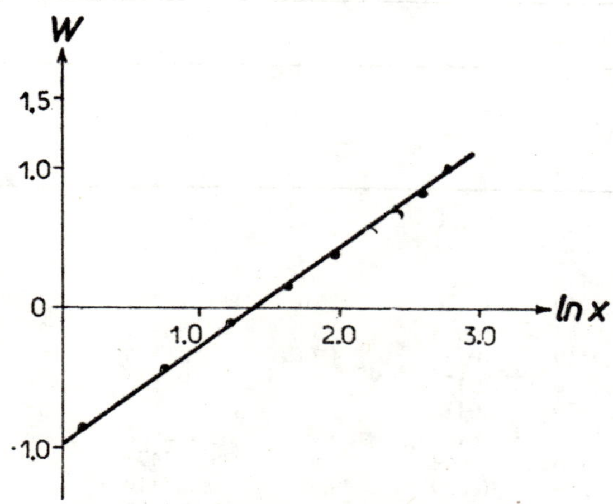

Figure 16. Probabilistic diagram of the Weibull distribution for parameter h_i.

Knowing the distribution as well as the values of its parameters we can estimate at the highest likelihood both of the mean and of the variance with the help of relations (44—45)

$$\bar{x}_w = 3.85\ \Gamma\left(\frac{1}{0.72} + 1\right) = 4.69\ \text{m}$$

$$\sigma_w^2 = 3.85^2\left[\Gamma\left(\frac{2}{0.72} + 1\right) - \Gamma^2\left(\frac{1}{0.72} + 1\right)\right] = 45.0\ \text{m}^2$$

$$\sigma_w = 6.7\ \text{m}$$

$$V_w = \frac{6.7}{4.69}\,100 = 143\%$$

Table 16

$a - b$	$x_i = \frac{a+b}{2}$	n_i	$f_i = \frac{n_i}{456}$	Cumulated relative frequency	1/2 jump	$F_n(x)$
0.2— 0.8	0.5	121	0.265	0.265	0.132	0.132
0.8— 1.6	1.2	73	0.160	0.425	0.080	0.345
1.6— 2.6	2.1	49	0.108	0.533	0.054	0.479
2.6— 4.0	3.3	47	0.103	0.636	0.051	0.584
4.0— 6.0	5.0	42	0.092	0.728	0.046	0.682
6.0— 8.0	7.0	35	0.077	0.805	0.039	0.767
8.0—10.0	9.0	18	0.039	0.844	0.020	0.825
10.0—12.0	11.0	17	0.037	0.881	0.018	0.862
12.0—14.0	13.0	17	0.037	0.918	0.018	0.899
14.0—18.0	16.0	16	0.036	0.954	0.018	0.936
18.0—23.0	20.5	15	0.033	0.987	0.017	0.971
23.0—29.0	26.0	6	0.013	1.000	0.007	0.994
		456	1.000			

Table 17

x_i	$\ln x_i$	$1 - F_n(x)$	$\frac{1}{1 - F_n(x)}$	$\ln\left[\frac{1}{1 - F_n(x)}\right]$	$W = \ln\ln\left[\frac{1}{1 - F_n(x)}\right]$
0.5	−0.6932	0.868	1.15	0.1398	−1.9676
1.2	+0.1823	0.655	1.53	0.4253	−0.8557
2.1	+0.7419	0.521	1.92	0.6523	−0.4277
3.3	+1.1939	0.416	2.40	0.8755	−0.1330
5.0	+1.6094	0.318	3.14	1.1442	+0.1347
7.0	+1.9459	0.233	4.29	1.4563	+0.3759
9.0	+2.1972	0.175	5.71	1.7422	+0.5551
11.0	+2.3979	0.138	7.25	1.9810	+0.6836
13.0	+2.5650	0.101	9.90	2.2925	+0.8296
16.0	+2.7726	0.064	15.62	2.7473	+1.0106
20.5	+3.0204	0.029	34.49	3.5410	+1.2644
26.0	+3.2581	0.006	166.66	5.1180	+1.6328

Except in the case of the mean ($\bar{x}_w = \bar{x}$), important errors appear in the case of the estimations made with the help of the empirical moments in spite of the very great volume of the sampling:

— for σ_w^2, $e = -30.2\%$
— for σ_w, $e = -16.4\%$
— for V_w, $e = -17.4\%$

Table 18

x_i	$\exp\left[-\left(\frac{x}{3.85}\right)\right]^{0.72}$	$F(x)$	$F_n(x)$	$d_n = \lvert F_n(x) - F(x)\rvert$
0.5	0.795	0.205	0.132	0.073
1.2	0.650	0.350	0.345	0.005
2.1	0.524	0.476	0.479	0.003
3.3	0.370	0.630	0.584	0.046
5.0	0.298	0.702	0.682	0.020
7.0	0.214	0.786	0.767	0.019
9.0	0.159	0.841	0.825	0.016
11.0	0.119	0.881	0.862	0.019
13.0	0.091	0.909	0.899	0.010
16.0	0.061	0.939	0.936	0.003
20.5	0.036	0.964	0.971	0.007
26.0	0.019	0.981	0.994	0.013

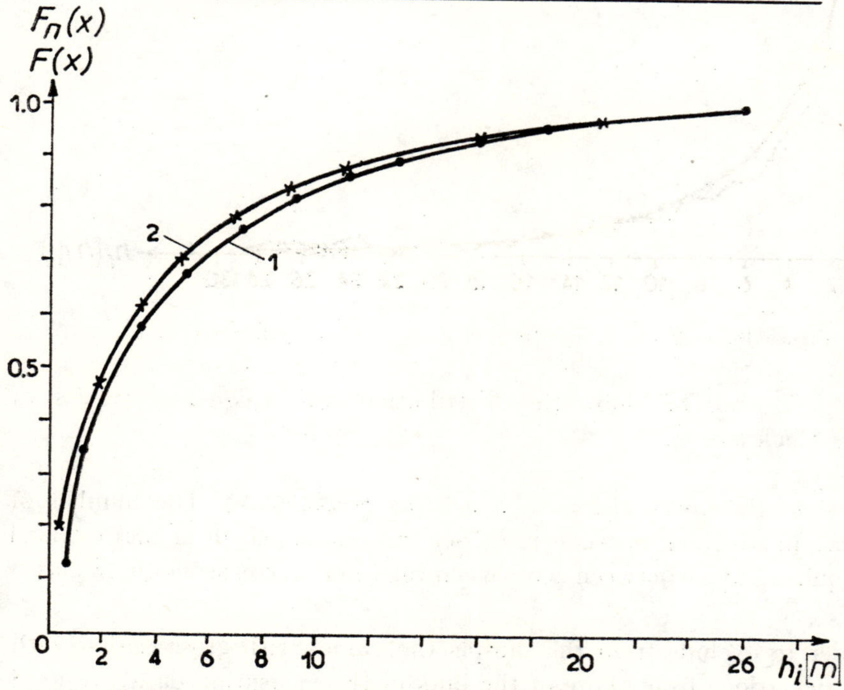

Figure 17. Polygon of the relative frequencies (*1*) and distribution curve (*2*) for parameter h_i.

The control of the correspondence between the empirical distribution and Weibull's model was made by Kolmogorov's test (Table 18). If equality $d_n = d_a = 0.073$ exists, the value of the significant level will be (see formula 44):

$$\lambda_q = d_n \sqrt{n} = 0.073 \sqrt{456} = 1.75; \quad q = 0.015$$

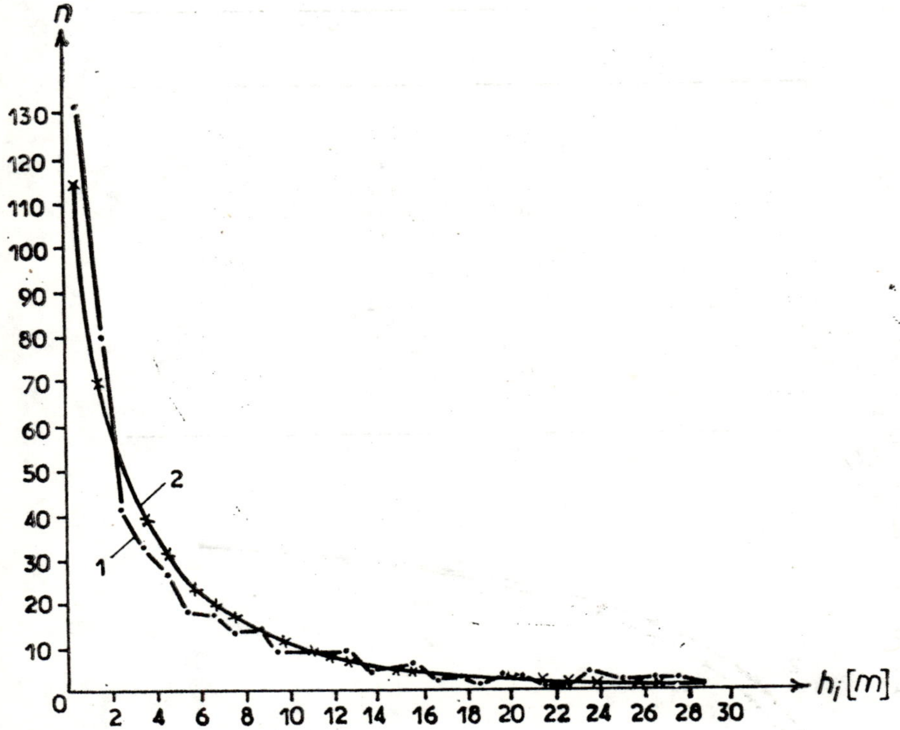

Figure 18. Polygon (*1*) and frequency curve (*2*) for parameter h_i.

Figures 17 and 18 show the distribution and frequency curves of elementary thickness h_i.

Number of the sandy bodies in a drilling borehole (N). The number of sandy bodies in a drilling borehole is a discrete variable which in coal interval II—III acquires values between zero (when the interval consists only in clays) and 12.

In order to exclude from the sample the values with gross errors (with important deviations from those of the bulk of the measuring data) we used the criterion supplied by relation (32) and found $x_{(n)P} = 12$; as a consequence all the values are included in the selection computation. Table 19 helped to compute:

$$N = \bar{x} = 2 + 0.5 = 2.5 \text{ (the mean number of sandy bodies in a drilling)}$$

$$S^2 = 5.38 - 0.5^2 - 5.1$$

$$S = \sqrt{5.1} = 2.3$$

$$V = \frac{2.3}{2.5} \cdot 100 = 92\,\%$$

Table 19

Number of sandy bodies x_i	Number of drillings n_i	$f_i(\%)$	α_i	$f_i\,\alpha_i^2$	$f_i\,\alpha_i^2$	$f_i\,\alpha_i^3$	$f_i\,\alpha_i^4$
0	22	13.8	− 2	−27.6	55.2	− 110.4	220.8
1	36	22.7	− 1	−22.7	22.7	− 22.7	22.7
2	44	27.7	0	0	0	0	0
3	24	15.1	+ 1	+15.1	15.1	+ 15.1	15.1
4	10	6.3	+ 2	+12.6	25.2	+ 50.4	100.8
5	7	4.4	+ 3	+13.2	39.6	+ 118.8	356.4
6	5	3.1	+ 4	+12.4	49.6	+ 198.4	793.6
7	4	2.5	+ 5	+12.5	62.5	+ 312.5	1 562.5
8	2	1.3	+ 6	+ 7.8	46.8	+ 280.8	1 684.8
9	1	0.6	+ 7	+ 4.2	29.4	+ 205.8	1 440.6
10	2	1.3	+ 8	+10.4	83.2	+ 665.6	5 324.8
11	1	0.6	+ 9	+ 5.4	48.6	+ 437.4	3 936.6
12	1	0.6	+10	+ 6.0	60.0	+ 600.0	6 000.0
	159	100.0		−50.3 +99.6	537.9 $V_2 = 5.38$	− 133.1 +2 884.8	21 458.7 $V_4 = 214.59$
				+49.3 $V_1 = +0.50$		+2 751.7 $V_3 = +27.52$	

So as to include this empirical distribution in Poisson's distribution it is necessary that the values of the mean, of the variance and of the centred moment of the third order (m_3) should be very near to one another; a great difference between the values of these characteristics leads to the rejection of this hypothesis. In this case, according to the values of $\bar{x}$ and S^2 it seems that the empirical distribution cannot be approximated by means of this theoretical distribution. For security's sake we applied Kolmogorov's criterion, by considering in addition to Poisson's distribution, the binomial one (Table 20).

In the case of Poisson's distribution, the probability function is:

$$P_x = \frac{\lambda^x}{x!} e^{-\lambda} \tag{55}$$

where parameter λ is equal to the variable mean and dispersion. Since these two characteristics estimated by means of the empirical moments have different values, we considered $\lambda = \bar{x} = 2.5$.

Table 20

x_i	f_i	$F_n(x)$	Poisson distribution		Binomial distribution		$d_n = \|F_n(x) - F'(x)\|$	$d_n = \|F_n(x) - F''(x)\|$
			P_x	$F'(x)$	$P_{n;x}$	$F''(x)$		
0	0.138	0.138	0.082	0.082	0.081	0.081	0.056	0.057
1	0.227	0.365	0.205	0.287	0.204	0.285	0.078	0.080
2	0.277	0.642	0.257	0.544	0.259	0.544	0.098	0.098
3	0.151	0.793	0.214	0.758	0.214	0.758	0.035	0.035
4	0.063	0.856	0.133	0.891	0.134	0.892	0.035	0.036
5	0.044	0.900	0.067	0.958	0.067	0.959	0.058	0.059
6	0.031	0.931	0.028	0.986	0.027	0.986	0.055	0.055
7	0.025	0.956	0.010	0.996	0.010	0.996	0.040	0.040
8	0.013	0.969	0.003	0.999	0.003	0.999	0.030	0.030
9	0.006	0.975	0.001	1.000	0.001	1.000	0.025	0.025
10	0.013	0.988	—	1.000	—	1.000	0.012	0.012
11	0.006	0.994	—	1.000	—	1.000	0.006	0.006
12	0.006	1.000	—	1.000	—	1.000	0	0

Poisson's distribution can be computed either with the help of the tables [62], or by the recurrence formula

$$P_{x+1} = P_x \cdot \frac{\lambda}{x+1} \tag{56}$$

By so doing, the computation is simplified; the first three terms are:

$$P_0 = \frac{2.5^\circ}{0!} e^{-2.5} = e^{-2.5} = 0.082$$

$$P_1 = 0.082 \frac{2.5}{1} = 0.205$$

$$P_2 = 0.205 \frac{2.5}{2} = 0.257$$

The probabilities of the binomial distribution are supplied by relation

$$P_{n;x} = C_n^x \cdot p^x \cdot q^{n-x} = \frac{n!}{x!\,(n-x)!} p^x \cdot q^{n-x} \tag{57}$$

The parameter of this distribution (p) is estimated with the help of the mean relation,

$$\bar{x} = np;\ p = \frac{\bar{x}}{n} = \frac{2.5}{159} = 0.0157$$

$$q = 1 - p = 0.9843$$

Owing to the computation difficulties in the case of high values of numbers n and x, the probabilities can be computed by recurrence formula

$$P_{n;x+1} = P_n(x) \cdot \frac{n-x}{x+1} \cdot \frac{p}{q} \qquad (58)$$

The first distribution terms of Table 20 were computed as follows:

$$P_{159;0} = \frac{159!}{0!(159-0)!} \cdot 0.0157^0 \times 0.9843^{159} = 0.9843^{159} = 0.081$$

$$P_{159;1} = 0.081 \cdot \frac{158}{1} \cdot \frac{0.0157}{0.9843} = 0.204$$

$$P_{159;2} = 0.204 \cdot \frac{157}{2} \cdot 0.01595 = 0.259$$

From the computations recorded in Table 20 there follows, for both the distributions, some interesting conclusions. Thus, the probabilities of the distributions possess practically the same values: this happens in the case of mean $np < 4$. Kolmogorov's criterion emphasized the possibility of approximating the empirical distribution by both the theoretical distributions taken as hypotheses since for $q = 0.05$, we obtain $d_n < d_a \left(d_a = \frac{1.36}{\sqrt{159}} = = 0.109\right)$. It is to be remarked here that according to the values of the sample mean and of the variance, the hypothesis of Poisson's distribution should have been rejected, which would not have been right.

In mathematical statistics [27] it is remarked that the so-called method of the empirical moments (proposed by C. Pearson) is the first general method of point estimation of the unknown parameters. It is widely used in practice owing to the much simpler computing procedures as compared to the more efficient methods of estimation which sometimes involve intricate computations. For this reason it is recommended that the estimations obtained with the help of the empirical moments should be used as first approximations by means of which the more efficient estimations could be determined. These considerations are especially valid in the case of abnormal distributions.

In case of a binomial distribution, the variance is supplied by relation

$$\sigma^2 = npq = 159 \times 0.0157 \times 0.9843 = 2.46 \qquad (59)$$

This value confirms once more the hypothesis of Poisson's distribution which is characterized by equation $\bar{x} = \sigma^2$. In this case, the true value of the coefficient of variation is

$$\sigma = \sqrt{2.46} = 1.57$$

$$V_b = \frac{1.51}{2.50} \cdot 100 = 63\%$$

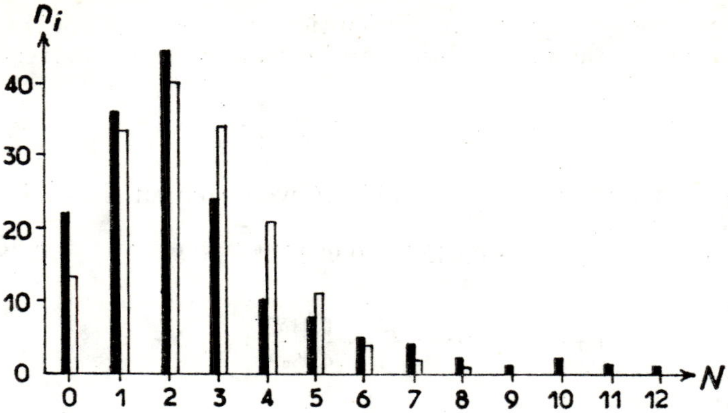

Figure 19. Distribution of parameter N: *black columns* = empirical distribution; *white columns* = binomial distribution.

A population of drillings, some of which meet with sandy bodies while others do not, can be considered as a statistical collectivity with qualitative indices; the dispersion can be used for the quantitative estimation of the macroheterogenity which is a most important element for appreciating the hydraulic connection between the aquifers. For this purpose, the population is considered to consist of two subpopulations, namely: the subpopulation of the drillings having opened sandy bodies (K) estimated at a unitary value and the subpopulation of the drillings that do not meet with sandy bodies (L) estimated at the zero value [16]. On this basis one can compute the arithmetic mean value:

$$\bar{x}_c = \frac{\Sigma x_i}{n} = \frac{1 \cdot n_1 + 0(n - n_1)}{n} = \frac{n_1}{n} = f \tag{60}$$

where n is the total number of drillings. Therefore the arithmetic mean of a population with qualitative indices is equal to the relative frequency of its index. The heterogeneity is expressed by the variance value (S_c^2). By developing the defining equation of the dispersion (S_c^2) we find

$$S_c^2 = \frac{\Sigma(x_i - \bar{x})^2}{n} = \frac{\left(1 - \frac{n_1}{n}\right)^2 n_1 + \left(0 - \frac{n_1}{n}\right)^2 (n - n_1)}{n} = \frac{n_1}{n}\left(1 - \frac{n_1}{n}\right) \tag{61}$$

Since $\left(1 - \frac{n_1}{n}\right)$ represents the frequency of the drillings which did not meet with sandy bodies (estimated to the zero value), it follows that the variance of the whole population is equal to the product of the frequencies of the two subpopulations possessed of opposite properties:

$$S_c^2 = f(1 - f) \tag{62}$$

The characteristic points of this function are:

— for $f = 1$ (when all the drillings have opened sandy bodies), $S^2 = 0$ whence the population is homogeneous;

— in interval $f = 0 - 0.5$, the function increases : $S^2 = 0 - 0.25$
— for $f > 0.5$, the function decreases : $S^2 < 0.25$.

To conclude, the heterogeneity of a population, whatever the qualitative indices may be, reaches a maximum where both the populations have the same number of terms. Two populations can only be compared from the standpoint of the heterogeneity (S^2) if in both cases terms with the same qualitative index are prevailing.

In the case of the lithological coal interval II—III:

$$\bar{x}_c = f = \frac{22}{155} = 0.14$$

$$S_c^2 = 0.14(1 - 0.14) = 0.12$$

For this value, a moderate heterogeneity of the distribution of both the subpopulations can be seen.

1.1.3 Lithological interval between coal layers I and II

In the preceding paragraph described in detail is the definition of theoretical distributions, the application of the tests of hypothesis and the computation methods for estimating the parameters of distributions and we shall now present only those elements which have something new to offer from the methodology standpoint. In fact, the main numerical characteristics of the distributions for all the lithological intervals of the carboniferous complex are recorded in Table 26.

The sampling distribution of the cumulated thickness of the sandy bodies (Σh_i) is adjusted with the probabilistic model of the Weibull distribution, as characterized by parameters $\eta = 1.4$ and $\sigma = 11.65$. The form of the frequency polygon (Figure 20) shows a complicated form of the empirical distribution the concordance of which with Weibull's law was verified by means of Kolmogorov's criterion.

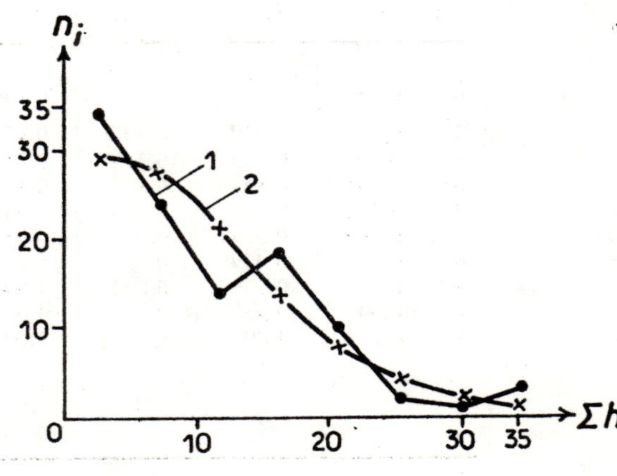

Figure 20. Polygon (*1*) and frequency curve (*2*) for parameter Σh_i in interval between I—II coal layers.

The population corresponding to the number of sandy bodies in a drilling (N) presents a rather peculiar situation due to the heterogeneity of the frequency distribution. From the total number of 15 intervals, only 6 of them have $N = 0$, which means that some of the drillings went only through marl-clays. Having rejected the interval with $N = 14$ (after verifying by relation 32), the following point estimation ensued:

$$\bar{x} = 1.77;\ S^2 = 3.03;\ S = 1.74;\ V = 98.3\%$$

The hypothesis of a concordance between the empirical distribution and the binomial one was verified with the help of Kolmogorov's criterion; for $q = 0.05$, we obtained the inequality $d_n < d_a$ $\left(\text{where } d_a = \frac{1.36}{\sqrt{142}} = 0.114\right)$ (Table 21). The following values

$$p = \frac{1.77}{142} = 0.0124$$

$$q = 1 - 0.0124 = 0.9876$$

$$\sigma^2 = \sqrt{142 \times 0.0124 \times 0.9876} = 1.74$$

$$\sigma = 1.32$$

$$V_b = \frac{1.32}{1.77} \cdot 100 = 74.5\%$$

are specific to the binomial distribution.

The probabilities of the binomial distribution (Table 22) were computed by relations (57, 58). The equality between the values of the mean and of the variance (σ^2) emphasizes the fact that this empirical distribution can also be adjusted by the Poisson's distribution.

Table 21

N (x_i)	Empirical distribution		Binomial distribution			$d_n = \lvert F_n(x) - F'(x) \rvert$
	f_i	$F_n(x)$	$P_{n;x}$	$F'(x)$	n'_i	
0	0.246	0.246	0.166	0.166	24	0.080
1	0.303	0.549	0.307	0.473	44	0.076
2	0.176	0.725	0.268	0.741	38	0.016
3	0.134	0.859	0.157	0.898	22	0.039
4	0.071	0.930	0.069	0.967	10	0.037
5	0.049	0.979	0.024	0.991	3	0.012
6	0	0.979	0.007	0.998	1	0.019
7	0	0.979	0.002	1.000	0	0.021
8	0.007	0.986	0	1.000		0.014
9	0.014	1.000	0	1.000		0
					142	

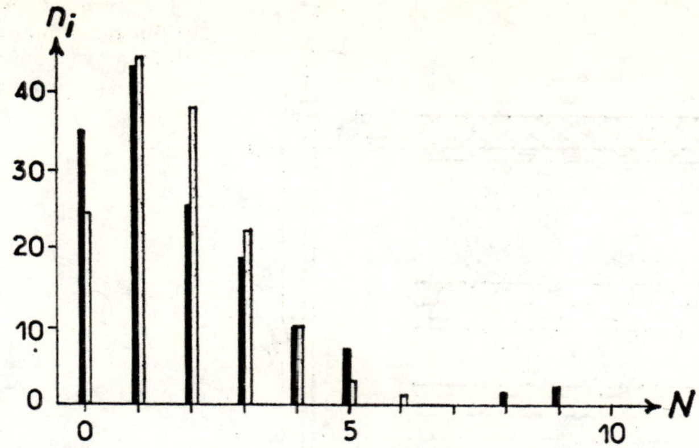

Figure 21. Distribution of parameter N (coal I—II interval): *black columns* = empirical distribution; *white columns* = = binomial distribution.

By admitting that the whole population consists of two qualitative subpopulations, the variance value (62) will be

$$S_c^2 = \frac{35}{142}\left(1 - \frac{35}{142}\right) = 0.185$$

At this value, an important heterogeneity can appear in the distribution of the two subpopulations.

1.1.4 The sandy horizon in the bottom of coal layer III

This interval is prevailingly sandy: in many drillings (31 from the total of 65) the marls and clays are missing. The differentiation of their interval (within the exploration area) and its consideration as a lithological population is imposed by the fact that it is separated from the rest of the early Dacian by a continuous marly horizon at least 10 m thick (Figure 22). Below this horizon, the development of the sands is subordinated to the marl-clays.

The lithological coefficient (C) (Table 26) presents a peculiarity; it has finite values for 34 wells while for the remaining 31 the values are infinite, owing to the fact that the horizon consists exclusively of sands. This being the case we can take into consideration a macroheterogeneity of the horizon assumed as a statistical collectivity with qualitative indices, the dispersion of which is

$$S_c^2 = f(1 - f) = \frac{31}{65}\left(1 - \frac{31}{65}\right) = 0.25$$

From this it follows that the non-homogeneity here reaches its highest value.

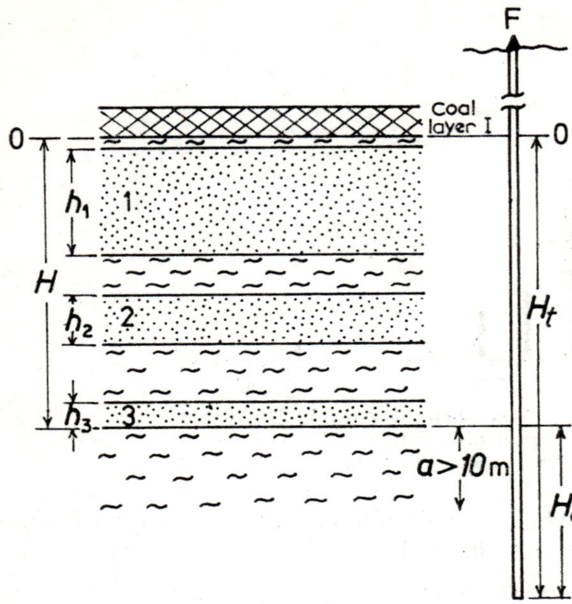

Figure 22. Scheme of the lithological conditions of the sandy horizon in the bottom of coal layer I.

1.1.5 The Dacian interval at the top of coal layer III

In certain zones, this lithological interval was not affected by erosion so that it was possible to consider 48 wells (Table 26).

The sampling distribution of parameter N (number of sandy bodies in a well) is characterized by a great amplitude, by small values of the intervals and by the existence of two peaks. Since the value of the second peak ($N = 15$) is nearing ($n=3$) those of the adjacent intervals, the distribution cannot be considered as binomial. The non-homogeneity of the lithological sequence results from the high mean value $\bar{x} = 8.5$. Since this is a discrete variable, attempts were made to make it tally with both the Poisson's distribution and the binomial one (Table 22), by using Kolmogorov's criterion for verifying the concordance. The probabilities of the binomial distribution were computed by recurrence formula (58), starting with the initial term

$$P_{45}^{(0)} = q^n = 0.81111^{45} = 0.00004592$$

$$p = \frac{8.5}{45} = 0.18889$$

$$q = 1 - 0.18889 = 0.81111$$

In the case of Poisson's distribution, the absolute value of the greatest difference ($d_n = 0.185$) is smaller than quantity $d_a = \frac{\lambda_q}{\sqrt{n}}$, at a significant

x_i	$F_n(x)$	Poisson distribution			Binomial distribution		$d_n = \lvert F_n(x) - F'(x) \rvert$	$d_n = \lvert F_n(x) - F''(x) \rvert$
		P_x	$F'(x)$	n_i	$P_{n;x}$	$F''(x)$		
0	0	0	0		0	0	0	
1	0.022	0.002	0.002	0.1	0.001	0.001	0.020	0.021
2	0.067	0.007	0.009	0.3	0.004	0.005	0.058	0.062
3	0.156	0.021	0.030	0.9	0.014	0.019	0.126	0.137
4	0.245	0.044	0.074	2.0	0.035	0.054	0.171	0.191
5	0.334	0.075	0.149	3.4	0.067	0.121	0.185	0.213
6	0.423	0.107	0.256	4.8	0.105	0.226	0.167	0.197
7	0.490	0.129	0.385	5.8	0.136	0.362	0.105	0.128
8	0.557	0.138	0.523	6.2	0.152	0.514	0.034	0.043
9	0.624	0.130	0.653	5.9	0.145	0.659	0.029	0.035
10	0.668	0.110	0.763	5.0	0.121	0.780	0.095	0.112
11	0.712	0.085	0.848	3.8	0.090	0.870	0.136	0.158
12	0.779	0.060	0.908	2.7	0.060	0.930	0.129	0.151
13	0.823	0.040	0.948	1.8	0.036	0.966	0.125	0.143
14	0.867	0.024	0.972	1.1	0.018	0.984	0.105	0.117
15	0.911	0.014	0.986	0.6	0.009	0.993	0.075	0.082
16	0.933	0.007	0.993	0.3	0.004	0.997	0.060	0.064
17	0.933	0.004	0.997	0.2	0.002	0.999	0.064	0.066
18	0.978	0.002	0.999	0.1	0.001	1.000	0.021	0.022
19	0.978	0.001	1.000	0	0	1.000	0.022	0.022
20	1.000	0	1.000		0	1.000	0	0
				45.0				

level $q = 0.05$, the value of it is $d_a = \frac{1.36}{\sqrt{45}} = 0.217$. In the case of the binomial distribution, this condition is fulfilled for $q = 0.03 \left(0.213 < \frac{1.45}{\sqrt{45}} = 0.217\right)$. Consequently Poisson's distribution is admitted. From Table 22 it follows that for both the distributions the value of the modulus is the same

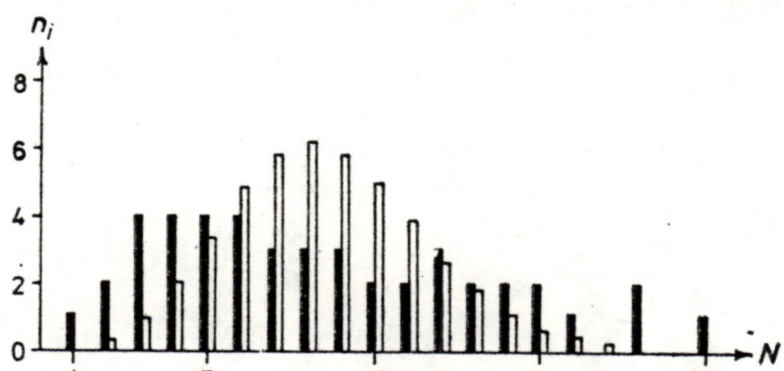

Figure 23. Distribution of parameter N in the rocks above coal layer III: *black columns* = empirical distribution; *white columns* = Poisson distribution.

($Mo = 8$); it can also be determined by means of the inequalities specific to each distribution:

— for Poisson's distribution

$$\lambda - 1 < Mo < \lambda; \qquad 7.5 < Mo < 8.5; \qquad Mo = 8$$

— for the binomial distribution

$$np - q < Mo < np + p; \qquad 7.69 < Mo < 8.69; \qquad Mo = 8$$

In the case of Poisson's distribution the estimated dispersion is

$$\sigma^2 = \lambda = 8.5; \qquad \sigma = \sqrt{8.5} = 2.9; \qquad V_p = \frac{2.9}{8.5} 100 = 34.1\%$$

1.1.6 The Levantine lithological complex

Among the rocks forming this complex, sands and gravels are permeable. The complex being to a great extent eroded it was only possible to ascertain the lithological coefficient C and the thickness of the elementary sandy bodies h_i the distributions of which were approximated as lognormal and respectively unilateral of the Weibull type (Table 26).

1.1.7 Coal layer II

As we noticed in § 1.1.1 the lignite layers are of hydrogeological importance because their thickness, structure and general distribution can determine the hydraulic isolation or connection of the aquifers along the vertical. For this reason the statistical processing was applied to the total thickness of the lignite layer II (including the small marly-clayey intercalations).

On the basis of the data given in Table 23, the following point estimator values were obtained (see relation 3—12):

$$\bar{x} = 5.30 \text{ m}; \; S^2 = 5.8 \text{ m}^2; \; S = 2.41 \text{ m}$$

$$V = \frac{2.41}{5.30} \cdot 100 = 45.5\%; \; m_3 = 5.47 \text{ m}^3; \; \alpha = \frac{5.47}{2.41^3} = 0.39$$

$$m_4 = 96.8 \text{ m}^4; \; \beta = \frac{96.8}{2.41^4} = 2.88; \; E = 2.88 - 3 = -0.12$$

Consequently, according to the observed data, the distribution of the total thickness of coal layer II is normal, with a small positive asymmetry and a slight flattening of the frequency curve. The method of the empirical moments supplies efficient estimations of the parameters defining the law of normal distribution only when the latter is practically symmetrical

Table 23

$a - b$	$\frac{a+b}{2}$	n_i	f_i	α_i	$f_i\alpha_i$	$f_i\alpha_i^2$	$f_i\alpha_i^3$	$f_i\alpha_i^4$
0.3— 1.8	1.05	12	7.8	−3	−23.4	70.2	−210.6	631.8
1.8— 3.3	2.55	18	11.7	−2	−23.4	46.8	− 93.6	187.2
3.3— 4.8	4.05	39	25.2	−1	−25.2	25.2	− 25.2	25.2
4.8— 6.3	5.55	38	24.7	0	0	0	0	0
6.3— 7.8	7.05	24	15.6	+1	+15.6	15.6	+ 15.6	15.6
7.8— 9.3	8.55	13	8.5	+2	+17.0	34.0	+ 68.0	136.0
9.3—10.8	10.05	6	3.9	+3	+11.7	35.1	+105.3	315.9
10.8—12.3	11.55	4	2.6	+4	+10.4	41.6	166.4	665.6
		154	100.0		−72.0 +54.7 −17.3 $V_1 =$ −0.17	268.5 $V_2 = 2.68$	−329.4 +355.3 + 25.9 $V_3 =$ +0.26	1 977.3 $V_4 = 19.77$

($\alpha = 0$—0.01) while the excess value tends to zero. In the contrary case, the mean and variance values (with all the subsequent implications) are affected by errors. When this is the case, the method of graphical estimations is recommended. This method can be applied to any distribution containing two unknown parameters.

The graphical method of determining the normal distribution (which will also be applied to lignite layer II) solves the following problems:

1. It verifies the normality of the empirical distribution.
2. It estimates parameters $\bar{x}$ and σ.
3. It establishes the adjusted theoretical frequencies.

When the empirical distribution is normal, the pair of values (x_i, z_i)[1)] considered as the co-ordinates of a point in plane Oxz (Figure 25), satisfies the straight-line equation deduced as follows:

$$z_i = \frac{x_i - \bar{x}}{\sigma} \tag{63}$$

$$z = \frac{1}{\sigma} x + \left(-\frac{\bar{x}}{\sigma}\right) \tag{64}$$

When the experimental points are collinear the adjusted straight-line equation becomes

$$z = a + bx \tag{65}$$

where a and b are determined coefficients. By comparing equations (64) and (65), we obtain

$$\sigma = \frac{1}{b}; \quad \bar{x} = -\frac{a}{b} \tag{66}$$

[1)] z_i represents the standard normal variable.

The computation of the variables z_i (Table 24) is made with the help of the distribution function and of the Laplace function:

$$F_n(x_i) = F(z_i) = 0.5 + \Phi(z_i) \tag{67}$$

$$\Phi(z_i) = F_n(x_i) — 0.5 \tag{68}$$

Knowing the value of Laplace function Φz_i, the variable value z_i is determined with the help of the table. There results from Figure 25 *a* "by

Table 24

x_i	n_i	Σn_i	$F_n(x_i)$	$\Phi(z_i)$	z_i	x_i^2	$x_i z_i$
1.05	12	12	0.07792	−0.42208	−1.420	1.10	−1.491
2.55	18	30	0.19480	−0.30520	−0.860	6.50	−2.193
4.05	39	69	0.44805	−0.05130	−0.123	16.40	−0.498
5.55	38	107	0.69481	0.19481	0.510	30.80	2.830
7.05	24	131	0.85065	0.35065	1.040	49.70	7.332
8.55	13	144	0.93507	0.43507	1.515	73.10	12.953
10.05	6	150	0.97402	0.47402	1.947	101.00	19.567
11.55	4	154	1.00000	0.50000	5.000	133.40	55.750
38.85 *)	154			Σ *)	+2.609	278.60	40.50

*) The sum is computed without the terms corresponding to $x_i = 11.55$

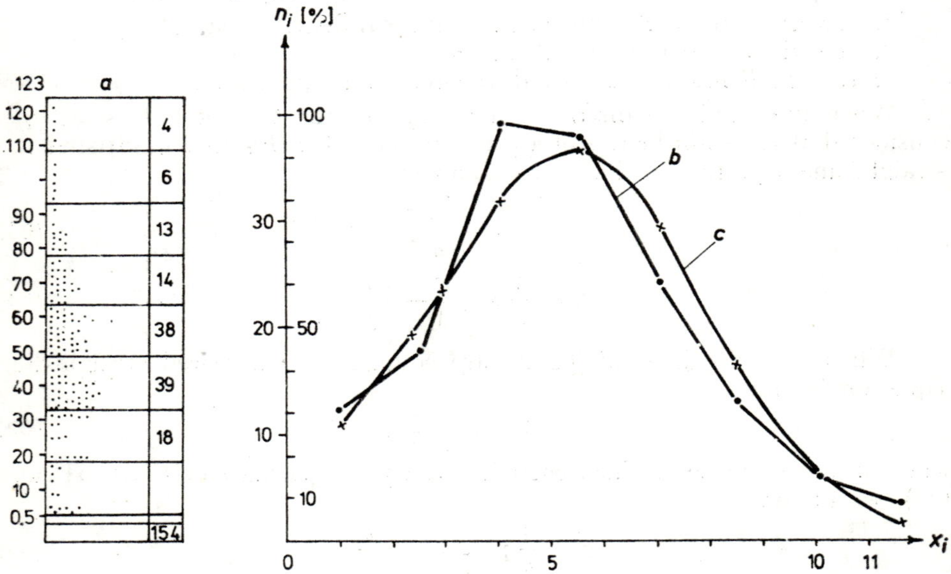

Fig. 24. Dispersion graph (*a*), polygon (*b*) and frequency curve (*c*) for the total thickness of coal layer II.

eye" verification of the distribution normality (the extreme interval $x_i = 11.55$ m was not taken into account).

So as to obtain an efficient estimation of the equation (65) parameters, the method of least squares was applied and the two following normal equations were found (Table 24):

$$Na + b\Sigma x_i = \Sigma z_i \tag{69}$$

$$a\Sigma x_i + b\Sigma x_i^2 = \Sigma x_i z_i$$

Without taking into account the extreme interval (because it cannot be included within the linear character of the distribution) the results were $a = -1.90$ and $b = 0.41$, then

$$z = -1.90 + 0.41\ x \tag{70}$$

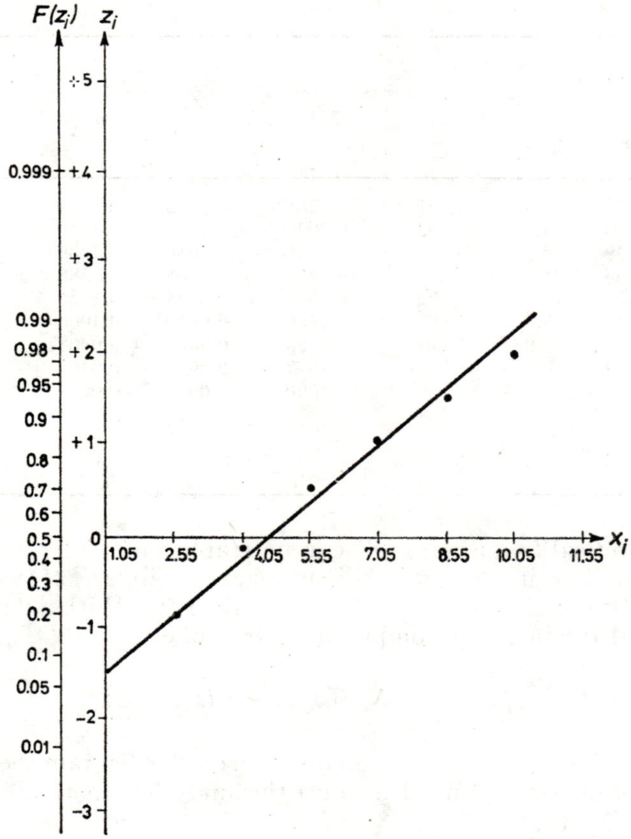

Figure 25. Normality verification (total thickness of coal layer II).

This equation permitted us to draw the straight line in Figure 25. The values of the distribution parameters are therefore:

$$\bar{x} = -\frac{-1.90}{0.41} = 4.63 \text{ m}$$

$$\sigma = \frac{1}{0.41} = 2.44 \text{ m}$$

$$V = \frac{2.44}{4.63} \cdot 100 = 52.7\%$$

Let us note that by estimating the mean with the help of the empirical moments an error $e = +15.1\%$ is introduced with regard to the efficient estimation $\bar{x} = 4.63$ m.

Table 25

x_i	Empirical frequencies		z_i'	$\Phi(z_i')$	$F(z_i')$	$F(z_{i-1}') - F(z_i')$	Theoretical frequencies	
	n_i	$f_i(\%)$					n_i'	$f_i'[\%]$
1.05	12	7.8	−1.467	−0.4288	0.0712	0.0712	11.0	7.1
2.55	18	11.7	−0.852	−0.3029	0.1971	0.1259	19.4	12.6
4.05	39	25.2	−0.238	−0.0920	0.4080	0.2109	32.5	21.2
5.55	38	24.7	+0.377	+0.1468	0.6468	0.2388	36.8	23.8
7.05	24	15.6	+0.992	+0.3394	0.8394	0.1926	29.7	19.3
8.55	13	8.5	+1.607	+0.4460	0.9460	0.1066	16.4	10.7
10.05	6	3.9	+2.221	+0.4868	0.9868	0.0408	6.3	4.1
11.55	4	2.6	+2.836	+0.4977	0.9977	0.0109	1.7	1.1
$+\infty$				+0.5000	1.000	0.0023	0.2	0.1
	154	100.0					154.0	100.0

The theoretical frequencies (n_i') can be graphically determined by means of the straight line in the probabilistic diagram (Figure 25) which, together with the axis Oz has a standardized normal scale $[F(z)]$. Within a given interval, the theoretical frequency is computed by formula

$$n_i' = N[F(x_{i-1}) - F(x_i)] \tag{71}$$

Because value of the distribution function on the diagram is only computed approximately, it is recommended to use the analytical method instead, where the normalized value is

$$z_i' = \frac{x_i - 4.63}{2.44}$$

The computation of the theoretical frequencies was achieved by means of Table 25, while the polygon and the frequency curve are represented by Figure 24.

1.2 The degree of certainty of point statistic estimations

All the numerical characteristics of the empirical and theoretical distributions previously analysed represent, as a matter of fact, point estimations with all the properties of the random variables. This constitutes an approximate method of supplying indications concerning the values of parameters behaving as random variables. These point estimations acquire a well-defined meaning when accompanied by the probability appreciation of the estimated parameter deviations from its true values. Thus, as a probabilistic measure of the certainty of a punctual estimation the *confidence interval* is used.

When the hydrogeological exploration is made by drillings, the parameter values determined by core-samples are affected, not only by errors of method and systematic errors but by analogical ones as well, which have a random character. These errors are due to the fact that the parameters of a geological structure cannot be determined continuously but only at isolated points. This is the reason for differences between the natural geological model and the conventional one. The problem is in estimating the characteristics of the target population on the basis of the point estimates of the sampled population. Owing to its random nature, the problem is solved by determining a probability P such that interval $[A, B]$ contains the point estimate of the considered parameter. A probability P such that the confidence interval could contain the unknown parameter value of the general population is called the *confidence probability*. Value $q = 1 - P$ is the significant level of the proposed estimation and represents the parameter probability of being outside the confidence limits (the critical interval).

When using the method of confidence intervals, the estimate of the mean target population, in the case of normal distribution, is supplied by relation

$$a = \bar{x} \pm t_q \frac{\sigma}{\sqrt{n}} \tag{72}$$

where the criterion t_q is, in fact, a reduced variable

$$t_q = \frac{\bar{x} - a}{\sigma_{\bar{x}}} = \frac{\bar{x} - a}{\frac{\sigma}{\sqrt{n}}} \tag{73}$$

where $\sigma_{\bar{x}}$ is the standard error of the mean distribution. The value of criterion t_q (also called Student-Fisher's variable) is to be found in tables [see ref. 54], as a function of the significant level and the degrees of freedom f. It follows from relations (72), (73) that the interval length is $I = 2\, t_q\, \sigma_x$. The smaller this interval and the larger (very near to unity) the probability P, the higher the degree of certainty of the respective point estimate will be. Among all the parameters involved in the computation of the confidence intervals, a keener attention should be paid to criterion t_q the value of which depends, in the case of large sample sizes, only on the value of the chosen significant level. A higher value of criterion t_q would result in a greater probability of containing random errors, and therefore a greater certainty in determining the values of these errors; on the other hand a lenthening occurs of interval I, which influences negatively the degree certainty of the point estimate.

The value of the significance level q (and implicitly of t_q) is chosen as a function of the activity area and of the precision required. In the case of industrial research, $q \leqslant 0.05$. In the case of mining exploration, R. Murard [49], proposes for q the following values:

— for certain reserves $\quad q \leqslant 10\%(P \geqslant 90\%)$

— for probable reserves $\quad q \leqslant 30\%(P \geqslant 70\%)$

— for possible reserves $\quad q \leqslant 70\%(P \geqslant 30\%)$

In the same problem, P. Ryjov [55] recommends for criterion t_q the following values:

— In the case of deposits where category A reserves are to be estimated $t_q = 1.7$ (rounded off) should be taken, which corresponds to $P = 90\%$ (for large volume selections);

— in the case of category B reserves, $t_q = 1.2$ must be taken which corresponds to $P = 75\%$.

Taking these values into account, it is recommended, for lignite reserves of category B in hydrogeological explorations, to consider $q = 0.20$, respectively $t_q = 1.25$. This significant level can be used for any detailed hydrogeological research (water supplies, irrigation, drainages, mineral waters etc.), as well as for small volume selections.

For determining the certainty degree of the point values, the theory of estimation offers several relations for the computation of random errors; these relations derive from the method of confidence intervals.

Further on, we shall give a minute analysis of the formulae of confidence intervals (implicitly, the problem of errors too) for the estimation of the normal distribution parameters. As a rule, these formulae are applied equally to non-normal distributions, when the population size volume is large (theoretically $n > 30$) since according to the law of great numbers the sample mean and variance tend to their theoretical values.

The value of the mean random error in a normal distribution is expressed by

$$\varepsilon_{\bar{x}} = \pm t_q \frac{\sigma}{\sqrt{n}} \tag{74}$$

As a matter of fact, this absolute statistical error determines the value of the confidence interval; the smaller this interval, the larger the determination precision of the respective parameter grows.

The relative error (or the index of precision) is supplied by relation

$$\varepsilon'_{\bar{x}} = \pm \frac{\varepsilon_{\bar{x}}}{\bar{x}} = \pm t_q \frac{\sigma}{\bar{x}\sqrt{n}} = \pm t_q \frac{V}{\sqrt{n}} \qquad (75)$$

This relation shows that $\varepsilon'_{\bar{x}}$ depends linearly on the coefficient of variation. In industrial research, the index of precision is limited to $\varepsilon'_{\bar{x}} \leqslant 5\%$. Taking into account the specific nature of geological research, it can be admitted that, for a hydrogeological exploration, $\varepsilon'_{\bar{x}} \leqslant 10\%$ is a good value. For instance in the case of the total thickness H of the coal II—III lithological interval (§ 1.1.2.1):

$$\varepsilon_{\bar{x}} = \pm\, 1.28 \frac{11.6}{\sqrt{160}} = \pm\, 1.2 \text{ m}$$

$$\varepsilon'_{\bar{x}} = \pm\, 1.28 \frac{33.5}{\sqrt{160}} = \pm\, 3.4\%$$

Therefore, the confidence limits are

$$a_1 = 34.6 - 1.2 = 33.4 \text{ m}; \quad a_2 = 34.6 + 1.2 = 35.8 \text{ m}.$$

If $\varepsilon'_{\bar{x}} = 10\%$, the significance level q (and respectively the confidence probability P) can be determined for the admissible value $\varepsilon'_{\bar{x}} = 10\%$,

$$t_q = \frac{10\sqrt{n}}{V(\%)} \qquad (76)$$

If $\varepsilon'_{\bar{x}} > 10\%$, it will be necessary to compute the minimum size of the sampled population corresponding to $\varepsilon'_{\bar{x}} = 10\%$ and $t_q = 1.28$ by means of the relation

$$n = t_q^2 \frac{V^2}{\varepsilon'^2_{\bar{x}}} = 1.28^2 \frac{V^2}{0.10^2} = 164\, V^2 \qquad (77)$$

The confidence interval of the sample standard deviation S can be computed with the help of the standard error σ_s which, in the case of a normal distribution, is expressed by

$$\sigma_s = \frac{S}{\sqrt{2n}} = 0.707\, \sigma_{\bar{x}} \qquad (78)$$

For parameter H of coal interval II—III, we shall have

$$\varepsilon_s = \pm t_q \frac{S}{\sqrt{2n}} = \pm 1.28 \frac{11.6}{\sqrt{2 \times 160}} = \pm 0.83 \text{ m}^2$$

The relative error of the root mean square deviation is

$$\varepsilon_s' = \pm \frac{\varepsilon_s}{S} \cdot 100 = \pm \frac{0.83}{11.6} \cdot 100 = \pm 7.2\%$$

The confidence limits are

$$\sigma_1 = 11.6 - 0.8 = 10.8 \text{ m}^2; \qquad \sigma_2 = 11.6 + 0.8 = 12.4 \text{ m}^2$$

To conclude, according to this parameter (σ), when the confidence probability $P = 80\%$, the minimum volume of the selection is exceeded.

The standard error of coefficient V of variation for a normal collectivity is supplied by relation

$$\sigma_v = \frac{V}{\sqrt{2n}} \sqrt{1 + \frac{2V^2}{10^4}}\,^{*)} \approx \frac{V}{\sqrt{2n}} \tag{79}$$

For parameter H of coal interval II—III, we have

$$\varepsilon_v = \pm t_q \frac{V}{\sqrt{2n}} = \pm 1.28 \frac{0.335}{\sqrt{2 \times 160}} = \pm 0.024$$

$$\varepsilon_v' = \frac{\varepsilon_v}{V} 100 = \pm \frac{0.024}{0.335} \cdot 100 = \pm 7.2\%$$

$$V_1 = 0.335 - 0.024 = 0.311; \quad V_2 = 0.335 + 0.024 = 0.359$$

Since the relative error is $\varepsilon_v' < 10\%$, it follows that the actual sample size exceeds the minimum size corresponding to probability $P = 80\%$.

In practice, the minimum sample size is usually evaluated on the basis of the mean value only; to this end, the estimation precision (the relative error) of the mean values should be computed. If the relative error exceeds the admissible value, we shall have to compute the minimum volume of the selection corresponding to the admissible values of the relative error and of criterion t_q. By this method, the degree of certainty of the mean values was

*) The value of $\frac{2V^2}{10^4}$ is practically negligible.

computed for all the parameters of the lithological intervals, for $\varepsilon'_{\bar{x}} = 10\%$ and $t_q = 1.28$. It follows from these computations (Table 26) that the relative errors are comprised within the limits 2.4—13.2%, the admissible value of 10% only being exceeded in 7 out of the 25 cases. Thence, the conclusion that most of the estimate mean values are significant, owing to the corresponding size of the samples.

A particular situation is that of parameter Σh_i (coal interval II—III) which, owing to the high variability and to the lognormal character of the distribution, has a smaller sample size than the minimum one corresponding to $P = 80\%$.

For coal layer II, though only the total thickness (the lignite banks + the included sterile intercalations between them) are considered, it is to be noted that the confidence probability has a high value which means a high degree of precision of the reserve estimate.

If the samples are of small size ($n < 30$), the value of the confidence intervals of the statistical parameters will be determined by formulae specific to the respective distributions [37].

The computed values of the random errors are only illustrative when the technical ones are of acceptable values. In case of the sandy-argillaceous lithological sequencies with variable lithofacies (as in the case of the Șotînga-Margineanca lignite deposit) it can be admitted that a mechanical coring is admissible from the hydrogeological standpoint as long as the measuring error does not exceed 10%, while the recovery ratio is at least 80%. Otherwise, the analogy error is affected by the technical one. To estimate the technical errors, the arithmetic means of the recovery ratio (R) of the sandy bodies from the drillings made in Șotînga and Cărbunaru fields were computed; it was found that the drillings with a good recovery ratio were those in series 4000 and the hydrogeological ones, while the drillings with a very variable recovery ratio belong to series 1000—6000—12000 (Table 27).

From Table 27, it follows that, in the drillings in series 1000—6000—12000, gross technical errors appear, which obviously influence the values of the random errors. The practical method of considering that the sections devoid of core-samples are sands is based on the fact that the lignite fields are of two principal lithological types: clay-marls which, owing to their cohesion, are not lost from the core-sample and sands, which can be lost. This method is not, however, correct since the variation of the lithofacies is important (many varieties of sands and clay-marls appear), while most of the sandy bodies are thinner (less than 2 m) than the lengths of unrecoverable sections. In this manner, it is possible to find a large unrecoverable section that could be considered as sand, though as a matter of fact it could equally include clay-marl banks, the final results being an artificial enrichment of the lithological interval with the sands.

Values of the statistical

Lithological interval and parameters	Variation interval $(a-b)$	Sample size (n)	Sample mean $\bar{x}$	Variance (σ^2)	Standard deviation σ	Theoretical limits of the sampling $X_{min} - X_{max}$	Measurements with gross errors
Levantine lithological complex							
Lithological coefficient (C)	0.10–0.76	32	0.35	0.03	0.17	–	0
Thickness of the elementary sandy bodies (h_i)	0.20–35.0 m	291	3.9 m	19.0 m²	4.35 m	X_{max} = 24.8 m	3
Dacian interval in the top of coal layer III							
Total thickness of interval (H)	50–270 m	48	116.3 m	2 314 m²	48.2 m	–	0
Cumulative thickness of the sandy bodies (Σh_i)	1–126 m	48	41.6 m	576.7 m²	24.0 m	–	0
Centre of gravity of the sandy bodies $(y_0\%)$	9–70.0	48	44.0	168.0	13.0	–	0
Lithological coefficient (C)	0.01–1.76	142	0.51	0.14	0.37	–	0
Thickness of the elementary sandy bodies (h_i)	0.1–37.0	1 462	3.0	17.6	4.2	X_{max} = 21.3 m	22
Number of sandy bodies per well (N)	1–20	45	8.5	8.5	2.9	–	0
Lithological interval between II–III coal layers							
Total thickness of interval (H)	10–74 m	160	34.6 m	135 m	11.6 m²	X_{max} = 69.4 m	1
Cumulated thickness of the sandy bodies (Σh_i)	0.16–40.0 m	137	13.3 m	258.3 m²	16.1 m	0.5–162 m	3
Centre of gravity of the sandy bodies $(y_0\%)$	10–82	136	45.9	206	14.4	–	0
Lithological coefficient C.	0–3.3	141	0.6	0.41	0.64	X_{max} = 3.01	1
Tickness of the elementary sandy bodies (h_i)	0.2–32.0 m	457	4.7 m	45.0 m²	6.7 m	X_{max} = 29.0	1

parameters

Coefficient of variation (V %)	Entropy H_S	Entropy H_{max}	Entropy $100\,H_r$	Empirical distribution type	Lew of theoretical distribution	Precision of the mean determination: Analogy error $\pm\varepsilon_{\bar{x}}$	Relative error $\pm\varepsilon'_{\bar{x}}$ (%)	Confidence interval ($q = 0.20$)	For $\varepsilon'_{\bar{x}} = 10\%$: t_q	For $\varepsilon'_{\bar{x}} = 10\%$: $P\%$	Value of n for $\varepsilon'_{\bar{x}} = 10\%$ and $P = 80\%$
49.5	1.40	1.80	77.8	Unimodal	Lognormal	0.04	12.2	0.31–0.39	1.07	70.0	42
111.0	1.85	2.40	77.1	Amodal, unilateral	Weibull	0.38 m	9.1	3.52–4.28 m	1.41	83.5	–
41.4	1 .45	1.94	74.7	Unimodal, asymmetrical	Lognormal	8.8 m	7.6	107.5–125.1 m	1.69	90.0	–
57.8	1.44	1.94	74.2	Unimodal, symmetrical	Normal	4.4 m	10.6	37.2–46.0 m	1.21	77.0	55
29.5	1.45	1.94	74.7	Unimodal, symmetrical	Normal	2.4	5.5	41.6–46.4	2.37	98.2	–
72.5	1.68	2.08	80.8	Unimodal, asymmetrical	Lognormal	0.04	7.8	0.47–0.55	1.64	90.0	–
140.0	1.61	2.48	64.9	Amodal, unilateral	Weibull	0.14 m	4.7	2.86–3.14 m	2.73	99.2	–
34.1	2.38	3.00	79.3	Dimodal	Poisson	0.9	11.0	7.6–9.4	1.19	75.6	54
33.5	1.76	2.08	84.5	Unimodal with marked asymmetry,	Normal	1.2 m	3.4	33.4–35.8 m	3.8	>99.9	–
121.0	1.76	2.08	84.5	Unimodal, asymmetrical	Lognormal	1.76 m	13.2	11.54–15.06 m	0.97	66.6	239
31.4	1.83	2.08	88. 0	Unimodal, small asymmetry, flattened form	Normal	1.6	3.5	44.3–47.5	3.7	>99.9	–
96.0	1.73	2.40	72.1	Amodal, unilateral	Weibull	0.07	10.4	0.60–0.74	1.24	82.5	151
143.0	1.34	2.30	58.2	Amodal, unilateral	Weibull	0.40	8.5	4.22–5.09	1.49	85.7	–

Lithological interval and parameter	Variation interval $(a-b)$	Sample size (n)	Sample mean $\bar{x}$	Variance (σ^2)	Standard deviation σ	Theoretical limits of the sampling $X_{min} - X_{max}$	Measurements with gross errors
Number of sandy bodies per drilling (N)	0 – 12	159	2.5	2.5	1.6	—	0
Lithological interval between I – II coal layers							
Total thickness of interval (H)	12.5 – 48.5 m	141	28.25 m	37.9 m²	6.2 m	—	0
Cumulative thickness of the sandy bodies (Σh_i)	0.5 – 36.5 m	107	10.6 m	58.9 m²	7.7 m	—	0
Centre of gravity of the sandy bodies $(y_0\%)$	10 – 100	107	46.7	269.0	16.4	$X_{max} = 94.2$	1
Lithological coefficient (C)	0.04 – 6.0	107	0.76	0.48	0.70	$X_{max} = 4.5$	2
Thickness of the elementary sandy bodies (h_i)	0.2 – 340 m	287	4.1 m	24.3 m	4.9 m	$X_{max} = 26.0$ m	2
Number of sandy bodies per drilling (N)	0 – 14	143	1.8	1.7	1.3	$X_{max} = 10$	1
Sandy horizon at the bottom of coal layer I							
Total horizon thickness (H)	4.6 – 56.4	65	19.5	150.6 m²	12.3 m	—	0
Cumulative thickness of the sandy bodies (Σh_i)	1.5 – 47.0 m	65	15.7 m	74.4 m²	8.6 m	—	0
Thickness of the elementary sandy bodies (h_i)	0.1 – 27.0	180	8.6 m	33.6 m²	5.8 m	—	0
Number of sandy bodies per drilling (N)	1 – 5	65	1.66	1.27	1.13	—	0
Coal layer II							
Total thickeness	0.3 – 12.3 m	154	4.6	6.0 m²	2.4 m	—	0

Table 26 (continued)

Coefficient of variation (%)	Entropy			Empirical distribution type	Law of theoretical distribution	Precision of the mean determination					Value of n for $\varepsilon'_{\bar{x}} = 10\%$ and $P = 80\%$
									For $\varepsilon'_{\bar{x}} = 10\%$		
	H_S	H_{max}	$100\,H_r$			Analogy error $\pm\varepsilon_{\bar{x}}$	Relative error $\pm\varepsilon'_{\bar{x}}$ (%)	Confidence interval ($q = 0.20$)	t_q	P %	
60.0	1.97	2.48	79.3	Unimodal, very asymmetrical	Poisson	0.16	6.4	2.34–2.66	2.0	95.0	–
21.9	1.68	2.08	80.7	Unimodal, symmetrical	Normal	0.67 m	2.4	27.6–28.9 m	5.45	>99.9	–
72.6	1.65	2.08	73.3	Amodal (unilateral)	Weibull	0.95 m	9.0	9.65–11.55 m	1.42	83.9	–
35.1	1.78	2.19	81.3	Unimodal, small asymmetry sharp from	Normal	2.0	4.3	44.7–48.7	2.96	99.4	–
92.1	1.17	2.30	50.9	Amodal, unilateral	Weibull	0.09	11.5	0.67–0.85	1.11	73	139
119.5	1.84	2.30	80.0	Amodal, unilateral	Weibull	0.37	9.0	3.73–4.47	1.41	82.6	–
74.5	1.65	2.71	60.9	Unimodal, very asymmetrical	Binomial	0.14	7.8	1.66–1.94	1.61	89.0	–
63.0	1.46	1.94	75.3	Unimodal, positive asymmetry	Lognormal	2.0 m	10.0	17.5–21.5	1.28	80.0	–
54.8	1.42	1.94	73.2	Unimodal, positive asymmetry	Lognormal	1.4 m	8.8	14.3–17.1	1.46	86.0	–
67.5	1.88	2.19	85.8	Amodal, unilateral	Weibull	0.55	6.4	8.05–9.15	1.99	95.0	–
68.0	1.03	1.61	64.0	Amodal, unilateral	Binomial	0.18	10.1	1.48–1.84	1.58	76.0	75
52.7	1.80	2.08	86.5	Unimodal, small asymmetry	Normal	0.25	5.5	4.35–4.85	2.35	97.9	–

Table 27

Drilling series	Variation limits (R)	Integral recovery ratio of core samples	Mean value of the recovery ratio	General appreciation of the recovery ratio
4 000	18 – 100%	50.6%	88%	Good recovery
1 000 – 6 000 – 12 000	0 – 100%	10.0%	48%	Reduced and very variable recovery ratio
Hydrogeological	46 – 100%	42.8%	88%	Good recovery

Naturally, the estimation of the influence of the technical errors upon the analogy errors should in each case be made as a function of the lithological sequence variability, of the physico-mechanical properties of the traversed rocks, as well as of several technical elements of the drilling process.

1.3 Estimation of the lithological complexity by means of the coefficient of variation

When practicing explorations, it is recommended to use as a measure of the geological complexity (variability) the *coefficient of variation.* In the case being studied, in order to be able to compare the variability of the lithofacies of the five lithological intervals (Table 28), the total coefficient of variation can be used, defined as the square root of the sum of the squared partial coefficients. The squared values are used to eliminate the influence of mathematical sign. In establishing a total coefficient of variation, parameters with various physical meanings should not be included, because this would lead to incorrect values. It is, for instance, a mistake to take both dimensional and grain size parameters into account at the same time. Since the indices (N, Σh_i, y_0, C, N, h_i) used previously are dimensional parameters of the lithofacies, they can define the total coefficient of variation. Because for certain lithological intervals it was impossible to determine all these coefficients, only three of them (Table 28) were computed. Thus, for lithological coal interval I—II, we have:

$$V_{t_1} = \sqrt{V_H^2 + V_{\Sigma h_i}^2 + V_{y_0}^2 + V_C^2 + V_{h_i}^2 + V_N^2} \qquad (80)$$

$$V_{t_1} = \sqrt{21.9^2 + 72.6^2 + 35.1^2 + 92.1^2 + 119.5^2 + 74.5^2} = 188\,\%$$

From the comparison between the computed values, it follows that the greatest complexity is reached in the lithological coal interval II—III and that above as well as below it this complexity diminishes. The sandy horizon in the bottom of coal bed I as well as the Levantine complex differ considerably from the other intervals, the values of their total coefficients of variation being much samller.

Table 28

Values of the coefficients of variation

Parameters used in computations	Lithological interval				
	Below coal layer I	Coal I–II	Coal II–III	Above coal layer III	Levantine
$[V_{t1}]$: $H.\ \Sigma h_i,\ y_0,\ C\ h_i.\ N$	—	188	224	179	—
$[V_{t2}]$: $H,\ \Sigma h_i,\ h_i,\ N$	127	160	200	161	—
$[V_{t3}]$: $C,\ h_i$	—	152	172	157	122

Since the complexity of the lithofacies is an essential factor on which depend the methods and the volume of the hydrogeological exploration works, it is necessary to use certain classifications according to the types of hydrogeological structures. Thus, in structures concerning the Romanian lignite fields, the following classification could be used:

— relatively constant lithofacies at $V_t \leqslant 70\%$
— variable lithofacies at $70 < V_t \leqslant 120\%$
— most variable lithofacies at $V_t > 120\%$.

By making use of data supplied by several hydrogeological structures of the same type, it is possible to establish similar classifications, which will prove very helpful when establishing the methods and the volume of the hydrogeological exploration works. These limits vary equally as a function of the number of the lithological parameters which will be used in computations.

1.4 Entropy used as a measure of the lithological complexity

In geology several parameters are used, each characterizing a distinct quantitative feature of an object or a phenomenon. Generally, the complexity, the non-homogeneity or the variability of the natural systems or structures means we must find information processing methods which will also highlight the global qualitative feature. This improvement in the processing methods must always be accompanied by an increase in the degree of objectivity, certainty and comprehension of the information.

As we have seen, the estimation of the hydrogeological structure non-homogeneity can be made with the help of the variance (σ^2), of the standard deviation (σ) and of the coefficient of variation ($V\%$). The statistical parameters σ^2 and σ are only correct in the case of symmetrical distributions with a single peak; in the case of asymmetrical distributions, the direct interpretation of these parameters is meaningless. Another deficiency of the standard deviation consists in the fact that its dimensions are identical with those of the measured parameter, which excludes the possibility of using it for comparing the non-homogeneities of the various objects or dimensional parameters. The coefficient of variation, being dimensionless, eliminates this deficiency while remaining nevertheless an approximate statistical measure of the non-homogeneity, owing to its dependence on the mean value. For instance, two normal populations having unequal mean values but the same standard deviations, have coefficients of variations of different values, which does not correspond to the fact that in reality both these population have the same homogeneity. In fact, from the value standpoint, the coefficient of variation represents the fraction (as a percentage) of the standard deviation of the mean value.

In qualitative information processing, certain notions and methods of the information theory can be used successfully by means of which the differentiation appears as a positive phenomenon. Such is for instance, the *probabilistic* or *informative entropy* which represents the mean value of the information of a finite probability field and can therefore be considered as a measure of the indetermination of our knowledge about this field. Since the complexity of a finite field of probability is determined by the amount of information necessary to describe this field, it follows that the entropy can be used as a measure of the complexity and of the non-homogeneity. In the case of a discrete field, the entropy can be computed by Shannon's formula:

$$H = -\sum_{i}^{n} p_i \ln p_i \tag{81}$$

where p_i is the probability of event i, and $\Sigma p_i = 1$. The entropy avoids the deficiencies of statistical measures σ and V used for estimating the non-homogeneity since: its value and dimensions depend only on the applied logarithm base; it is not determined by the mean value; it characterizes the asymmetrical and complex forms of distributions. As a function of the applied logarithm base, the following units of entropy are known:

— base 2 logarithm, $\log_2 2 = 1$ bit
— $\ln e = 1$ natural unit $= 1$ nit
— $\log 10 = 1$ decimal unit $= 1$ dit.

The notion of entropy can be applied to systems formed by several components as well as to any experiment or random variable that brings into relief a probability field; we can therefore speak of the entropy of an experiment or of a random variable.

The high values of the entropy correspond to a great complexity, and a small informative capacity of a system; such situations occur when the system is formed by a large number of component elements or when the probabilities of these elements are close to one another. The entropy reaches a maximum (at a given n and $\Sigma p_i = 1$) when $p_1 = p_2, \ldots = p = \frac{1}{n}$, i.e., when all the events or component elements n have the same probability, so that:

$$H_{max} = -n \frac{1}{n} \ln \frac{1}{n} = \ln n \tag{82}$$

Taking all this into account, it follows that an objective estimate of the lithological complexity can be made by means of the entropy of parameters H, Σh_i, C, y_0 and N, analysed in the previous paragraphs. The entropy can be easily determined by using the tables computed for the three logarithms; these tables supply the values of expression $-p \ln p$ for various probabilities p while the entropy is supplied by the summation of the tabular values (Table 29).

Table 29

Entropy values

(in nit units)

H_{max} – maximum entropy

Number of components	H_{max}
3	1.0986
4	1.3863
5	1.608
6	1.795
7	1.943
8	2.076
9	2.186
10	2.303
11	2.397

p	$-p \cdot \ln p$	p	$-p \cdot \ln p$	p	$-p \cdot \ln p$	p	$-p \cdot \ln p$
0.001	0.007	0.19	0.315	0.46	0.356	0.73	0.230
0.002	0.012	0.20	0.322	0.47	0.355	0.74	0.223
0.003	0.017	0.21	0.328	0.48	0.354	0.75	0.216
0.004	0.022	0.22	0.333	0.49	0.350	0.76	0.208
0.005	0.027	0.23	0.340	0.50	0.346	0.77	0.200
0.006	0.031	0.24	0.342	0.51	0.344	0.78	0.194
0.007	0.035	0.25	0.346	0.52	0.340	0.79	0.186
0.008	0.038	0.26	0.350	0.53	0.336	0.80	0.178
0.009	0.042	0.27	0.353	0.54	0.334	0.81	0.170
0.01	0.046	0.28	0.357	0.55	0.329	0.82	0.162
0.02	0.084	0.29	0.359	0.56	0.325	0.83	0.152
0.03	0.106	0.30	0.361	0.57	0.320	0.84	0.147
0.04	0.129	0.31	0.364	0.58	0.316	0.85	0.139
0.05	0.150	0.32	0.365	0.59	0.312	0.86	0.126
0.06	0.168	0.33	0.366	0.60	0.307	0.87	0.121
0.07	0.186	0.34	0.366	0.61	0.302	0.88	0.113
0.08	0.202	0.35	0.3675	0.62	0.297	0.89	0.102
0.09	0.216	0.36	0.368	0.63	0.291	0.90	0.094
0.10	0.230	0.37	0.370	0.64	0.286	0.91	0.086
0.11	0.243	0.38	0.368	0.65	0.280	0.92	0.077
0.12	0.254	0.39	0.3675	0.66	0.274	0.93	0.067
0.13	0.265	0.40	0.366	0.67	0.268	0.94	0.058
0.14	0.275	0.41	0.366	0.68	0.261	0.95	0.050
0.15	0.285	0.42	0.364	0.69	0.256	0.96	0.039
0.16	0.293	0.43	0.364	0.70	0.250	0.97	0.029
0.17	0.301	0.44	0.361	0.71	0.243	0.98	0.019
0.18	0.309	0.45	0.360	0.72	0.236	0.99	0.010

H (m)	$a-b$	**10—18**	**18—26**	**26—34**	**34—42**
	p_i	0.069	0.181	0.244	0.256
	$-p_i \ln p_i$	0.184	0.310	0.344	0.348
Σh_i (m)	$a-b$	**0—5**	**5—10**	**10—15**	**15—20**
	p_i	0.240	0.270	0.190	0.117
	$-p_i \ln p_i$	0.342	0.353	0.315	0.250
y_0(%)	$a-b$	**10—19**	**19—28**	**28—37**	**37—46**
	p_i	0.015	0.103	0.177	0.192
	$-p_i \ln p_i$	0.060	0.233	0.307	0.316
C	$a-b$	**0—0.3**	**0.3—0.6**	**0.6—0.9**	**0.9—1.2**
	p_i	0.319	0.243	0.212	0.078
	$-p_i \ln p_i$	0.365	0.343	0.329	0.199
	$a-b$	**2.4—2.7**	**2.7—3.0**	**3.0—3.3**	
	P_i	0.014	0.07	0.007	
	$-p_i \ln p_i$	0.051	0.035	0.035	
h_i[m]	$a-b$	**0.2—3.4**	**3.4—6.6**	**6.6—9.8**	**9.8—13**
	p_i	0.606	0.149	0.079	0.059
	$-p_i \ln p_i$	0.304	0.284	0.200	0.166
	$a-b$	**25.8—29**	**29—32.2**		
	P_i	0.013	0.002		
	$-p_i \ln p_i$	0.057	0.012		
N	x_i	**0**	**1**	**2**	**3**
	p_i	0.138	0.227	0.277	0.151
	$-p_i \ln p_i$	0.273	0.338	0.356	0.286
	x_i	**8**	**9**	**10**	**11**
	p_i	0.013	0.006	0.013	0.006
	$-p_i \ln p_i$	0.057	0.031	0.057	0.031

So as to be able to compare the non-homogeneities of the various lithological parameters, the sampled population of the continuous variables must be divided into intervals of standard values, which can be obtained by applying H. Sturges's formula (2). The entropy of such a population is slightly lower than that of a discrete variable according to formula

$$H_s = H - \frac{n-1}{N}$$

where H_s is the entropy of the continuous variable, H is the entropy of a discrete type variable computed by relation (81), n is the number of intervals and N the sample size.

Table 30

				H / H_s	H_{max} / $100\ H_r$
42—50 0.156 0.290	**50—58** 0.062 0.172	**58—66** 0.019 0.080	**66—74** 0.013 0.056	1.78/1.76	2.08/84.5
20—25 0.080 0.202	**25—30** 0.059 0.166	**30—35** 0.037 0.122	**35—40** 0.007 0.035	1.79/1.76	2.08/84.5
46—55 0.257 0.349	**55—64** 0.154 0.288	**64—73** 0.051 0.152	**73—82** 0.051 0.152	1.86/1.83	2.08/88.0
1.2—1.5 0.043 0.135	**1.5—1.8** 0.035 0.118	**1.8—2.1** 0.028 0.102	**2.1—2.4** 0.014 0.051	1.76/1.73	2.40/72.1
13—16.2 0.046 0.142	**16.2—19.4** 0.024 0.093	**19.4—22.6** 0.022 0.088	**22.6—25.8** 0 0	1.35/1.34	2.30/58.2
4 0.063 0.173 **12** 0.006 0.031	**5** 0.044 0.137	**6** 0.031 0.108	**7** 0.025 0.095	1.97	2.48/79.3

In the case of a normal distribution with zero values of the asymmetry degree (α) and of excess (E), by using the natural logarithms, the entropy as a function of the standard deviation is supplied by formula

$$H = 0.916 + \ln \frac{\sigma}{c} \tag{83}$$

where c is the interval size.

As in the case of the coefficient of variation, the total entropy (H_t) of a lithological interval can be computed by summing up the elementary entropies of each lithological parameter. This global entropy permits us to make comparisons between the complexity of the lithological intervals and to back up conclusions of a practical and scientific nature (genesis, sedimentation conditions, lithogenesis, etc.).

To estimate the complexity of the systems with several components, C. R. Pelto introduced the notion of relative entropy H_r as the ratio between the measured value of the entropy and its maximum values (H_{max}), which can be obtained for a given number of components n [48]:

$$100\,H_r = \frac{-100\sum_{1}^{n} p_i \ln p_i}{H_{max}} \qquad (84)$$

In this form, a high relative entropy (or a low informative capacity), occurs when the contents of the components are near enough to one another; the maximum entropy of 100% corresponds to equal contents while the maximum of information, and therefore $H_r = 0$, occurs in the case of a single component. The notion of relative entropy can moreover be used in statistical distributions of the random variables, the components corresponding then to the sample intervals.

Table 30 presents in detail the computation of the entropies for the lithological parameters of coal interval II—III. For the other lithological intervals the entropies (sample, maximum and relative) are recorded in Table 26.

As in the case of the coefficient of variation, the values of the total entropies $H_t = \Sigma H_s$ (Table 31) were compared with one another it follows from this table that intervals I—II, II—III and the top of Dacian coal interval III present practically the same degree of lithological complexity. The sandy horizon at the bottom of coal bed I is remarkable by its higher informative capacity ($H_{t2} = 5.8$), and therefore by a smaller lithological complexity. The total entropy of the Levantine lithological complex has no significant meaning because of its low number of parameters C and h_i.

Table 31

Values of total entropy

Parameters taken in the computation	Lithological interval				
	Below coal layer I	Coal I–II	Coal II–III	Above coal layer III	Levantine
$[H_{t1}]$: $H, \Sigma h_i, y_0, C, h_i, N$	—	9.8	10.4	10.0	—
$[H_{t2}]$: $H, \Sigma h_i, h_i, N$	5.8	6.8	6.8	6.9	—
$[H_{t3}]$: C, h_i	—	3.0	3.1	3.3	3.2

The differentiation of the three lithological intervals can however be made with the help of the relative entropy, the variation limits of which are

$H_r = 50.9 — 88.0\%$. The arithmetic mean of the relative entropy of each interval is as follows:

— Dacian top of coal layer III $H_{rm} = 74.8\%$
— Interval II—III coal layer $H_{rm} = 77.7\%$
— Interval I—II coal layer $H_{rm} = 72.2\%$

Despite these low differences, the conclusion that the global lithological complexity of coal interval II—III is greater than that of the other intervals can be accepted.

As in the case of the total coefficient of variation, it is possible to establish classifications of the degrees of the lithological complexity according to the types of hydrogeological structures as a function of the total entropy value or of its mean relative one.

1.5 Global lithological considerations

From the synthesis of the statistical parameters (Table 26), there follows a qualitative and quantitative global characteristic of the distributions of the sandy bodies and generally of the lithofacies for each lithological interval.

The continuous increase of the sandy bodies' weight towards the bottom of the lignite deposit results from the values of the lithological coefficient (C) according to which the lithological intervals may be characterized as follows:

— the Levantine complex is clayey-sandy;

— the horizon in the top of coal layer III, and the intervals between II—III and I—II coal layers are sandy-clayey;

— the horizon in the bottom of coal layer I is prevailingly sandy.

Since the mean values of the sandy bodies' in a well (N) diminishes from the bottom of coal bed III downwards, it follows that along them elementary sandy bodies (h_i) increase in thickness, which means a better hydraulic connection between them and consequently better natural hydrogeological conditions of dewatering.

The homogeneity of a geological population is moreover determined by the number and characteristics of the laws of theoretical distribution of both the discrete and continuous variables. From this standpoint, the lithofacies of the Şotînga Mărgineanca field is characterized by a high degree of complexity because the empirical distributions of the lithological parameters are represented by five probabilistic models corresponding to: the normal distributions, the lognormal and amodal-unilateral distributions of Weibull type for the continuous variables and the Poisson and binomial distributions for discrete variable N. The wide range of the distributions — from the symmetrical unimodal to the unilateral-amodal ones with an extreme asymmetry — is a consequence of the interaction of the lithological factors in the sedimentation process, which determined the appearance of a heterogeneous population. At first sight, owing to the absence of the multimodal distributions and of the integration of the sampling distributions among the classical models, it would seem that these populations are homogeneous. The geochemists dealt with the mathematical, formal feature of the asymmetrical modal distributions. Thus, L. N. Ahrens introduced the multinormal distribution model by considering that the distribution with a positive or negative asymmetry is composed of two or three normal distributions, truncated and adjacent that can be graphically individualised. If the means of the component distributions are nearly equal, the heterogeneity may be hidden and the result is the appearance of a symmetrical associated distribution. If the values of the means are different, a heterogeneous population will result which could, in an extreme case, be a unilateral amodal distribution. Distributions of the Weibull type prevail among the continuous lithological variables of the Şotînga-Mărgineanca field (Table 32). This probabilitistic model, however, contains only two lithological parameters: the thickness of the elementary sandy bodies (h_i) and, partially, the lithological coefficient (C). For the rest, the centre of gravity of the sandy bodies (y_0) has a normal distribution while the other parameters vary between the normal and lognormal law of distribution. W. C. Krumbein [37] considers that when applying the various probabilistic models to the study of geological populations, the most important after the normal and lognormal distributions is the gamma distribution. This distribution is applied when the logarithmic transformation of the variable values does not altogether eliminate the positive asymmetry; it is widely applied because of its various forms. The density function of the gamma distribution is given in the works of mathematical statistics in several forms; one of them [37] is:

$$f(x) = \frac{x^{r-1} e^{-\frac{x}{\beta}}}{\Gamma_{(r)} \cdot \beta^r} \quad ; \quad x \geqslant 0 \tag{85}$$

where Γ is a gamma function.

The values of both the parameters ($r > 0$ and $\beta > 0$) are estimated by means of observation data. The estimation of these parameters is difficult because of the impossibility of linearizing the density function (the probabilistic diagram method). Certain publications supply simplifying diagrams with the help of which efficient estimations can be obtained for parameters r and β as well as for the mean, the variance, the coefficient of variation, etc. A simpler but only approximate method is based upon the use of the moments of the lower order, which are estimated by means of the empirical moments of selection. In this way, the mean and variance expressions are used, specific to gamma law:

$$\begin{aligned} \bar{x} &= r\,\beta \\ S^2 &= r\,\beta^2 \end{aligned} \tag{86}$$

If sample mean values $\bar{x}$ and S^2 are known, parameters r and β will be estimated by means of these equations.

Table 32

Lithological parameter	Lithological interval	Theoretical distribution				
		Normal	Lognormal	Weibull	Poisson	Binomial
Total thickness of interval H[m]	Top of coal III (Dacian)		×			
	Between II and III coal layers (Dacian)	×				
	Between coal layers I–II (Dacian)	×				
	Sandy horizon in the bottom of coal layer I		×			
Cumulated thickness of the sandy bodies Σh_i[m]	Top of coal III (Dacian)	×				
	Between II and III coal layers (Dacian)		×			
	Between coal layers I–II (Dacian)			×		
	Sandy horizon in the bottom of coal layer I		×			
Centre of gravity of the sandy bodies y_0 (%)	Top of coal III (Dacian)	×				
	Between II and III coal layers (Dacian)	×				
	Between coal layers I–II (Dacian)	×				
Lithological coefficient C	Levantine complex		×			
	Top of coal III (Dacian)		×			
	Between II and III coal layers (Dacian)			×		
	Between coal layers I–II (Dacian)			×		
Thickness of the elementary sandy bodies h_i[m]	Levantine complex			×		
	Top of coal III (Dacian)			×		
	Between II and III coal layers (Dacian)			×		
	Between coal layers I–II (Dacian)			×		
	Sandy horizon in the bottom of coal layer I			×		
Number of sandy bodies per drilling N	Top of coal III (Dacian)				×	
	Between II and III coal layers (Dacian)				×	
	Between coal layers I–II (Dacian)					×
	Sandy horizon in the bottom of coal layer I					×
TOTAL		6	6	8	2	2

1.6 Trend surface analysis of maps

In the preceding paragraphs the statistical analysis of the lithological parameters had a global character because the elements of the respective populations were not localized in the space. Such an analysis especially in the case of large areas or and of a variable lithofacies should be completed by a "space" statistical processing and synthesis of the population which on the primary maps can be represented as very intricate cartographic images, which could not be easily interpreted for scientific or practical purposes.

For the statistical analysis of these models, a start is made with the concept of "response surface", introduced by G. E. Box for the chemical reactions where the unsystematic fluctuations of this surface are a consequence of the experimental errors. W. C. Krumbein [36] pointed out that the adoption of this concept by the geological sciences too is possible, since each geological cartographic image (analogous to the "response surface") is the cumulated effect of two groups of factors: some of which operate on a large scale and others with relatively local effects. The first group contains the regional factors which control the sedimentation process together with the environmental tectonics, which affects the entire sedimentation area, including all the structural and erosional changes occurring after the deposition. The effect of these factors is cumulated in the regional component. The residual (local) component is the effect of the fluctuations on a small scale, associated with the local influences (local erosion areas, sand barriers, etc.).

The terms *regional* and *local* are relative since they depend on the scale of the map considered; indeed, situations may appear when a regional component at scale 1 : 500 can become a local one at scale 1 : 50,000. In order to eliminate such a confusion, the term "regional component" should be considered equivalent to the term "trend"; the residual component is a non-systematic error. By so doing the confusions due to the various scales in use are avoided.

The trend has an immediate geometric correspondent and is defined as a regular surface with relatively stable values presenting systematic variations from a point to another. The residual component is the relatively unstable part of the observed value which varies more or less irregularly and does not contain significant gradients.

The theoretical and practical importance of both these types of maps of trends and of the residual component — is underlined in several works [36, 37, 38, 46] which emphasize the wide area of application of this method in all the geological disciplines. For instance, D. R. Matuszak [46] analyses the tectonic-sedimentary significance of these two components and emphasizes the reciprocal connections between the sedimentation process and the tectonic activity as well as some paleogeographical consideration. Likewise, in economic geology problems, the method does not need further recommendation.

The inclusion of this method among the important hydrogeological documentations facilitates the interpretation and the possibility of quantitative appreaciation of the hydrogeological conditions in certain areas, a problem that cannot yet be dealt with because of the high degree of complexity of the maps achieved on the basis of the measured values.

1.6.1 Methods of trends separation

As previously shown, the notions of "trend" and "residual component" are only relative since their authentic separation would be unthinkable; this situation is however in agreement with the pairs of antonyms which casually correspond to them (in this case the geological processes with their "regional" and "local" character). In gravimetry, the separation process of the regional anomaly (trend) is also called *smoothing* because its values are deduced by the graphical, numerical or analytical levelling of the observed values of the gravity. The separation of the two components may be influenced by certain errors owing to the arbitrariness of the smoothing (especially in case of the graphical one), to the conditions of the smoothing process (choice of the relative size of the grid), to the premises considered in case of the analytical methods and to interpolation between the measured values. The choice of the smoothing methods depends both on the cartographic image complexity and on the accuracy required. Obviously, analytical methods supply the most accurate smoothing of the trend.

The geophysicists have developed a great many methods of the trend separation some of which can be directly applied to contour-type maps. Among the graphical methods, those of the isoline smoothing and of geological profile smoothing made at equal distances over the whole map surface are usually used. These quick methods lack a physical and mathematical basis and are arbitrary. In case of an intricate lithofacies, these practical methods cannot be applied.

The numerical methods offer the advantage of a higher degree of objectivity than the graphical ones and can be automatically applied. The method most used in practice, is that of the *mobile mean* which requires a knowledge of the values selected from a grid of squares having side d. The trend value at every grid point is supplied by the arithmetic mean of the following nine values: the value at the given point, four values at neighbouring points along the sides of the squares, and four values at neighbouring points along the diagonals [11].

The use of the analytical smoothing methods with a geometrical support are based on the trend, being assimilated with a regular surface (a plane, a surface of the II-er degree and even of the V-th or VI-th degree) in regard to which the residual components appear as random deviations. In this manner, the general linear model of the mathematical statistics is found, which offers the advantage that the independent variables can be studied under various forms (at various powers, logarithmic, etc.). The values of trend (t) can be expressed by a polynomial of the (p, q) degree [37]:

$$t(x,\ y) = a_{00} + a_{10}x + a_{01}x + a_{01}y + \dots + a_{pq}x^p y^q \tag{87}$$

where $a_{00}, \dots, a_{pq}$ are the polynomial coefficients of the general linear model.

When the selected points are situated at equal distances in a rectangular grid, we can apply the methods based on the analysis of the orthogonal polynomials which offer the advantages of standardization and of quick computation. In this case, expression (87) can be written in the following equivalent form:

$$t(x,\ y) = b'_{00} + b'_{10}\,\zeta'_1(x) + b'_{01}\,\eta'_1(y) + \dots + b'_{pq}\,\zeta'_p(x)\,\eta'_q(y) \tag{88}$$

where

$\zeta'_p(x)$ is the value of orthogonal polynom of degree p along x direction;

$\eta'_q(y)$ is the value of the orthogonal polynom of degree q, along a y direction;

$b'_{00} \dots b_{pq}$ are the estimations of coefficients $a_{00} \dots a_{pq}$.

Lastly, to conclude, function $t(x, y)$ is deduced from $T(x, y)$ (measured values) which gives the best estimation of the trend for the highest possible number of observations. The difference between $T(x, y)$ and $t(x, y)$ for each observation point represents the residual component $\varepsilon(x, y)$ which varies from one point to another; its probable mean for the whole investigated area is equal to zero.

The above considerations belong to the *method of complete regression*, the best mathematically elaborated method which gives moreover a higher accuracy. This accuracy is due to the fact that the orientation of the observation grid is independent of the structure of the isolines as well as to the consideration of all the direct and cross-product polynomial terms.

Function $t(x, y)$ can be computed by the grid nodes by using the tables of values ζ' and η' and the computed polynomial coefficients b'_{rs}. This method implies an arduous computation as well as the use of a computer for the computation of the matrices. That is why W. C. Krumbein recommends, in the case of lithofacies maps, simplified analytical methods that, applied within certain limits, can supply satisfactory results; among them we shall exemplify hereafter the abbreviated regression method and the "expected value" method.

1.6.2 The abbreviated regression method

Generally in the regression method the data are schematically arranged as in Figure 26. The importance of the cross-product polynomial terms is more important when the grid orientation does not coincide with the principal direction of the isolines. If however the grid orientation is concordant with the trend gradients, the cross-product polynomial terms tend to have small values and can be neglected.

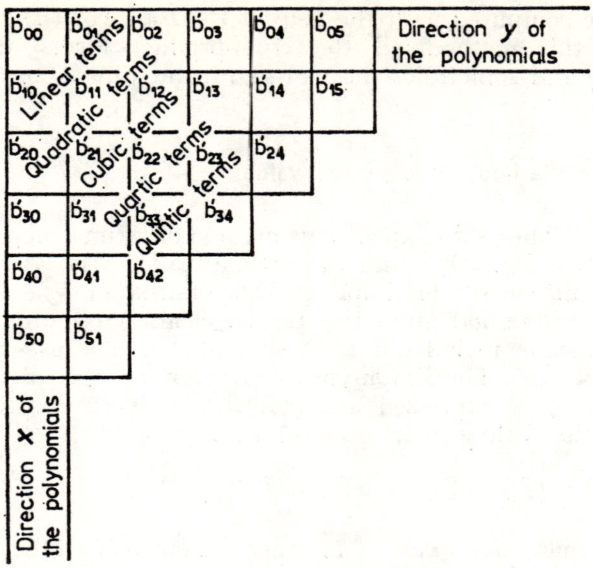

Figure 26. Arrangement of the polynomial terms.

In the case of a grid where the equidistance between points is assumed to be equal to one, the independent variables can be replaced by positive integers 0, 1, 2, ... such that the computation of direct polynomial terms of the matrix can be made by substituting the map, for a table with columns and lines. The observed value T_{ij} at a grid point can then be expressed by relation

$$T_{ij} = \mu + \alpha_i + \beta_j + \varepsilon_{ij} \tag{89}$$

where T_{ij} is the measured value in i-th row and j-th column; μ is the population mean; α_i the contribution from the i-th row; β_j the contribution of the j-th column and ε_{ij} the residual component.

If the polynomial terms along direction ox are located on the line component, while those along oy are on the column component, the quantities in expression (89) can be written clear as follows:

$$\mu = b'_{00}$$

$$\alpha_x = b'_{10} \cdot \zeta'_1(x) + b'_{20}\, \zeta'_2(x) + b'_{30}\, \zeta'_3(x) + \ldots$$

$$\beta_y = b'_{01}\, \zeta'_1(y) + b'_{02} \cdot \zeta'_2(y) + b'_{03}\, \zeta'_3(y) + \ldots \tag{90}$$

$$\varepsilon_{xy} = b'_{11}\, \zeta'_1(x)\, \zeta'_1(y) + b'_{21}\, \zeta'_2(x)\, \zeta'_1(y) + b'_{12}\, \zeta'_1(x)\, \zeta'_2(y) + \ldots$$

By substituting expressions (90) in equation (89) a simple model (used in the mathematical statistics) is obtained for the conventional rows-columns analyses of the variance with significant gradients which is based on the expansion in series of the orthogonal polynomials [36]. The polynomial coefficients are computed with the help of the totals of the rows and of the columns. By this method all the cross-product terms are included in component ε_{xy} and sometimes the non-significant direct ones too.

1.6.3 Methods of expected value

This is an expeditious method of estimating the trend that is sometimes only qualitative and approximate. That is why it only gives satisfactory results in rather simple geological situations, because when these are complex the method gives rise to significant cross-product polynomial terms that cannot be included in the trend. This method uses a small number of rows and columns. The simplifying character of the method is because the equation (88) is expressed as an algebraic identity based on the means of the observation values in the rows and columns [39]:

$$T_{ij} = \bar{T}.. + (\bar{T}_{i}. - \bar{T}..) + (\bar{T}._{j} - \bar{T}..) + (\bar{T}_{ij} - \bar{T}_{i}. - \bar{T}._{j} + \bar{T}..) \quad (91)$$

where T_{ij} is the measured value; $\bar{T}..$ the general mean of all the measurements; $(\bar{T}_i - \bar{T}..)$ is the difference between the row mean and the grand mean which represents component α_i; $(\bar{T}._j - \bar{T}..)$ is the difference between the column mean and the grand mean equivalent to the column component β_j; the expression in the last paranthesis is the "error" equivalent to ε_{ij} (the residual component).

If the residual component is neglected and the reductions are effected, the trend is supplied by relation

$$t_{ij} = \bar{T}_{i}. + \bar{T}._{j} - \bar{T}.. \quad (92)$$

In the case of a matrix of the observation data with a few elements, the trend can be estimated by means of a I and II degree surface; the method of the "expected value" gives then satisfactory results.

1.6.4 Arrangement of the observation data

Since the geological and hydrogeological exploration by drillings does not represent a systematic sampling, because of the partial inclusion of a series sampling, the observation data are not to be found in the nodes of a geometric grid, an absolutely necessary condition when using orthogonal polynomials. In the mathematical statistics, there are also methods [37] for estimating the trends even in the case of irregularly spaced points of observation data by using the non-orthogonal polynomials. In this case however, the computation is very arduous and the simplified methods cannot be applied. That is why it is customary to elaborate a new map with interpolated observation data, situated in a rectangular grid. Since the interpolated values introduce a variable component liable to distort the actual image, the density of the observations on the original map should be greater than grid cell made for establishing the selection interval. R. L. Miller [48] advises the use of certain means proposed by J. T. Curtis for ecological researches, to wit

$$\text{— density } D = \frac{\textit{Total number of points}}{\textit{Total area}} \qquad [\text{points/km}^2]$$

$$\text{— area of the best square } S = \frac{2 \times \textit{the total area}}{\textit{Total number of points}}$$

In this manner, the area of the best square depends both on the investigated area and on the number of points. When choosing the area, the "marginal effect" should also be taken into account since sometimes it can be rather important. This is due to the fact that the examined perimeter belongs to a larger area, which leads to a low harmonization of the data at the confines of the map. This drawback can be avoided by considering a larger area.

In connection with the grid orientation when using the simplified methods of estimating the trend, it is necessary to rotate it so that the axes should be parallel with the isolines of the trend. By so doing, the importance of the interaction effect is diminished since the cross-product polynomial terms can be neglected.

1.6.5 Estimation of the trends and of the residual components in the lithological interval between coal layers II and III

To exemplify the methods described above, a perimeter of $2.5 \times 1.75 = 4.4$ km² was chosen south of fault F_1 that comprises most of the Cărbunaru and Gheboeni mining fields and the northern part of Șotînga mining field (Figure 2). The square side (the cell) is estimated according to J. T. Curtis' criterion at 250 m because the most frequent sizes of the exploration grid are smaller than 200 m.

Total thickness of the sandy bodies (Σh_i). The processing of this parameter was minutely made, with a view to giving the computation method a basis which is also valid for the other parameters of the lithofacies.

The abbreviated regression method. Although the cartographic image of the measured parameter is rather intricate (Figure 27) it is possible to see that the isopachytes trend is parrallel to the geographical co-ordinates. Since the geometric grid is uniform, a suitable origin can be chosen and to the independent variables positive integers 0, 1, 2... can be substituted. In this way, the orthogonal polynomials acquire a standard form and are conventionally represented by ξ'. Among the orthogonal polynomials, the most accessible are those of the Tchebyshev type, because their values are

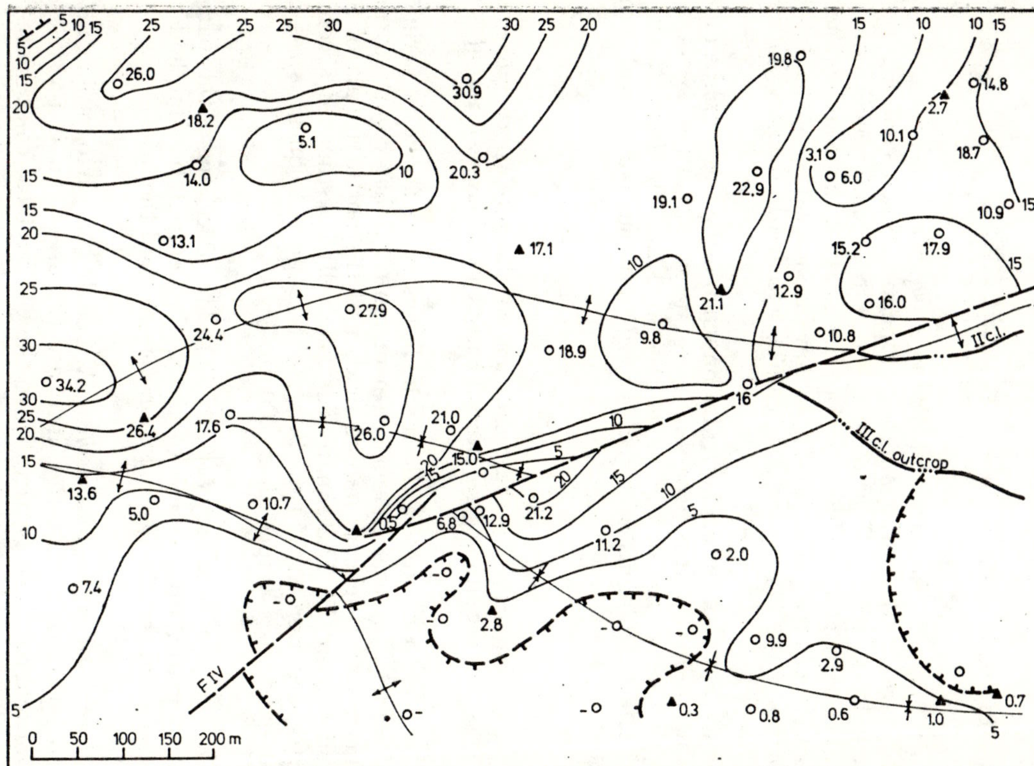

Figure 27. Isopachyte map of the sandy bodies in coal interval II–III (Σh_i).

expressed by integers. The values of these orthogonal polynomials are recorded in Table 33 in the cases where the numbers of rows and columns are equal to 3—12 [15].

Table 34 represents the computation of the polynomial coefficient by means of orthogonal polynomials ζ'. To this end, we first estimate by linear interpolation the values of Σh_i in the nodes of a 250 m-spacing grid; this operation is effected by means of the isopachyte map (Figure 27). We thus obtain an arrangement of the observation data by rows and columns corresponding to directions Ox and Oy. Both the sums (of the rows and columns) are multiplied by polynomial terms after which the summation is made by columns (*1—7*) and by rows (*8—17*). On each column and row of polynomials the sum of their squares is recorded (SS) and this value is recorded too. Values SS on the columns are multiplied by the number of rows while those on the rows are multiplied by the number of columns which gives the products noted by 18—34. Lastly, the polynomial coefficients of various orders are computed along directions Ox (marked by index l) and Oy (index c) by dividing values 1—17 by values 18—34. These coefficients represent the estimations of coefficients "b" from equations (90).

The polynomial coefficients are also used for defining the various elementary areas and choosing the most important of them, i.e. those that control the trend. The trend estimation by an area of a certain degree is also made with the help of some conventional row-columns analyses. Table 35 shows the trend computation approximated by a quadratic surface. The first column of the table records the arithmetic mean of all observations. The contribution of the I order terms along Ox (column 2) is supplied by multiplying polynomial coefficient I_l by each orthogonal polynomial of the first column, e.g.: $-1.335 \times (-7) = +9.34$; $-1.335 \times (-5) = +6.67$. Column 3 is obtained by multiplying coefficient II_l by each orthogonal polynomial of the second column (for instance: $-0.427 \times (-7) = +2.99$, etc.). In the following column the mean algebraic summation is made with the two terms of the I and II degree.

The contribution of the terms along Oy is similarly computed, though by taking into account both the I_c and II_c polynomials coefficients as well as the orthogonal polynomials of the I and II degree (the first two rows). For instance: $-0.903 \times (-5) = +4.51$; $0.903 \times (-4) = +3.61$ and so on. The last value results from the addition of the two sums corresponding to a row and a column that determine the position of the point in the observation grid.

The problem of estimating the approximation polynomial of the trend can be solved by a series of statistical tests from which the analysis of variance was chosen. By means of this test the contribution of the surfaces and of the terms of various degrees is successively computed, this operation also being known as the estimation of the trend strength. This strength refers to the total sum of the squared observation data deviations, a parameter calculated by relation

$$SSY^2 = \sum_i \sum_j Y_{ij}^2 - \frac{1}{n} (\Sigma \Sigma Y_{ij})^2 \qquad (93)$$

Table 33

Values of the orthogonal polynomials of Tchebyshev type

$n = 3$

	ζ_1'	ζ_2'
	−1	+1
	0	−2
	+1	+1
SS	2	6

$n = 4$

	ζ_1'	ζ_2'	ζ_3'
	−3	+1	−1
	−1	−1	+3
	+1	−1	−3
	+3	+1	+1
SS	20	4	20

$n = 5$

	ζ_1'	ζ_2'	ζ_3'	ζ_4'
	−2	+2	−1	+1
	−1	−1	+2	−4
	0	−2	0	+6
	+1	−1	−2	−4
	+2	+2	+1	+1
SS	10	14	10	70

$n = 6$

	ζ_1'	ζ_2'	ζ_3'	ζ_4'	ζ_5'
	−5	+5	−5	+1	− 1
	−3	−1	+7	−3	+ 5
	−1	−4	−4	+2	− 10
	+1	−4	−4	+2	+ 10
	+3	−1	−1	−3	− 5
	+5	+5	+5	+1	+ 1
SS	70	84	180	28	252

$n = 7$

	ζ_1'	ζ_2'	ζ_3'	ζ_4'	ζ_5'	ζ_6'
	−3	+5	−1	+3	−1	+1
	−2	0	+1	−7	+4	−6
	−1	−3	+1	+1	−5	+15
	0	−4	0	+6	0	−20
	+1	−3	−1	+1	+5	+15
	+2	0	−1	−7	−4	−6
	+3	+5	+1	+3	+1	+1
SS	28	84	6	154	84	924

$n = 8$

	ζ_1'	ζ_2'	ζ_3'	ζ_4'	ζ_5'	ζ_6'	ζ_7'
	−7	+7	−7	+7	−7	+1	−1
	−5	+1	+5	−13	+23	−5	+7
	−3	−3	+7	−3	−17	+9	−21
	−1	−5	+3	+9	−15	−5	+35
	+1	−5	−3	+9	+15	−5	−35
	+3	−3	−7	−3	+17	+9	+21
	+5	+1	−5	−13	−23	−5	−7
	+7	+7	+7	+7	+7	+1	+1
SS	168	168	264	616	2 184	264	3 432

$n = 9$

	ζ_1'	ζ_2'	ζ_3'	ζ_4'	ζ_5'	ζ_6'	ζ_7'	ζ_8'
	0	−20	0	+18	0	−20	0	+70
	+1	−17	−9	+9	+9	+1	−14	−56
	+2	−8	−13	−11	+4	+22	+14	+28
	+3	+7	−7	−21	−11	−17	−6	−8
	+4	28	+14	+14	+4	+4	+1	+1
SS	60	2 772	990	2 002	468	1 980	858	12 870

$n = 10$

	ζ'_1	ζ'_2	ζ'_3	ζ'_4	ζ'_5	ζ'_6	ζ'_7	ζ'_8	ζ'_9
	+1	−4	−12	+18	+6	−8	−56	+14	+126
	+3	−3	−31	+3	+11	+6	−42	−28	−84
	+5	−1	−35	−17	+1	+10	+86	+20	+36
	+7	+2	−14	−22	−14	−11	−47	−7	−9
	+9	+6	+42	+18	+6	+3	+9	+1	+1
SS	330	132	8 580	2 860	780	660	29 172	2 860	48 620

$n = 11$

	ζ'_1	ζ'_2	ζ'_3	ζ'_4	ζ'_5	ζ'_6	ζ'_7	ζ'_8	ζ'_9	ζ'_{10}
	0	−10	0	+6	0	−40	0	+70	0	−252
	+1	−9	−14	+4	+4	−12	−28	−14	+42	+210
	+2	−6	−23	−1	+4	+36	−2	−68	−48	−120
	+3	−1	−22	−6	−1	+29	+33	+73	+27	+45
	+4	−6	−6	−6	−6	−48	−23	−31	−8	−10
	+5	+15	+30	+6	+3	+15	+5	+5	+1	+1
SS	110	858	4 290	286	156	11 220	4 862	27 170	9 724	184 756

$n = 12$

	ζ'_1	ζ'_2	ζ'_3	ζ'_4	ζ'_5	ζ'_6	ζ'_7	ζ'_8	ζ'_9	ζ'_{10}	ζ'_{11}
	+1	−35	−7	+28	+20	−20	−140	+70	+210	−42	−462
	+3	−29	−19	+12	+44	+4	−204	−74	+102	+90	+330
	+5	−17	−25	−13	+29	+25	+83	−65	−303	−75	−165
	+7	+1	−21	−33	−21	+11	+251	+119	+227	+35	+55
	+9	+25	−3	−27	−57	−31	−225	−61	−79	−9	−11
	+11	+55	+33	+33	+33	+11	+55	+11	+11	+1	+1
SS	572	12 012	5 148	8 008	15 912	4 488	369 512	65 208	408 408	33 592	705 432

Observations: The value of polynomial ζ_0 (first column) is equal to +1 in all the cases; this polynomial is not present in the computation by the method of the abbreviated regression. For $n = 9 \ldots 12$, only the lower part of the table is given; the values of the terms in the upper part result by symmetry with the same sign for the even ζ and with the opposite sign for the odd ζ.

	Arrangement of the observation data											Sums
Measured values	0	25	25	30	32	27	15	17	17	10	18	216
	20	20	12	8	20	20	17	20	13	11	18	179
	20	15	14	18	17	18	16	21	12	17	12	180
	28	27	25	27	24	20	13	16	11	15	15	221
	16	26	15	23	22	16	15	14	10	3	0	160
	11	5	8	17	6	17	10	4	3	0	0	81
	7	4	1	0	0	1	0	4	4	0	0	21
	3	3	2	0	0	0	0	1	7	2	9	27
Sums	105	125	102	123	121	119	86	97	77	58	72	1085
Polynomial terms for $n = 11$	−5	−4	−3	−2	−1	0	1	2	3	4	5	110
	15	6	−1	−6	−9	−10	−9	−6	−1	6	15	858
	−30	6	22	23	14	0	−14	−23	−22	−6	30	4290
	6	−6	−6	−1	4	6	4	−1	−6	−6	6	286
	−3	6	1	−4	−4	0	4	4	−1	−6	3	156
	15	−48	29	36	−12	−40	−12	36	29	−48	15	11 220
	−5	23	−33	2	28	0	−28	−2	33	−23	5	4862
	5	−31	73	−68	−14	70	−14	−68	73	−31	5	27 170
	−1	8	−27	48	−42	0	42	−48	27	−8	1	9724
	1	−10	45	−120	210	−252	210	−120	45	−10	1	184 765
Column sums × polynomial terms	−525	−500	−306	−446	−121	0	86	194	231	232	360	−795
	1575	750	−102	−738	−1089	−1190	−774	−582	−77	348	1080	−799
	−3150	750	2244	2829	1694	0	−1204	−2231	−1694	−348	2160	+1050
	630	−750	−612	−123	484	714	344	−97	−462	−348	432	+212
	−315	750	102	−492	−484	0	344	388	−77	−348	216	+84
	1575	−6000	2958	4428	−1452	−4760	1032	3492	2233	−2784	1080	−262
	−525	2875	−3366	446	3388	0	−2408	−194	2541	−1334	360	+1783
	525	−3875	7446	−8364	−1694	8330	−1204	−6596	5621	−1796	360	−247
	−105	1000	−2754	5904	−5082	0	3612	−4656	2079	−464	72	−1394
	105	1250	−4590	−14 760	25 410	−29 988	18 060	11 640	3 465	−580	72	+19 264

the abbreviated method of regression for Σh_i

Polynomial terms for $n = 8$							Rows sum × polynomial terms						
−7	7	−7	7	−7	1	−1	−1512	1512	−1512	1512	−1512	216	−216
−5	1	5	−13	23	−5	7	−895	179	895	−2327	4117	−895	1253
−3	−3	7	−3	−17	9	−21	−540	−540	1260	−540	−3060	1620	−3780
−1	−5	3	9	−15	−5	35	−221	−1105	663	1989	−3315	−1105	7735
1	−5	−3	9	15	−5	−35	160	−800	−480	1440	2400	−800	−5600
3	−3	−7	−3	17	9	21	246	−246	−574	−246	1394	738	1722
5	1	−5	−13	−23	−5	−7	105	21	−105	−273	− 483	−105	−147
7	7	7	7	7	1	1	189	189	189	189	189	27	27
168	168	264	616	2184	264	−3432	−2468	−790	+336	+1744	−270	−304	+994
							(1)	(2)	(3)	(4)	(5)	(6)	(7)

Sum of the orthogonal polynomial squares

	168 × 11	1848	(18)	110 × 8	880	(25)
	163 × 11	1848	(19)	858 × 8	6864	(26)
	264 × 11	2904	(20)	4 290 × 8	34 320	(27)
	616 × 11	6776	(21)	286 × 8	2288	(28)
	2164 × 11	24 024	(22)	156 × 8	1248	(29)
	261 × 11	2904	(23)	11 220 × 8	89 760	(30)
	3432 × 11	37 752	(24)	4862 × 8	38 896	(31)
				27 170 × 8	217 360	(32)
(8)				9 724 × 8	77 792	(33)
(9)				184 756 × 8	1 478 048	(34)

(10)	Polynomial coefficients							
(11)	I_l	1/18	−2468/1848	−1.335	I_c	8/25	−795/880	−0.903
(12)	II_l	2/19	−790/1848	−0.427	II_c	9/26	−799/6,864	−0.116
(13)	III_l	3/20	+336/2904	0.116	III_c	10/27	+1050/34 320	0.030
(14)	IV_l	4/21	+1744/6776	0.257	IV_c	11/28	+212/2288	0.093
(15)	V_l	5/22	−270/24 024	−0.011	V_c	12/29	+84/1248	0.067
(16)	VI_l	6/23	− 304/2904	−0.105	VI_c	13/30	−262/89 760	−0.002
(17)	VII_l	7/24	994/37 752	0.026	VII_c	14/31	+1783/38 896	0.046

Table 35

Trend estimation by a quadratic surface for Σh_i (the abreviated regression method)

Contribution of the terms along ta Oy / Contribution of the terms along to Ox			Degree I Degree II	+4.51 −1.74	+3.61 −0.70	+2.71 +0.12	−1.81 +0.70	+0.90 +1.04	0.00 +1.16	−0.90 +1.04	−1.81 +0.70	−2.71 +0.12	−3.61 −0.70	−4.51 −1.74
Mean	Degree I = linear	Degree II = quadratic	Sum	+2.77	+2.91	+2.83	+2.51	+1.94	+1.16	+0.14	−1.11	−2.59	−4.31	−6.25
12.33	+9.34	−2.99	18.68	21.5	21.6	21.5	21.2	20.6	19.8	18.8	18.6	16.1	14.4	12.4
12.33	+6.67	−0.43	18.57	21.3	21.5	21.4	21.1	20.5	19.7	18.7	17.5	16.0	14.3	12.3
12.33	+4.00	+1.28	17.61	20.4	20.5	20.4	20.1	19.6	18.8	17.8	16.5	15.0	13.3	11.4
12.33	+1.33	+2.14	15.80	18.6	18.7	18.6	18.3	17.7	17.0	15.9	14.7	13.2	11.5	9.6
12.33	−1.33	+2.14	13.14	15.9	16.0	16.0	15.6	15.2	14.3	13.3	12.0	10.6	8.8	6.9
12.33	−4.00	+1.28	9.61	12.4	12.5	12.4	12.1	11.6	10.8	9.8	8.5	7.0	5.3	3.4
12.33	−6.67	−0.43	5.23	8.0	8.1	8.1	7.7	7.2	6.4	5.4	4.1	2.6	0.9	0
12.33	−9.34	−2.99	0	2.8	2.9	2.8	2.5	1.9	1.2	0.1	0	0	0	0

Table 36

Point number	Initial data Y_{ij}	Y^2_{ij}	Linearly harmonized values Y'_{ij}	Residual values (linear) $Y_{ij} - Y'_{ij}$	$(Y_{ij}-Y'_{ij})^2$	Quadratically harmonized values Y''_{ij}	Residual values (quadratic) $Y_{ij} - Y''_{ij}$	$(Y_{ij}-Y''_{ij})^2$
1	0	0	10.5	−10.5	110.25	11.1	−11.1	123.21
2	15	225	13.6	1.4	1.96	14.0	1.0	1.00
3	10	100	12.5	− 2.5	6.25	13.8	− 3.8	14.44
4	25	625	13.0	12.0	144.00	14.5	10.5	110.25
5	10	100	15.5	− 5.5	30.25	15.3	− 5.3	28.09
6	20	400	15.0	5.0	25.00	15.0	5.0	25.00
7	15	225	14.5	0.5	0.25	14.0	1.0	1.00
8	21	441	14.0	7.0	49.00	13.0	8.0	64.00
9	3	9	12.5	− 9.5	90.25	11.5	8.5	72.25
10	11	121	12.0	− 1.0	1.00	11.0	0	0
11	8	64	13.6	− 5.6	31.36	14.1	− 6.1	37.21
12	12	144	14.0	− 2.0	4.00	14.0	− 2.0	4.00
Σ	150	2 454			493.57			480.45

Sum total of the squared deviations:

$$SSY^2 = 2.454 - \frac{(150)^2}{12} = 5.79$$

Deviation from the linear surface:

$$SS(Y_{ij} - Y'_{ij})^2 = 493.57$$

Contribution of the linear surface:

$$SSY^2 - SS(Y_{ij} - Y'_{ij})^2 = 579 - 493.57 = 85.43$$

Deviation from the quadratic surface:

$$SS(Y - Y''_{ij})^2 = 480.45$$

Contribution of the quadratic surface:

$$SSY^2 - SS(Y_{ij} - Y''_{ij})^2 = 579 - 480.45 = 98.55$$

Contribution of the quadratic terms:

$$98.55 - 85.43 = 13.12$$

For a better understanding of the computation technique of this test, Table 36 offers a conclusive example with values of parameter Σh_i. Table 37 shows the results of the dispersional analysis for the trends of various degrees of parameter Σh_i. Thus, the deviation from the quadratic surface is:

$$SS(Y_{ij} - Y''_{ij})^2 = 2\,586$$

The difference supplies the sum of the squares attributed to the quadratic surface:

$$SSY^2 - SS(Y_{ij} - Y''_{ij})^2 = 6\,896 - 2\,586 = 4\,310.$$

The estimation of the trend for parameter Σh_i was made up to the quintic surface (Table 38). The problem of the "upward" analytical continuation of the trend can be solved with the help of the polynomial coefficients; Table 34 shows that the values of the polynomial coefficients are appreciable up to the IV degree but, afterwards, they diminish abruptly. Although the sum of the squares is large for ζ'_5 (2 184) the contribution of the V degree terms is very small (0.06%) which indicates the conclusion that the general (total) trend can be estimated by a quartic surface.

Table 37

Results of the analysis of variance for Σh_i

Variation source	Sum of the squared SS	%
Contribution of linear surface	3712	53.81
Deviation from the linear surface	3184	46.19
Contribution of quadratic surface	4310	62.50
Contribution of quadratic terms	599	8.69
Deviation from quadratic surface	2586	37.50
Contribution of cubic surface	4316	62.59
Contribution of the cubic terms	6	0.09
Deviation from the cubic surface	2580	37.41
Contribution of the quartic surface	4766	69.11
Contribution of the quartic terms	450	6.53
Deviation from the quartic surface	2130	30.89
Contribution of the quintic surface	4770	69.17
Contribution of the quintic terms	4	0.06
Deviation from the quintic surface	2126	30.83
Total:	6896	100.00

Whitten [37] pointed out that on the colour index map (the percentage of the melanocratic minerals) of the "older granite" of Donegal (Ireland) systematic effects of degree higher than II and III can be brought into relief. Under such conditions, when the contribution of quadratic or cubic surface (expressed by the sum of squares) represents less than 90% of the corrected total sum of the squares, Whitten recommends the separation of the secondary component of the trend by subtracting every value of the principal trend from every one of those corresponding to the overall trend. In this way, it can be considered that the principal trend represents the major geological structure while the secondary trend is the result of geological processes of smaller intensity and action. Let us note that while the secondary trend has a systematic distribution allowing us to establish certain connections with the geological structure, the residual component has only a random character.

The cartographic images of the trends present some characteristics. Thus, the quadratic trend (Figure 28) shows a slow and continuous increase of the isopachytes towards the north-west; for the cubic trend (Figure 29) an axis of maximum is outlined in the northern part of the area while the rest of the cartographic image remains approximately unchanged. The quartic trend (Figure 30) offers a quite different image in regard to the preceding one; especially the axis of maximum is displaced towards the south. For the quintic trend (Figure 31) no elements can be observed distinct from the ones of the IV degree. The comparison between the gradients along the two principal directions emphasizes the fact that along the south-north direction their values are smaller than those along the east-west direction; it is also to be remarked that the sandless area ($\Sigma h_i = 0$) increases concomitantly with the trend.

Table 38

Trend estimation by a quintic surface, for Σh_i (abbreviated regression method)

Contribution of the terms along Ox / Contribution of the terms along Oy						Degree I	+4.51	+3.61	+2.71	+1.81	+0.90	0.00	−0.90	−1.81	−2.71	−3.61	−4.61
						Degree II	−1.74	−0.70	+0.12	+0.70	−1.04	+1.16	+1.04	+0.70	+0.12	−0.70	−1.74
						Degree III	−0.90	+0.18	+0.66	+0.69	+0.42	0	−0.42	−0.69	−0.66	−0.18	−0.90
						Degree IV	+0.56	−0.56	−0.56	−0.09	+0.37	+0.56	+0.37	−0.09	−0.56	−0.56	+0.56
						Degree V	−0.20	+0.40	+0.07	−0.27	−0.27	0	+0.27	+0.27	−0.07	−0.40	+0.20
Mean	Degree I	Degree II	Degree III	Degree IV	Degree V	Sum	+2.23	+2.93	+3.00	+2.84	+2.46	+1.62	+0.36	+1.72	−3.88	−5.45	−6.39
12.33	+9.34	−2.99	−0.81	+1.80	+0.08	+19.75	22.0	22.7	22.8	22.6	22.0	21.6	20.1	18.1	15.9	14.3	13.4
12.33	+6.67	−0.43	+0.58	−3.34	−0.25	+15.56	17.8	18.5	18.6	18.4	18.0	17.3	15.9	13.9	11.7	10.1	9.2
12.33	+4.00	+1.28	+0.81	−0.77	+0.19	+17.84	20.1	20.8	20.8	20.8	20.7	19.6	19.2	19.2	14.0	12.4	11.1
12.33	+1.33	+2.14	+0.35	+2.31	+0.17	+18.63	20.9	21.6	21.6	21.5	21.1	20.4	19.0	17.0	14.8	13.2	12.2
12.33	−1.33	+2.14	−0.35	+2.31	−0.17	−14.93	17.9	17.9	17.9	17.8	17.4	16.7	15.3	13.3	11.0	9.5	8.5
12.33	−4.00	+1.28	−0.81	−0.77	−0.19	+ 7.84	10.5	10.8	10.8	10.7	10.3	19.6	8.2	6.2	4.0	2.4	1.4
12.33	−6.67	−0.43	−0.58	−3.34	+0.25	+ 1.56	3.8	4.5	4.6	4.4	4.0	3.3	1.9	0	0	0	0
12.22	−9.34	−2.99	+0.81	+1.80	−0.08	+ 2.53	4.8	5.5	5.5	5.4	5.0	4.3	2.9	0,9	0	0	0

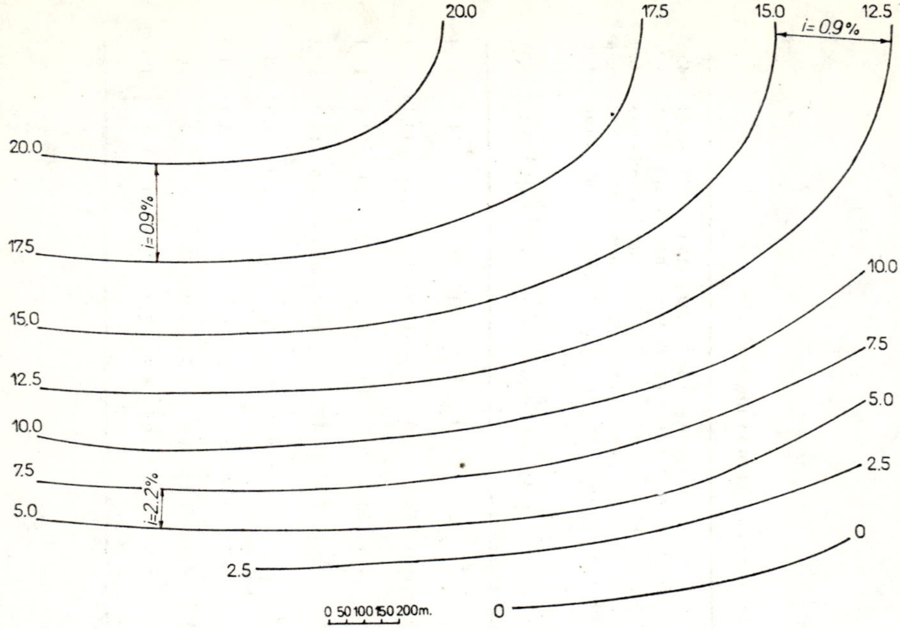

Figure. 28. Map of the quadratic trend of parameter Σh_i.

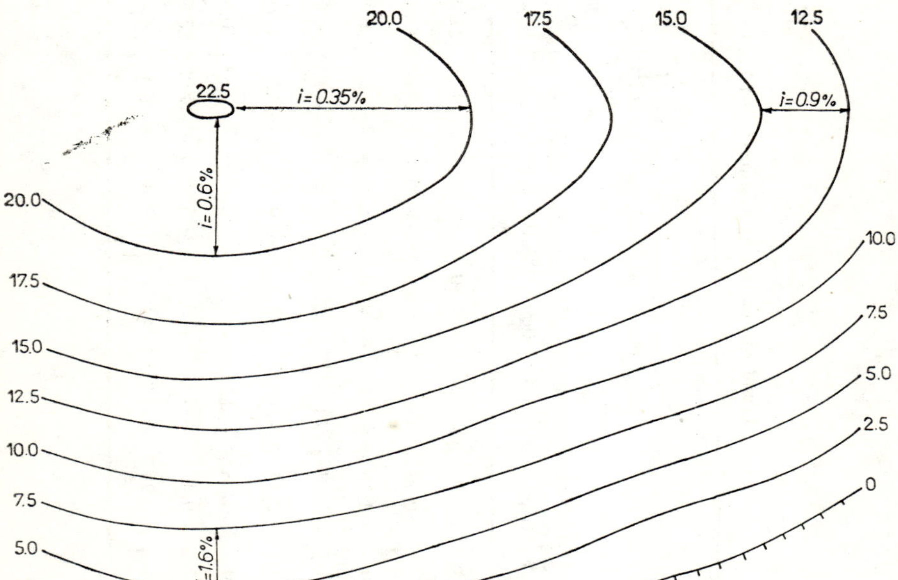

Figure 29. Map of the cubic trend of parameter Σh_i.

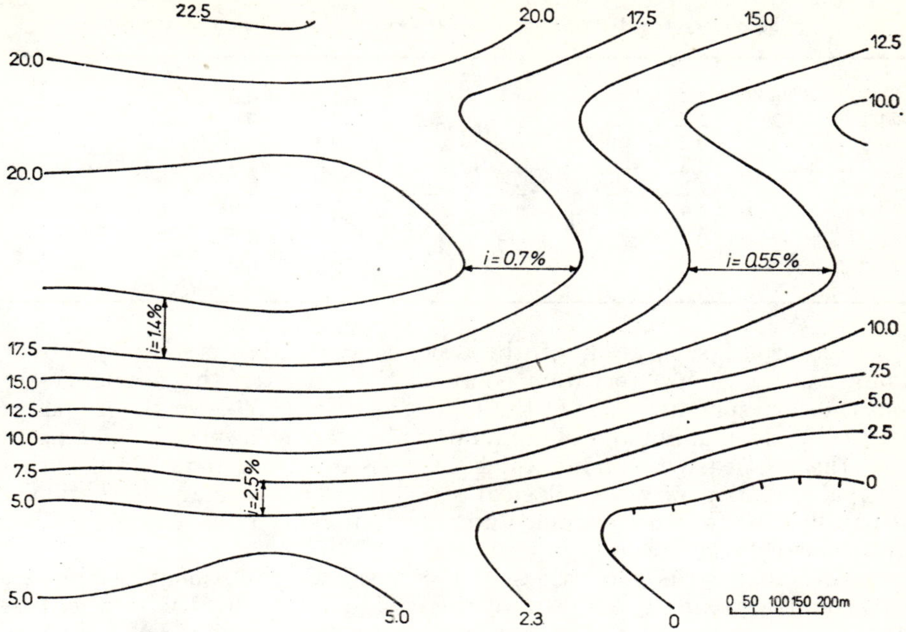

Figure 30. Map of the total trend (of the IV degree) of parameter Σh_i.

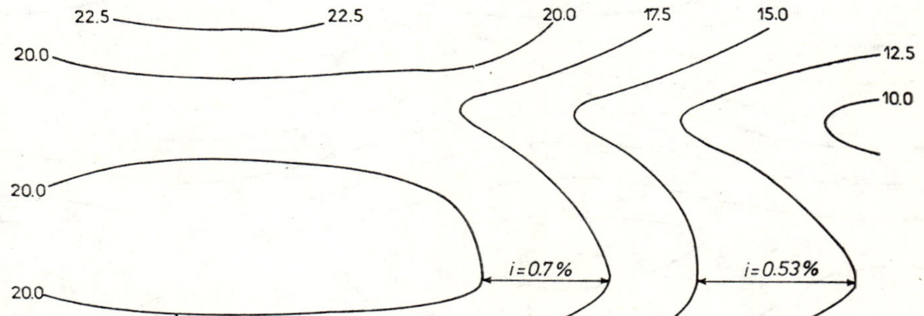

Figure 31. Map of the quintic trend of parameter Σh_i.

Table 39

Secondary trend (IV—II) for Σh_i (abbreviated regression method)

+0.6	+0.6	+1.1	+1.6	+1.8	+1.6	+1.0	+0.2	−0.2	+0.2	+0.7
−3.1	−3.2	−2.7	−2.2	−2.0	−2.2	−2.8	−3.6	−4.0	−3,5	−3.0
−0.3	−0.3	+0.2	+0.7	+0.8	+0.6	−0.1	−0.7	−1.2	−0.7	−0.3
+2.3	+2.3	+2.8	+3.3	+3.5	+3.2	+3.8	+1.9	+1.4	+1.9	+2.3
−1.6	+1.6	+2.0	+2.6	+2.7	+2.5	+0.7	+1.2	+0.7	+1.2	+1.6
−1.9	−1.9	−1.4	−1.0	−0.8	−1.0	−1.7	−2.4	−2.8	−2.3	−2.0
−4.3	−4.3	−3.9	−3.3	−3.2	−3.4	−4.0	−4.1	−2.6	−0.9	0
+2.2	+2.2	+2.7	+3.2	+3.4	+3.1	+2.6	+0.7	0	0	0

The formal separation of the secondary trend was obtained by the diminution, at each observation point, of the values of the secondary trend from the corresponding ones of the IV degree (Table 39). On the proper map (Figure 32) the maximum and minimum values are concentrated and oriented along the west-east directions which suggests a certain parallelism with the principal direction of the geological structure. The regular alternation and the parallelism of these maximum and minimum axes emphasizes the geological significance of the secondary trend.

Table 40 records the values of the residual component with regard to the quartic trend. The mean of the residuals is slightly different from

Figure 32. Map of the secondary trend component for Σh_i.

Table 40

Residual component for Σh_i (abbreviated regression method)

−33.1	+2.8	+2.4	+ 7.2	+9.6	+5.6	−4.8	−0.8	+1.1	−4.6	+4.9	1.3
− 1.8	+1.7	−6.7	−10.9	+1.5	+2.5	+1.1	+6.1	+1.0	+0.2	+8.8	+ 7.1
− 0.1	−5.2	−6.6	− 2.8	−3.4	−1.4	−1.7	+5.2	−1.8	+4.4	+0.3	−12.5
+ 7.1	+6.0	+3.6	− 5.4	+2.8	−0.2	−5.5	−0.6	−3.6	+1.6	+3.1	+19.7
− 1.5	+8.4	−3.0	+ 4.8	+4.2	−0.8	−0.2	+0.8	−1.3	−7.0	−8.5	− 4.1
+ 0.5	−5.6	−3.0	+ 5.9	−4.8	+7.2	+1.9	−2.1	−1.2	−3.0	−1.4	− 5.6
+ 3.3	+0.2	−3.2	− 1.4	−4.0	−2.0	−1.4	+4.0	+4.0	0	0	− 3.5
− 2.0	−2.1	−3.5	− 5.7	−5.3	−4.3	−2.7	+0.3	+7.0	+2.0	+9.0	− 7.3

zero (—4.9/88 = —0.056) owing to the rounding of the computation results. The intensity, area distribution and morphology of this component are presented in Figure 33; the intensities are widely variable but the values of their gradients are for the most part of the perimeters close to those of the trends.

Method of the "expected value". According to Table 41, the observed values are arranged in rows and columns while the last two columns give the sums and the mean values. Table 42 gives the previously computed means of the rows and columns; each element of this table is completed by the trend value resulting from subtracting the general mean from the sum of means of rows

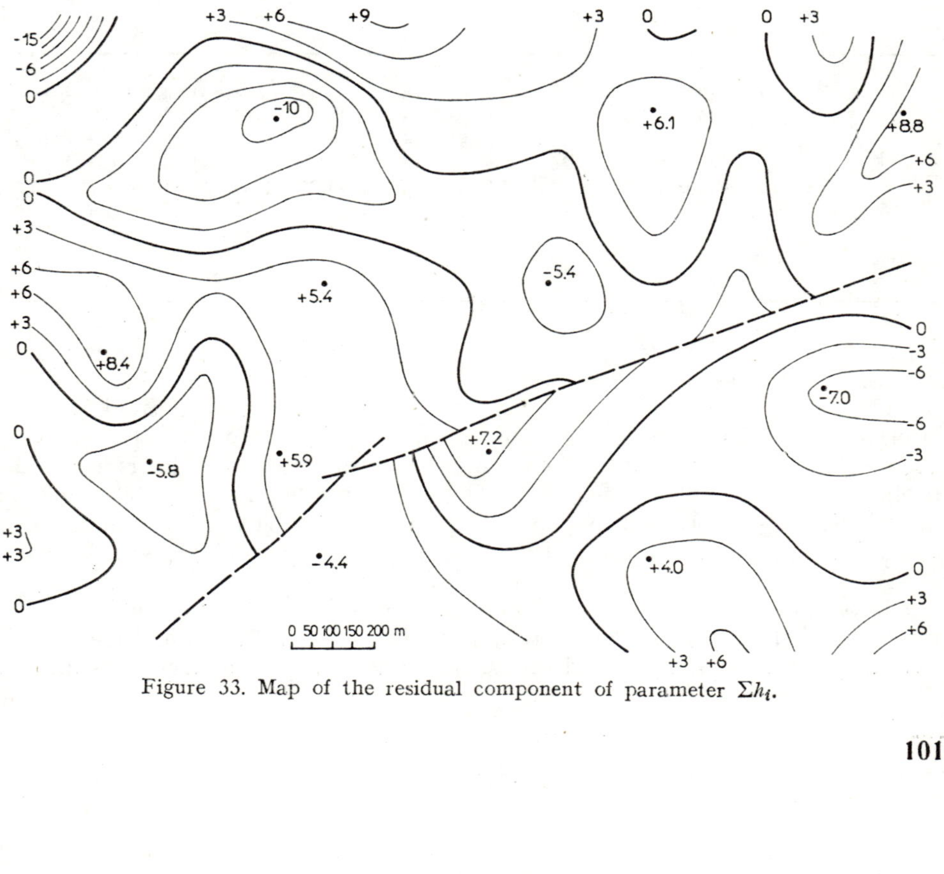

Figure 33. Map of the residual component of parameter Σh_i.

Application of the expected value method for Σh_i

Table 41

	Measured values											Row sums	Row means
	0	25	25	30	32	27	15	17	17	10	18	216	19.6
	20	20	12	8	20	20	17	20	13	11	18	179	16.3
	20	15	14	18	17	18	16	21	12	17	12	180	16.4
	28	27	25	27	24	20	13	16	11	15	15	221	20.1
	16	26	15	23	22	16	15	14	10	3	0	160	14.5
	11	5	8	17	6	17	10	4	3	0	0	81	7.4
	7	4	1	0	0	1	0	4	4	0	0	21	1.9
	3	3	2	0	0	0	0	1	7	2	9	27	2.5
Column sums	105	125	102	123	121	119	86	97	77	58	72	1085*	
Column means	13.1	15.6	12.7	15.4	15.1	14.9	10.8	12.1	3.6	7.2	9.0	—	12.3**

* Grand total;
** Grand mean

Table 42

Trend estimation for Σh_i (expected value method)

Column means	13.1	15.6	12.7	15.4	15.1	14.9	10.8	12.1	9.6	7.2	9.0	Row means
	20.4	22.9	20.0	22.7	22.4	22.2	18.1	19.4	16.9	14.5	16.3	19.6
	17.1	19.6	16.7	19.4	19.1	18.9	14.8	16.1	13.6	11.2	13.0	16.3
	17.2	19.7	16.8	19.5	19.2	19.0	14.9	16.2	13.7	11.3	13.1	16.4
	20.9	23.4	20.5	23.2	22.9	22.7	18.6	19.9	17.4	15.0	16.8	20.1
	15.3	17.8	14.9	17.6	17.3	17.1	13.0	14.3	11.8	9.4	11.2	14.5
	8.2	10.7	7.8	10.5	10.2	10.0	5.9	7.2	4.7	2.3	4.1	7.4
	2.7	5.2	2.3	5.0	4.7	4.5	0.4	1.7	0	0	0	1.9
	3.3	5.8	2.9	5.6	5.3	5.1	1.0	2.3	0	0	0	2.5

and lines. The residual component is obtained by subtracting the trend from the observed values.

The cartographic image of the trend estimated by this method is quite different from that supplied by the method of the abbreviated regression (Figure 34); the form and the intricate connections of the isopachytes emphasize the drawback of this method when it is applied to intricate geological conditions. In support of this conclusion we have moreover the value of the residual mean, which is quite different from zero (—33.3/88 = —0.38).

Method of the mobile means. The cartographic image (Figure 35) of the trend supplied by this method (by also adding a slight graphical smoothing of the isopachytes) can be placed among the cubic and quartic trends estimated

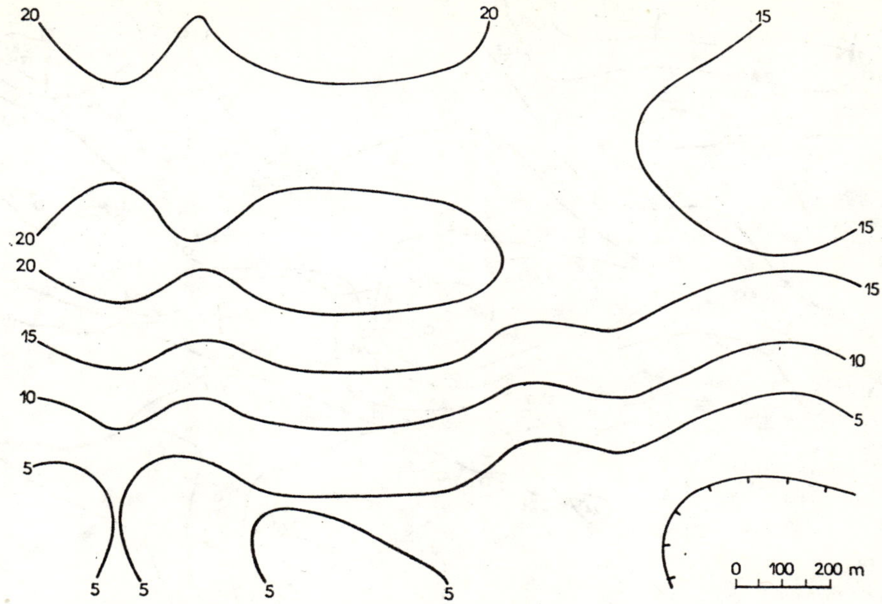

Figure 34. Map of the trend for Σh_i by the method of "expected value".

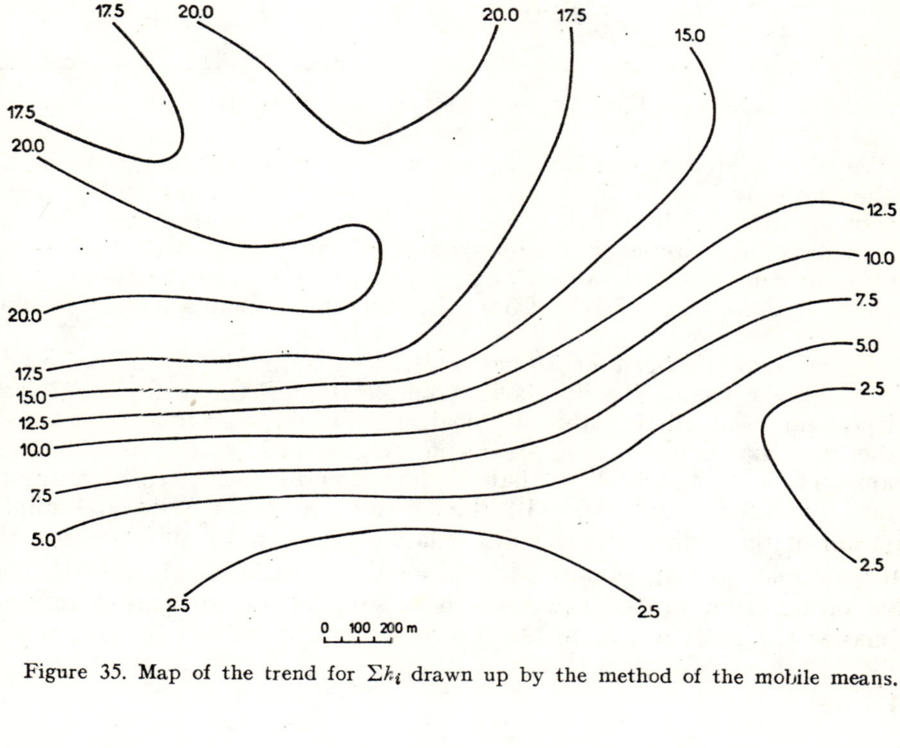

Figure 35. Map of the trend for Σh_i drawn up by the method of the mobile means.

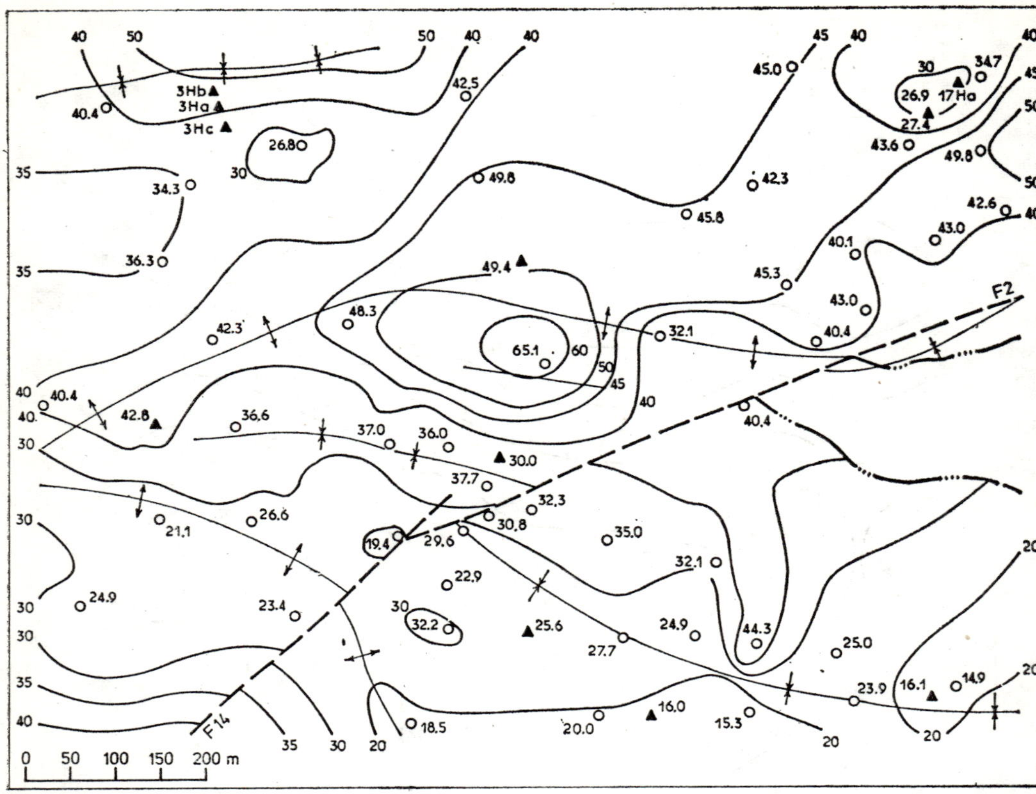

Figure 36. Isopachyte map for coal interval II—III (H).

by the abbreviated method of regression. Obvious changes appear in the south-eastern and north-western areas of the perimeter. Thus, in the south-eastern area the sandless area is missing while in the north-western one the 17.5 m isopachyte appears. These conspicuous differences are due to the insufficient number of data and this imposed to estimate the trend at the considered points on the basis of only six observation points instead of nine.

Estimation of the trend for parameters H, C, y_0, N and isobaths of coal bed II. Taking into account the lithofacies complexity of the complex, it was indispensable that all the other lithological parameters be treated by means of the abbreviated method of regression. A general characteristic of these parameters (H, C, y_0 and N) is that the total trend is estimated by a quintic surface (Figures 37, 39, 41 and 43), which is imposed by the quite non-uniform variation of the values of the polynomial coefficients. In all the cases, the principal trend was approximated by a quadratic surface owing to the high value of the sum of the squares. The results of the analysis of variance emphasize the great weight of the unsystematic fluctuations in the case of

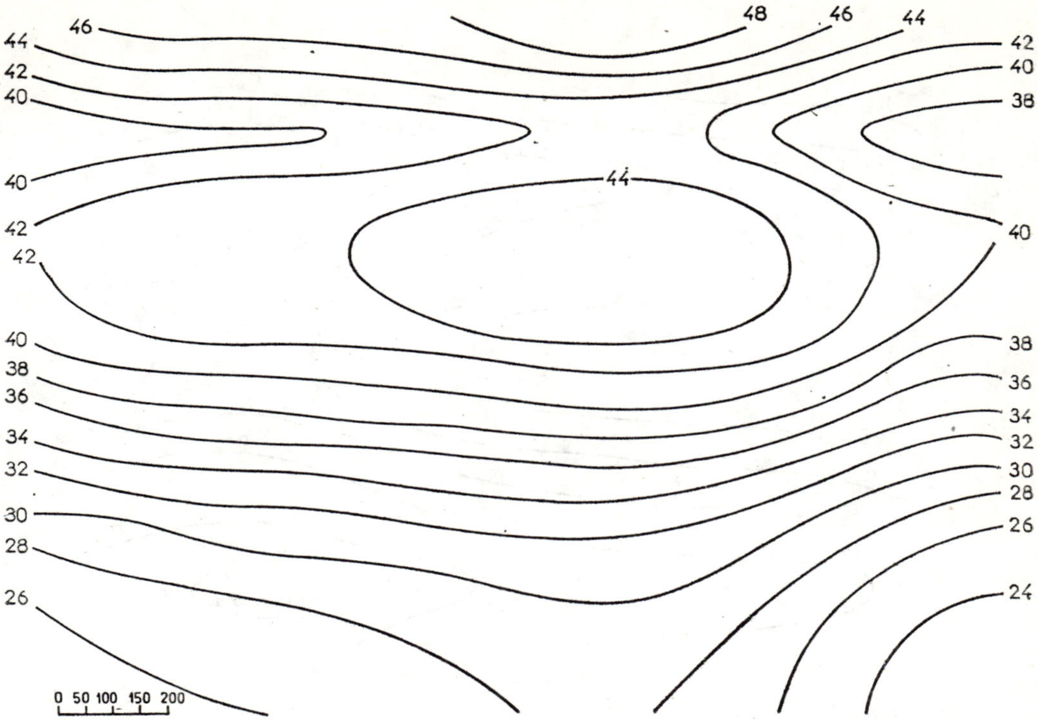

Figure. 37. Total trend map of parameter *H*.

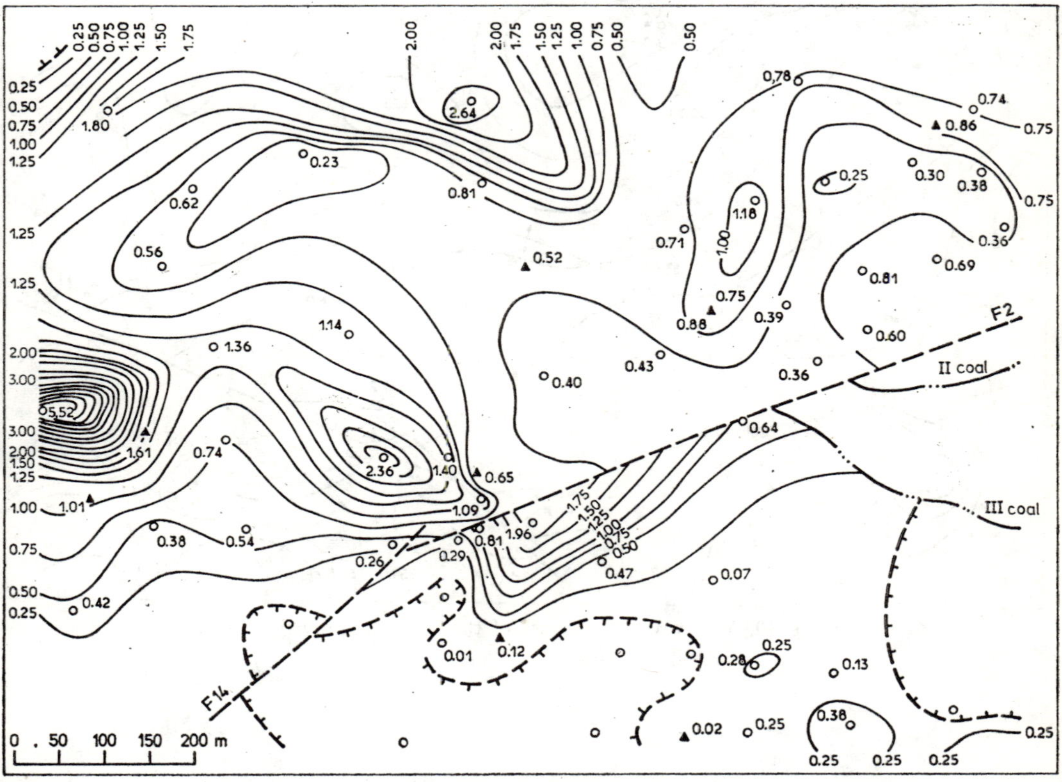

Figure 38. Map with the isolines of lithological coefficient *C*.

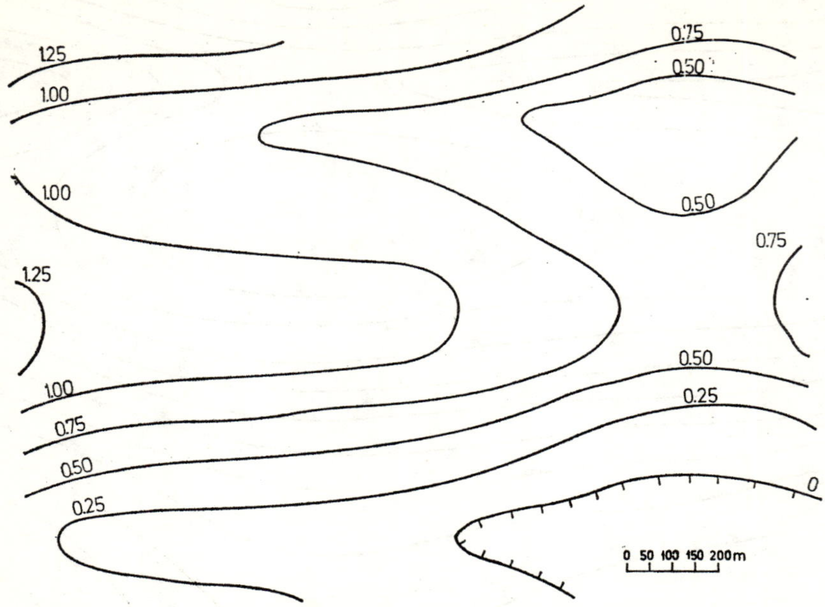

Figure 39. Total trend map (V degree) of lithological coefficient *C*.

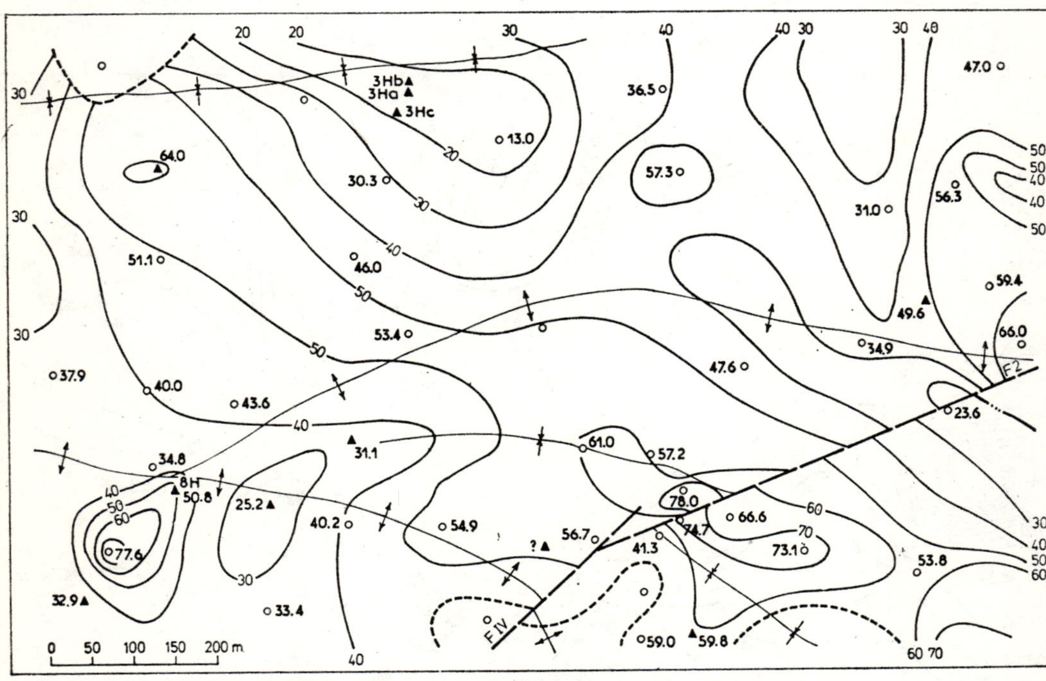

Figure 40. Map with the isolines of the centre of gravity y_0 (%).

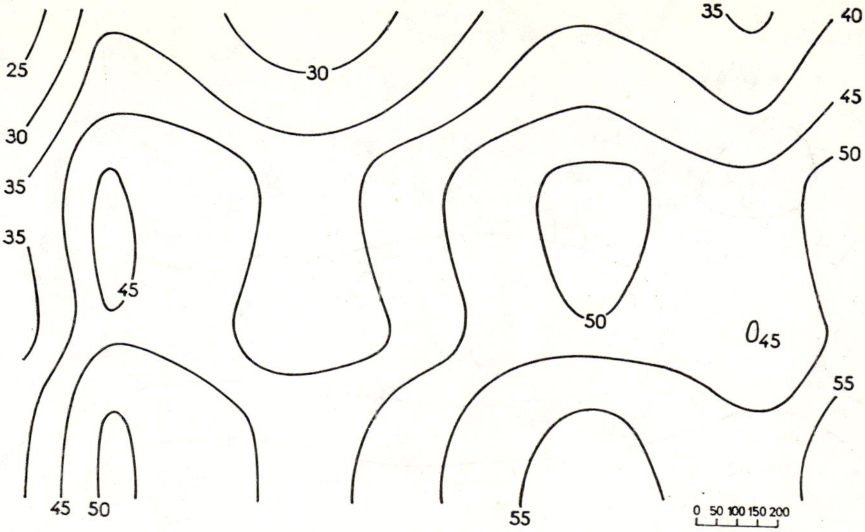

Figure 41. Total trend map of the centre of gravity y_0 (%).

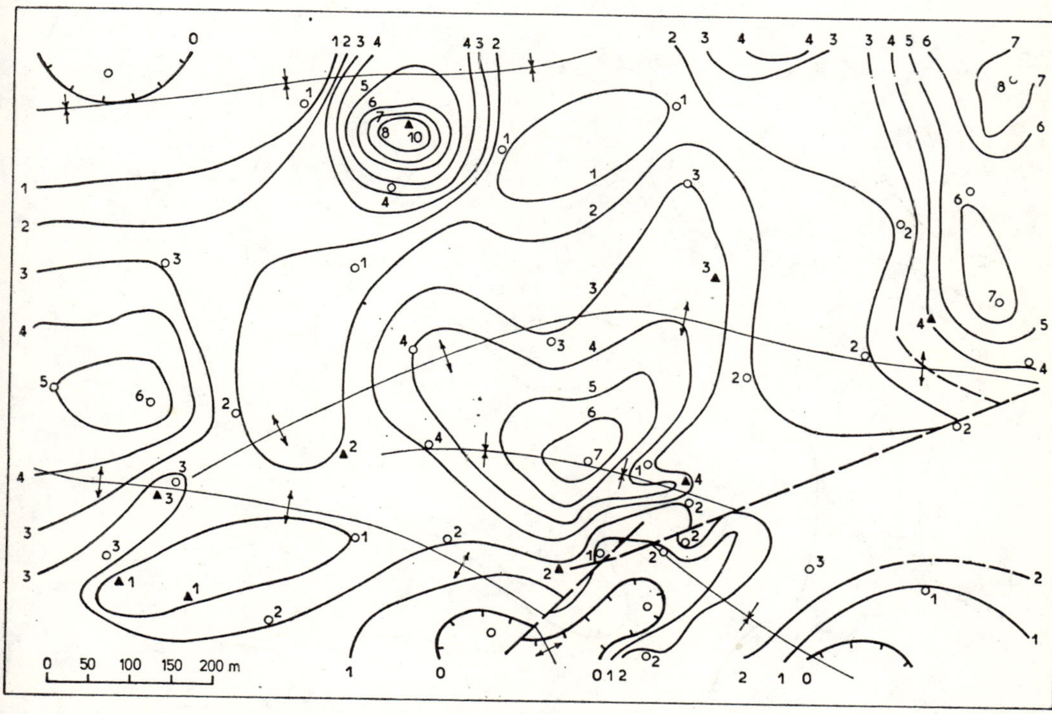

Figure 42. Map of the isolines representing the number of sandy bodies (N) by drilling.

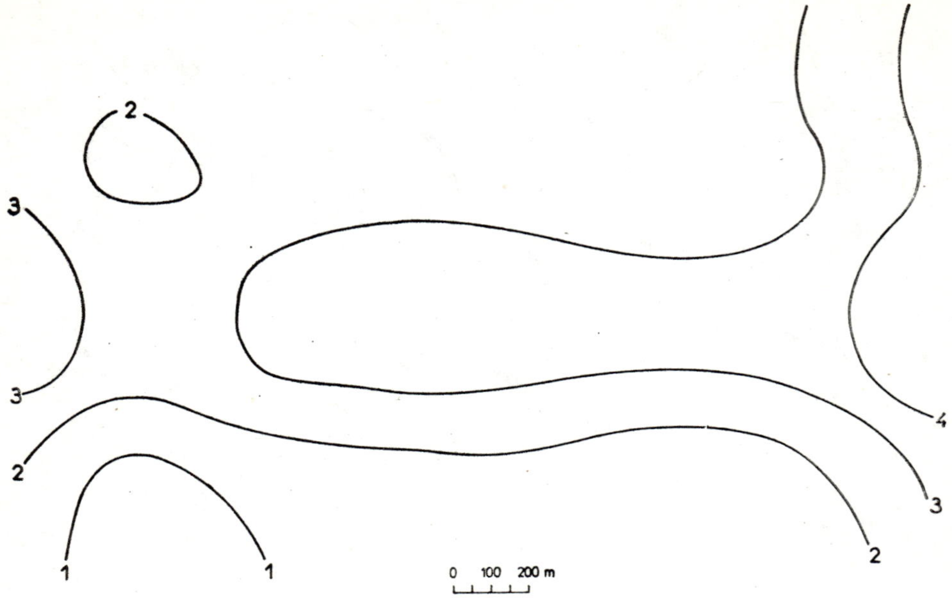

Figure 43. Total trend map of parameter *N*.

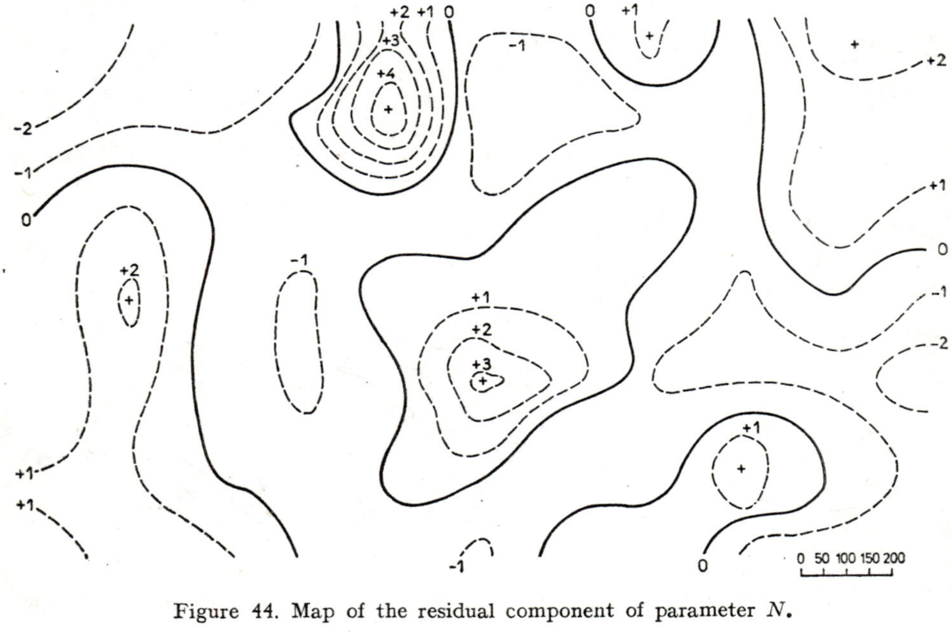

Figure 44. Map of the residual component of parameter *N*.

parameters y_0 and N, which shows that the value of the residual component is higher than that of the total trend. Then, in the case of parameter N,the maximum values of the total trend and of the residual component must be equal ($N = 4$).

The isobaths of coal bed II drawn on the basis of the observation data present a lower degree of variability than the above lithological parameters.

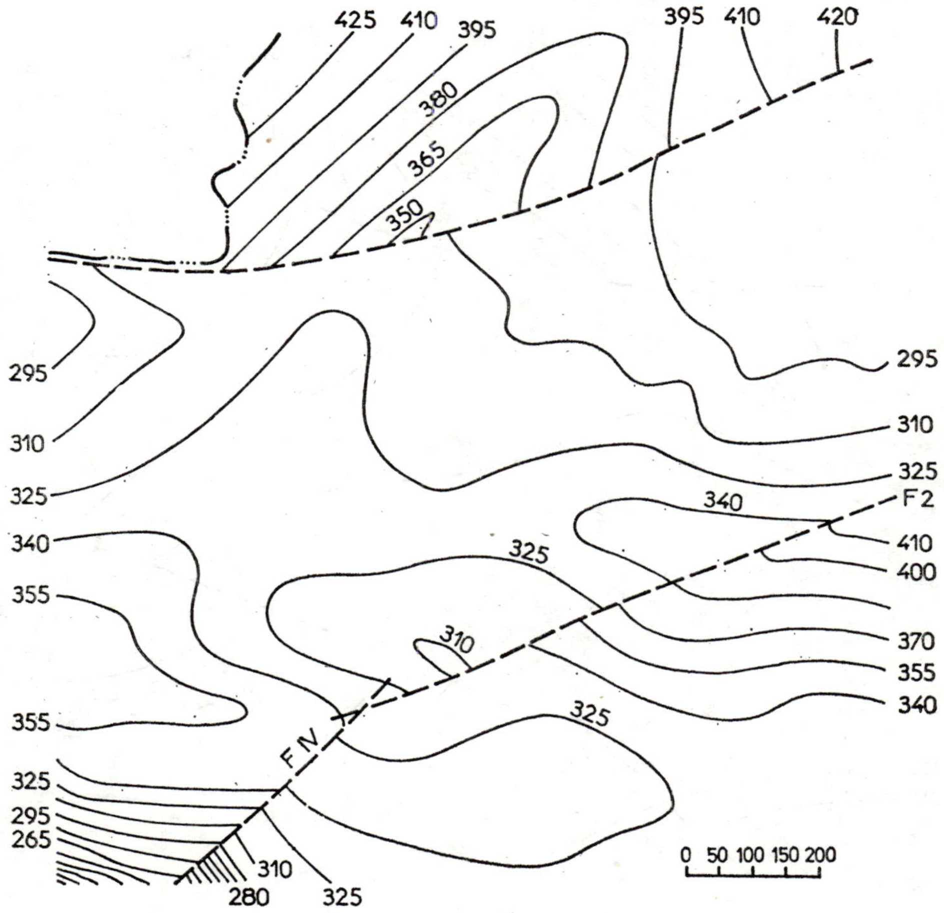

Figure 45. Isobath map at the level of coal layer II.

Although the structural map (Figure 45) shows clearly the trends of each of the tectonic compartment, we proceeded to smooth them with the help of the mobile means. Such a map (Figure 46) — of the trend of isobaths — is very useful to hydrogeologists for appreciating the regional dynamics of the aquifer. With its help, the principal directions of the undisturbed ground water flow can be established; when establishing the drainage scheme it is necessary to know the ratio of the draining flow to the bed slope.

When approximating the trend corresponding to the lithological parameters, fault F_2 was not taken into consideration, the parameters being assumed as continuous variables depending only on co-ordinates x and y. In the case of the isobaths, the situation is different since the incline reaches its maximum value of 80 m. When this is the case, the delimitation of the

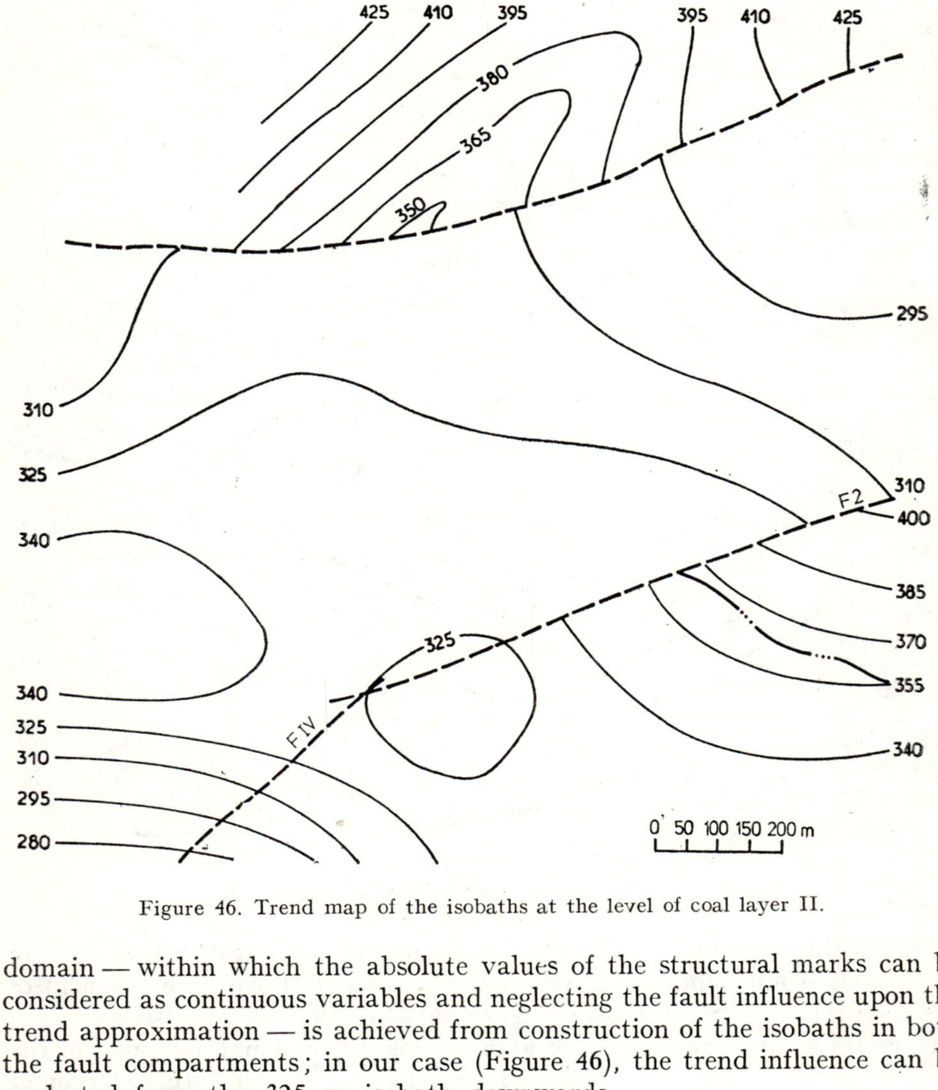

Figure 46. Trend map of the isobaths at the level of coal layer II.

domain — within which the absolute values of the structural marks can be considered as continuous variables and neglecting the fault influence upon the trend approximation — is achieved from construction of the isobaths in both the fault compartments; in our case (Figure 46), the trend influence can be neglected from the 325 m isobath downwards.

1.6.6 Hydrogeological importance of the trend and of the residual component

To appreciate the hydrogeological conditions, parametric maps are used that permit us to obtain an overall view of the distribution of the sandy bodies (or of other geological bodies constituting the permeable lithofacies) and also to emphasize certain quantitative relations necessary for estimating the complexity and difficulty of the hydrogeological conditions. In most cases however the cartographic image of the parametric maps is too complicated to permit a suitable interpretation. The drawing of the trend map as well as of those with residual components offers a convenient and effective method of hydrogeological interpretation of the distribution of the permeable bodies which finally may lead to a zoning of the hydrogeological conditions.

From the computation made for perimeters Cărbunaru-Gheboeni in the coal interval II—III a marked diversity resulted from the trend residual component relationships of the various parameters, which makes the global interpretation more labourious (Table 43).

Table 43

Comparative results of the analysis of variance

Significane	Σh_i	H	C	y_0	N	Isobaths
Principal trend (%)	62.50	53.44	43.36	32.51	21.80	—
Secondary trend (%)	6.61	4.68	8.06	7.90	6.40	—
Total trend (%)	69.11	58.12	51.42	40.41	28.20	76.5
Residual component (%)	30.89	41.88	48.58	59.59	71.80	23.5
Residual mean	−0.056	−0.03	−0.022	−0.008	+0.04	+1.71

In the geological literature, examples of the trend analysis of quite variable lithofacies are rare. A similar case was solved by Matuszak by means of an electronic computer for a clayey-sandy horizon (middle Pennsylvanian age) in southern Oklahoma U.S.A. [46]. The thickness of this horizon varies from 90—150 m. By the complete regression method, Matuszak found that the total trend confirms 53% of the squares sum in the case of the structural data, 65% in the case of the isopachytes of the horizon and 30% in the case of the sand/clay coefficient which means in general a large residual variability just as in the case of the perimeter considered above.

Generally, the practical significance of the trend maps is more important when the trend contribution (as materialized by the percentage of the squares sum) is larger. In intricate geological and hydrogeological conditions, the trend contribution diminishes, which leads to an even greater weight of the residual component than in the case of parameters y_0 and N. Evidently, when this is the case, the complexity of the hydrogeological conditions is very great and so as to be able to appreciate it, we also must appeal to the residual maps.

By a separate interpretation of the trend maps, one may affirm, for instance, that the hydrogeological conditions in the top of a coal layer (underground worked) become more difficult and intricate, as the values of parameters Σh_i, C and N are higher, while the value of y_0 is lower. From the complex interpretation of the maps of principal trend, drawn for the particular perimeter, it follows that the intricacy of the hydrogeological conditions grows higher particularly when going from south towards the north. The principal change introduced by the secondary trend into the image of the total trend is the appearance of certain areas with small parameters Σh_i, H and C and of an area with maximum values of y_0. These maps are very useful for estimating the hydrogeological conditions of the mining working and moreover for their progress schedule (as long as no other more important criteria occur). In this way it is possible to estimate the volume of dewatering work and the duration of dewatering by panels, both of which influence the working tehnology as well as the cost price.

The trend maps of the permeable facies are useful for any hydrogeological problems of an economic or technical nature, when the lithological structure is complex. For example the capture perimeter of an aquifer can also be determined by the high values of parameters Σh_i, N and C as well as by the small value of the centre of gravity ($y_0\%$) with regard to the upper or lower limit of the aquifer; under these conditions, for a uniform permeability of the aquifer the considered perimeter offers the best hydraulic and economic capture conditions. Obviously, as a function of the structure peculiarities of the water bearing formation, other parameters of the lithofacies may be taken into consideration.

1.6.7 Application conditions of the methods of trend separation

In connection with the use of the simplified methods, which obviously offer the advantage of standard computations, the problem arises of the conditions under which they can be used with risk that their results would be affected by gross errors in regard to those supplied by the more elaborate method of complete regression. W. C. Krumbein [36] distinguishes in the problems of lithofacies three situations:

1. The lithofacies characteristics are simple enough, so that all cross-product polynomial terms can be disregarded while the tendency can be estimated by a quadratic surface. If the number of rows and columns is taken to be small (to avoid the terms of higher degree), the "expected value" method can be applied by orienting the network along the variation direction of the observation data. As a consequence, the negative result appearing in the case of parameter Σh_i is justified because the characteristics of the lithofacies are rather intricate.

2. If the relations between the grid orientation and the lithofacies gradient agree enough to permit the discarding of the cross-product polynomial terms

it will be possible to use the method of the abbreviated regression which permits us to compute all the direct polynomial terms as well as to select the most important ones. The use of the "expected value" method would lead in such cases to a trend surface with many undulations (Figure 34).

3. If the characteristics of the lithofacies are very intricate, making it impossible to appreciate the principal variation directions, it will be necessary to take into account the cross-product polynomial terms, which can be accomplished by the method of the complete regression.

It should also be stated that the method of the mobile means can approximate the greatest part of the total trend; the isolines so obtained should however be submitted to a graphical smoothing.

1.7 Estimation of the linear correlation between the lithological parameters

The correlation method analyses the connections and the interdependence of the various factors and expresses in the form of synthetic indicators the strength and the direction of the bond as well as their degree of influence. In the quantitative analysis of the geological problems one must take into account the existence of the complex relations, which suggest we should consider, together with the dependent variable, several independent ones. As a consequence, the computation of the correlation exceeds the frame work of the pairs of variables and includes moreover other independent variables which have a significant influence upon the dependent variable under study. By so doing we find the multiple linear correlation which possesses several specific indicators. The correlation between x and y for instance may be completely or partially due to the strong bond between x and y on the one hand and by a third random value z on the other hand. Information concerning the degree of dependence between dependent variable x and a single independent variable y with the exclusion of the other independent variable z is found with the help of the partial correlation coefficient [62]

$$r_{xy(z)} = \frac{r_{xy} - r_{xz} \cdot r_{yz}}{\sqrt{(1 - r_{xz}^2)(1 - r_{yx}^2)}} \tag{94}$$

$$r_{xz(y)} = \frac{r_{xz} - r_{xy} \cdot r_{yz}}{\sqrt{(1 - r_{xy}^2)(1 - r_{yz}^2)}} \tag{95}$$

$$r_{yz(x)} = \frac{r_{yz} - r_{xy} \cdot r_{xz}}{\sqrt{(1 - r_{xy}^2)(1 - r_{xz}^2)}} \tag{96}$$

where r_{xy}, r_{xz} and r_{yz} are the estimations of the linear-correlation coefficient determined by formula

$$r = \frac{n\Sigma xy - \Sigma x \Sigma y}{\sqrt{[n\Sigma x^2 - (\Sigma x)^2][n\Sigma y^2 - (\Sigma y)^2]}} \tag{97}$$

Since the correlation is computed on the basis of the sampled population, its probabilistic estimation is necessary for characterizing the relation between the parameters involved. The criteria are the same as in the case of a simple correlation.

Hereafer, the correlation between parameters Σh_i, H, C, y_0 and N for coal interval II—III (Șotînga-Mărgineanca deposit) will be estimated in two forms: the values observed in the drillings and the values corresponding to the total trend.

Correlation between the parameters representing the effective values. Taking into consideration the Cărbunaru mining field with 44 drillings, the linear-correlation coefficients of simple relation (Table 44) were estimated which emphasized the low intensity of the bonds between $\Sigma h_i — C$ and $H — y_0$. For examining the significance of correlation coefficients r, in the case of small sample size ($n < 50$) criterion

$$\varkappa = \frac{Z}{\sigma_Z}; \quad \sigma_z = \frac{1}{\sqrt{n-3}} \tag{98}$$

may be used, where $Z = \frac{1}{2} \ln \frac{1+r}{1-r}$ is Fisher's criterion while σ_z is the mean standard deviation. When $\varkappa \geqslant 3$, the coefficient of correlation being tested is significant for the target population ($P = 99.73\%$). When $\varkappa < 3$, whatever the absolute value of r, the linear correlation between the respective parameters is considered as non-significant. In our case, excepting linear correlation $\Sigma h_i—C$, all the others are not significant (Table 44).

Table 45 shows that the values of the coefficient of partial correlation are quite close to those of the coefficient of linear correlation which makes it

Table 44

Values of the linear-correlation coefficients

Parameter	H		y_0		C		N	
	r	$\varkappa$	r	$\varkappa$	r	$\varkappa$	r	$\varkappa$
Σh_i	−0.014	0.09	+0.10	0.64	+0.54	3.87	+0.11	0.71
H	—	—	−0.34	2.27	−0.25	1.64	+0.12	0.77
y_0	—	—	—	—	+0.07	0.45	+0.17	1.10
C	—	—	—	—	—	—	+0.12	0.77

Table 45

Values of the partial-correlation coefficients

Parameter	H		y_0		N	
	r	$\varkappa$	r	$\varkappa$	r	$\varkappa$
$\Sigma h_i - C$	+0.53	3.78	+0.55	3.96	+0.53	3.78

impossible to emphasize the influence of each of the factors and so to exclude the influence of the other ones.

A qualitative analysis of the linear correlation can be made with the help of the correlation field built in the system of logarithmic co-ordinates. A notable strength of the linear relation corresponds to a correlation field in the form of a very elongated domain or of an ellipse, as in the case of correlation $\log(\Sigma h_i) - \log C$ (Figure 47). When the correlation fields have a nearly circular form $[\log(\Sigma h_i) - \log Z;\ \log C - \log Z]$ or a very elongated form on the horizontal or on the vertical, the linear correlation practically vanishes.

Correlation between the parameters representing the values of the total trend. Taking into account the separation of each of the measured values into its two components — trend and residual—the problem arises as to whether a stronger linear correlation exists between the trend values.

With the help of relation (97) we computed the coefficients of linear correlation coefficients, including the probabilistic estimation (table 46.)

Table 46

Values of linear-correlation coefficients (total trend)

Parameters	H		y_0		C		N	
	r	$\varkappa$	r	$\varkappa$	r	$\varkappa$	r	$\varkappa$
Σh_i	+0.82	22.77	−0.63	8.28	+0.91	50.56	+0.35	3.15
H	—	—	−0.56	6.43	+0.71	13.15	+0.18	1.48
y_0	—	—	—	—	−0.65	10.48	−0.12	1.07
C	—	—	—	—	—	—	+0.33	2.94

The linear correlation between the structure Z and the other parameters was not the object of the computation since the absence of this dependence results from the correlation fields (Figure 47).

The values of r and χ, recorded in Table 46 prove the existence of a significant linear correlation between parameters $\Sigma h_i - H - y_0 - C$ and $|r| = 0.56 - 0.91$. The linear correlation with N is generally weak but it is significant for Σh_i and C. In order to eliminate these influences, the partial correlation coefficients (Table 47) were also computed (Table 47).

In connection with the importance of the multiple linear correlation, P. I. Sharapov points out that in the general case of relations between variables x, y and z, the simple correlation between x and y can be very strong while the partial correlation resulting from the elimination of the z influence can vanish or even change of sign. This is due to the fact that the strength and the character of the connection are determined on the one hand by x and z and on the other hand by z and y, from which it follows that no direct connection exists. That is why a multilateral analysis of the connection between the parameters is

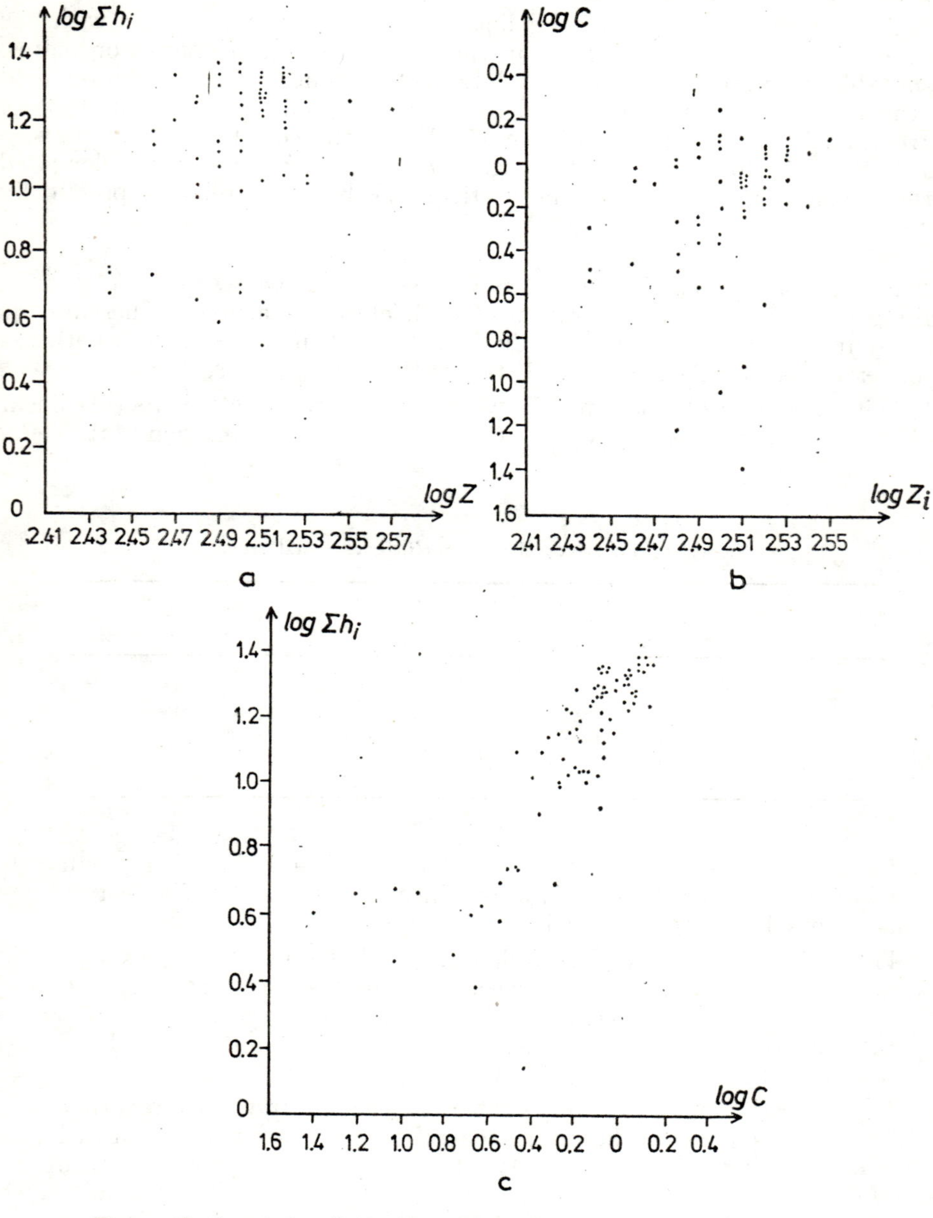

Figure 47. Correlation fields (Z — elevation at the top of the coal layer II).

Values of the partial-correlation coefficients (total trend)

Parameters	$\sum h_i$		H		y_0		C		N	
	r	$\varkappa$	r	$\varkappa$	r	$\varkappa$	r	$\varkappa$	r	$\varkappa$
$\Sigma h_i - H$	—	—	—	—	+0.73	14.60	+0.59	8.55	+0.82	22.77
$\Sigma h_i - y_0$	—	—	−0.35	3.15	—	—	−0.13	1.04	−0.63	8.28
$\Sigma h_i - C$	—	—	+0.80	21.05	+0.85	28.33	—	—	+0.91	50.55
$\Sigma h_i - N$	—	—	+0.35	3.15	+0.35	3.15	+0.13	1.04	—	—
$H - y_0$	−0.09	0.72	—	—	—	—	−0.19	1.57	−0.55	6.40
$H - C$	−0.15	1.44	—	—	+0.55	7.33	—	—	+0.70	13.10
$y_0 - C$	−0.25	2.12	−0.42	4.04	—	—	—	—	−0.65	10.48
$C - N$	+0.03	0.24	+0.29	2.52	+0.33	2.94	—	—	—	—

necessary, the linear correlation methods being insufficient to estimate the reciprocal influence between several parameters.

In our case, the general diminution of the intensity degree of the partial correlation with regard to that of the linear one should be noted; the only exception to this rule is parameter N the influence of which is equal to zero because the values of both the series are practically constant. As a function of the degree of influence the parameters could even be arranged in a hierarchical system. Let us then note the very strong influence of Σh_i on $H - y_0$, $H - C$, $y_0 - C$ and $C - N$; if the influence of Σh_i is eliminated, these correlations become very weak and insignificant. A special situation is presented by correlation $H - C$ the sign of which changes into negative. The influence of H is strong upon correlations $\Sigma h_i - y_0$ and $y_0 - C$ the intensity of which diminishes considerably. The influence of C upon correlations $\Sigma h_i - y_0$ and $H - y_0$ appears to be likewise. The influence of parameters y_0 and N are small upon the correlations.

From the analysis of the linear and partial correlations it can be deduced that parameter Σh_i has a very strong influence upon all correlations, while H and C have only a limited action. It is therefore to be expected that any change in Σh_i value would be followed by variations of the other parameters. The influences of y_0 and N upon the correlations is practically insignificant.

From a theoretical standpoint we must emphasize the fact that by eliminating the residual component, the dependency between the values of the total trend has become much greater with regard to the correlations of the effective values. This also speaks in favour of the random character of the residual component. The hydrogeological importance of the linear correlation is that it shows the complexity of hydrogeological conditions: the higher the number of these correlations, the simpler the hydrogeological conditions become. In our case, the number and the intensity of the correlations grows with the values of the total trend while for the effective values a single linear correlation is obvious.

The partial correlation can bring into relief the parameters the contributions of which are important for the complexity of the hydrogeological conditions; in our case they are in order of importance: Σh_i, H and C.

1.8 Sedimentological considerations of a hydrogeological importance

The trend and residual maps combined with the indicators of the linear correlations can be used for emphasizing certain reciprocal ratios between the sedimentary processes and the development of the tectonic structure syncronous with the sediment accumulation. Evidently so as to be able to explain the influence of the tectonic movements upon the primary variations of the layer thickness, maps of trends with a regional character should be necessary but do not exist in the case of the Şotînga-Mărgineanca deposit. Nevertheless, some interpretations could be made and would be very useful for guiding the hydrogeological researches. It has already been pointed out that between the structures Z and H on the one hand and C on the other hand no significant linear correlation exists; this attempt was made for the tectonic compartment between fault F_1 in the north and F_2 in the south, up to the area where the throw becomes insignificant. This does not exclude however the existence of certain non-linear relations between these parameters. Between the principal trend of H and the coefficient C a very close parallel exists, the respective values growing from south to north which means that the areas possesing the thicker sediments are richer in sands; the same remark is valid for the total trend. Comparing the structure trend on the one hand, with the principal trend of parameters C and H on the other hand, no parallelism appears. However, by taking into account the total trend of parameters C and H we find:

— The western area of structural maximum ($Z > 340$ m) corresponds to the highest values of coefficient C (0.75—1.25) while the north-eastern area of structural minimum corresponds to small values of C (< 0.50);

— On the northern slope of the anticline, the C values increase while on its southern slope they diminish;

— The central area that shows the highest values of H has no significant relations with the structure.

This information shows that the sand sedimentation was partially controlled by the simultaneous development of the structure. Additional information about this problem can be obtained from residual maps. Indeed, the areas presenting positive anomalies of thickness H correspond to negative anomalies of coefficient C and vice versa. This constitutes an additional argument for the simultaneity of tectonic activity with sedimentation, when the higher areas — the more energetic ones — have a greater specific content of sandy sediments.

The relations between the structure and coefficient C on the residual maps are just as inconclusive as the trends are, due to the fact that the structurally elevated areas possess negative anomalies of coefficient C. This can be explained by the raising of the area above the basic level of the sedimentation so that the respective sediments could have been subjected to erosion. In this way a new redistribution of the sedimentation conditions could have occurred in which the previous areas of minimum energy were changed into areas of high energy with high values of coefficient C.

These general sedimentological considerations underline once more the complicated relations between the quantitative parameters of the lithofacies distribution, which determine a high complexity of the hydrogeological deposit conditions.

2

HYDRODYNAMICAL INVESTIGATION OF AQUIFERS AND WELLS

In designing capture and dewatering systems of groundwater, besides schematizing the natural hydrogeological conditions, it is important to know the hydrogeological parameters. Their determination constitutes one of the principal problems of hydrogeology, due to the various degrees of accuracy of the methods employed and of the expense of field test. Under these conditions, the methods of determining the hydrogeological parameters must be improved by dropping some of the classical methods in favour of those of greater technical and economic efficiency.

Hydrogeological parameters are divided into two groups:

The first group includes the parameters characterizing the experimental drilling and its adjacent area, namely the rate of discharge Q and the drawdown S which permit us to draw the indication diagram $Q = f(S)$, the well loss Δh and the seepage resistance ζ_0 of the well.

The second group includes the parameters characterizing the aquifer proper in areas more or less remote from the experimental well, to wit the hydraulic conductivity k, the transmisivity kM or kH, the storage coefficient, the hydraulic diffusivity a and the capillary rise h_c. Naturally to the second group could be added the aquifer thickness M or H, the position of the piezometric surface or of the water table as well as other geometrical characteristics of the aquifer (scattering area, slope, shape of its boundaries in the horizontal and in the vertical sections); but these parameters do not involve particular experimental works since they are determined by hydrostatic methods.

The hydrodynamic investigation of the aquifers by pumping, frequently followed by the recovery of the water level in the experimental well constitutes an investigation technique indispensable to any kind of hydrogeological research.

The greatest influence on the ground-water régime belongs to the structure of the aquifer system and to its boundary conditions both in the plane and in cross-section. That is why a typification was proposed of the hydrogeological conditions, in which each aquifer type is characterized by certain features of the ground water régime, i.e. by certain laws of the variation of the dynamic level as a function of time:

1. "Infinite" aquifers homogeneous as to the permeability and isolated both at their top and their bottom by practically impermeable beds:

a. Confined aquifers, within detrital rocks, with an interstitial permeability.

b. Unconfined aquifers, within detrital rocks with interstitial permeability.

c. Unconfined and confined aquifers, in fissured and karstic rocks.

2. Aquifers (two or more) situated in complexes of detrital strata, constituting the so-called leaky systems:

a. Aquifers separated from one another by very thick semipervious layers (usually clayly).

b. Leaky system consisting in a sequence of not too thick aquiferous and semi-pervious layers.

3. Limited aquifers, the depression cone of which created by pumping is extended up to the barrier or recharge boundary.

a. Aquifers hydraulically connected to water courses or to lakes.

b. Limited aquifers connected to one or more boundaries. As a function of their number and of their geometry they can be divided into;
— Semi-infinite aquifer in the case of a straight boundary;
— Rectangular aquifer in the case of two intersecting boundaries at an approximately right angle;
— Linear aquifers determined by two parallel straight boundaries;
— Closed aquifers determined by the presence of several boundaries closing a more or less circular surface.

c. Aquifers formed by several areas with various hydraulic properties, separated by linear or circular or more complicated boundaries.

d. Aquifer areas (groups) representing local foci of recharge or discharge.

4. Aquifers characterized by a great non-homogeneity and anisotropy of their parameters so that their models can only be established by statistical processes.

The certainty degree of the tested aquifer parameter depends on the importance that the design and the accomplishment of the hydrodynamic test granted to the features of the hydrogeological conditions as well as to theoretical achievements in this field.

This chapter presents the solution by analytical or graphoanalytical methods, of a few actual hydrogeological cases belonging to the abovementioned scheme.

2.1 Hydrogeological parameters

2.1.1 Yield-drawdown diagram

Unlike the hydrodynamic methods, based on equations of mathematical physics and of theoretical hydrodynamics, the hydraulic method uses empirical relations based directly on the experimental data. Among these hydraulic methods, an important role is played by the yield-drawdown diagram together with its characteristic equation, which shows the correlation between the rate of discharge and the drawdown produced in the experimental well. It can be assumed that the yield-drawdown diagram is a global characteristic since it is quantitatively determined by both the hydrogeological parameters of both the aquifer and the experimental well. That is why the construction of the yield-drawdown diagram of an experimental well (requiring the achievement of the equilibrium pumping test) is only necessary when the catchment or drainage of the respective aquifer will be realized by wells similar to those of the experimental ones, or else when the so-called exploration-exploitation wells are used. When only the determination of the filtration properties of the aquifer is intended or when the drainage is to be made by special works (as in the case of subsurface drainage for the lignite mine working), it is quite reasonable that in each experimental well a single experimental pumping should be made. The form and the equation of a yield-drawdown diagram are determined by the flow régime of the aquifer (by gravity for the unconfined aquifer, by water drive and elasticity for the confined aquifer) by the filtration régime, by the non-permanent processes in the adjacent area of the experimental well (suffusion, clogging), by the hydraulic resistances of the well screen and casing.

Specific capacity q, the parameter of the well productivity, expresses the discharge rate per unit of drawdown.

$$q = \frac{Q}{S_0} \quad [\mathrm{l/s.m}]$$

Its values permit us to ascertain the correctness of the pumping test under equilibrium conditions and to estimate the type of correlation between Q and S_0 (Figure 48). The increasing specific capacity with the drawdown (curve 1) means that the pumping data are not correct and that probably the equilibrium condition has not been fulfilled. The upholding of a constant value of the specific capacity (curve *II*) corresponds to a laminar régime with water drive or elastic and with negligible hydraulic resistances of the well. If the oscillations of the specific capacity are small around a mean value (curve *III*), the latter can be computed as an arithmetic mean $q = \frac{q_1 + q_2 + q_3}{3}$;

For the ratios $\frac{q - q_1}{q}$, $\frac{q - q_2}{q}$ and $\frac{q - q_3}{q}$ are greater than 3%, Altovski recommends that the pumping should be considered unsatisfactory; if error ε is smaller than 3%, the specific capacity can be considered constant and equal to the mean value q. Curve *IV* corresponds to a parabolic correlation between Q and S_0. The diminishing of the specific capacity according to curve *V* indicates a logarithmic or exponential dependence between Q and S_0. As a function of the hydraulic peculiarities, cases *IV* and *V* can occur in unconfined aquifers and cases *II, III, IV* and *V* in confined aquifers.

In establishing the equations of the yield-drawdown diagrams, it should be taken into account that they are not functional relations but probabilistic correlations containing one or two empirical parameters whose values can be

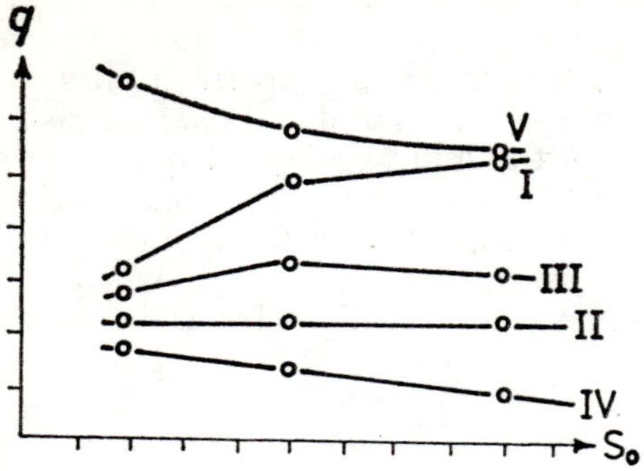

Figure 48. Various dependency cases of drawdown S_0 on specific yield q.

determined by means of the error analysis. According to the required accuracy, the parameter estimation can be made graphically (Figure 48) or analytically by using either the method of the compensating errors or the method of least squares. The most proper one, in cases where the number of the parameters is less than that of the pumpings, it is the last method that supplies efficient estimates of the parameters.

The analytic expression of the yield-drawdown diagram of a perfect experimental well, which penetrates up to the bottom of an unconfined aquifer, is derived from Dupuit's equation

$$Q = 1.366\, k \frac{(2H - S_0)\, S_0}{\log \dfrac{R}{r_0}} \tag{99}$$

where R is the radius of influence and r_0 is the radius of the experimental well

$$Q = C(2HS_0 - S_0^2) \tag{100}$$

$$Q_{max} = C(2H^2 - H^2) = CH^2 \tag{101}$$

$$\frac{Q}{Q_{max}} = \frac{2HS_0 - S_0^2}{H^2} \tag{102}$$

$$\frac{Q}{Q_{max}} = H_0^2 = 2HS - S_0^2 \tag{103}$$

$$H^2 - \frac{Q}{Q_{max}} H^2 = H^2 - 2HS_0 + S_0^2 \tag{104}$$

$$(H - S_0)^2 = \frac{H^2}{Q_{max}} (Q_{max} - Q) \tag{105}$$

$$(H - S_0)^2 = \alpha(Q_{max} - Q) \tag{106}$$

The value of Q_{max} (theoretical maximum rate of discharge corresponding to $S_0 = H$) is estimated by the method of the least squares

$$Q_{max} = \frac{\Sigma H^4 Q - \Sigma H^2 Q (H - S_0)^2}{\Sigma [H - (H - S_0)^2]^2} \tag{107}$$

For confined aquifers, the equation of the yield-drawdown diagram can be linear, parabolic, logarithmic or exponential.

Hereafter, the basic equations are presented as well as the estimation relations of the parameters resulting from the application of the method of least squares.

The linear correlation (Figure 49) is supplied by formula

$$Q = qS_0 \tag{108}$$

where the specific capacity can be obtained by relation

$$q = \frac{\Sigma S_0 Q}{\Sigma S_0^2} \tag{109}$$

This correlation represents an ideal case when the hydraulic properties in the vicinity of the experimental well are not changed and the seepage resistance of the well is non-significant.

In case that within the depression cone appears moreover the turbulent flow, the correlation between Q and S_0 can be parabolic:

$$S_0 = aQ + bQ^2 \tag{110}$$

where a and b are coefficients characterizing the laminar and turbulent flow.

To verify if a situation is of the parabolic correlation type, the equation will be linearized by dividing all its terms by Q:

$$\frac{S_0}{Q} = s = a + bQ \tag{111}$$

In this case, s is the specific drawdown, i.e. the drawdown required for obtaining a unit of yield:

$$b = \frac{\Sigma s\, \Sigma Q - N\Sigma sQ}{(\Sigma Q)^2 - N\Sigma Q^2} \tag{112}$$

$$a = \frac{\Sigma s - b\Sigma Q}{N} \tag{113}$$

where $N > 2$ is the number of the pumping test. The logarithmic correlation is expressed by

$$Q = a + b \log S_0 \tag{114}$$

which represents the straight-line equation (Figure 49). The parameters of this equation are estimated by relations

$$b = \frac{N\Sigma Q \log S_0 - \Sigma Q\, \Sigma \log S_0}{N\Sigma(\log S_0)^2 - (\Sigma \log S_0)^2} \tag{115}$$

$$a = \frac{\Sigma Q - b\Sigma \log S_0}{N} \tag{116}$$

FORMULA	DIAGRAM	
	basic	transformed
Linear correlation $Q = qS_0$ $q = tg.\alpha$	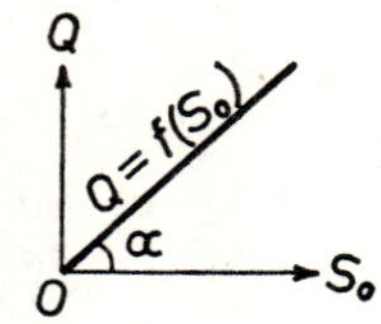	
Parabolic correlation $S_0 = aQ + bQ^2$ $Q = \frac{\sqrt{a^2 + 4bS_0} - a}{2b}$ $b = tg.\alpha$ $s = a + bQ$	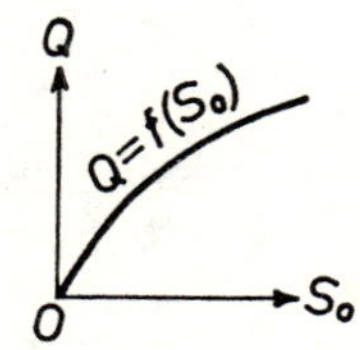	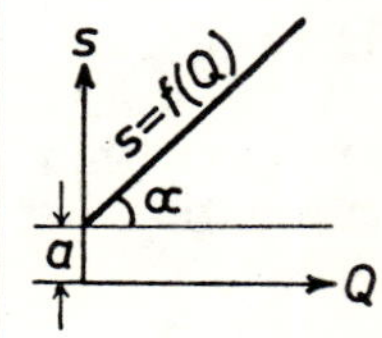
Logarithmic correlation $Q = a + b\log S_0$ $\log S_0 = \frac{Q - a}{b}$ $b = tg.\alpha$	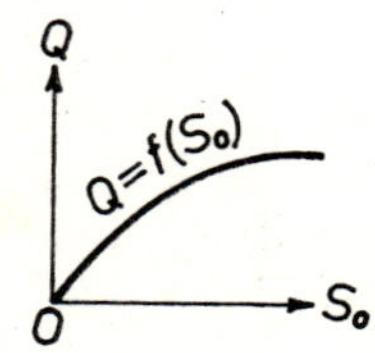	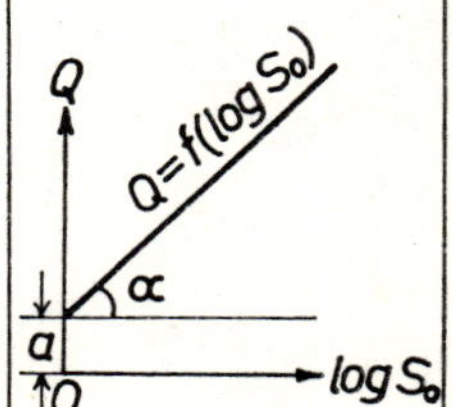
Exponential correlation $Q = q_0\sqrt[m]{S_0}$ $S_0 = \left(\frac{Q}{q_0}\right)^m$ $\log Q = \log q_0 + \frac{1}{m}\log S_0$ $\frac{1}{m} = tg.\alpha$	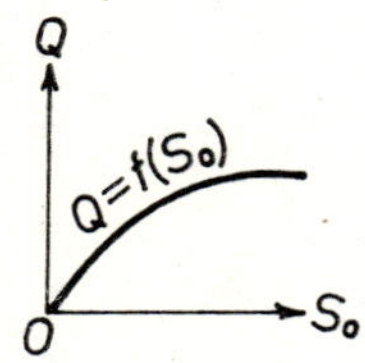	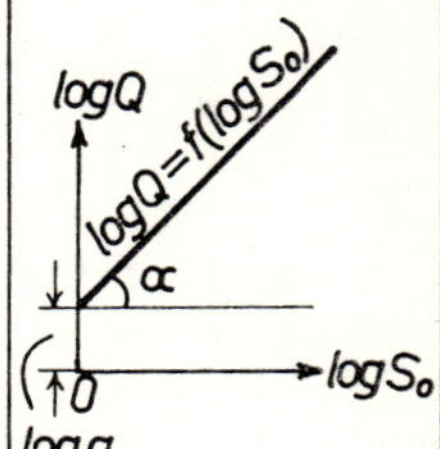

Figure 49. Yield-drawdown diagrams.

Smreker's equation expresses an exponential interdependence between Q and S_0

$$Q = q_0 \sqrt[m]{S_0} \tag{117}$$

where m can assume values going from 1 (in case of the laminar régime hence of a linear correlation $Q = q\,S_0$) to 2 which leads to Krasnopolski's equation that is valid for the turbulent régime

$$Q = q_0 \sqrt{S_0} \tag{118}$$

The linearization of equation (117) permits one to verify the possibility of including in this type of correlation (Figure 49)

$$\log Q = \log q_0 + \frac{1}{m} \log S_0 \tag{119}$$

When $N > 2$, this equation leads to the expression of the following parameters:

$$\frac{1}{m} = \frac{N\Sigma \log S_0 - \Sigma \log S_0 \, \Sigma \log Q}{N\Sigma(\log S_0)^2 - (\Sigma \log S_0)^2} \tag{120}$$

$$\log q_0 = \frac{\Sigma \log Q - \frac{1}{m} \Sigma \log S_0}{N} \tag{121}$$

From the above correlations, only (105) and (108) are expressed by theoretical formulae: they represent the ideal cases when there is no "skin-effect", the filtration régime is laminar and the hydraulic resistance of the borehole is insignificant. These ideal yield-drawdown diagrams can be made conspicuous by the linear dependence between q and S_0 in case of the unconfined aquifer and by the constant value of q in case of a confined aquifer. Deviations from the theoretical situation could also appear in the case of unconfined aquifers, due to the appearance of the turbulent régime, to the clogging effect or poor natural development around the well, or to the imperfection of the latter; in these cases (curve V in Figure 48), all three parabolic, logarithmic or exponential types of correlations that were initially elaborated for the confined aquifer should be tried. The choice of a correlation from these three possibilities is graphically made by means of linearization diagrams; the best collinearity of the experimental points define the real type of correlation.

An important problem of the yield-drawdown diagrams is that concerning their extrapolation, *i.e.* the establishment of the maximum limit of the drawdown S_M up to which the exponential equation can be applied, which implies the upholding of the hydraulic régime corresponding to the experimental pumpings. On the basis of the practical results and under the conditions of a relative homogeneity of the aquifer, the following extrapolation limits are admitted as a function of the maximum drawdown realized by the experimental pumping S_{max}:

- for the theoretical correlations (105) and (108) $S_M = 1.5$ to $2S_{max}$
- for the parabolic correlation (110) $S_M = 1.75$ to $2S_{max}$
- for the logarithmic correlation (114) $S_M = 2$ to $3S_{max}$
- for the exponential correlation (117) $S_M = 1.5$ to $1.75 \; S_{max}$

While for the theoretical correlations at least two experimental pumpings are required, for the empirical correlations at least three pumpings are necessary. If the problem has nevertheless to be solved by two pumpings ($N = 2$), then the extrapolation should be done for all three correlations (parabolic, logarithmic and exponential), the most embracing one being chosen. The equations of the yield-drawdown diagrams can be expressed either with yield as a function and the drawdown as an argument or inversely.

Between the form of the yield-drawdown diagram and certain qualitative characteristics of the aquifer (variation of the hydraulic conductivity on the vertical and on the horizontal, the existence of a recharge or impermeable boundary, etc.) there are relations of correspondence that can be made conspicuous especially in the case of the unconfined aquifers.

2.1.2 Total seepage resistance of a well

In an experimental or a production well, two forms of imperfection can be distinguished: 1) according to the degree of the opening (the active filter length l_0 is smaller than the aquifer thickness M); 2) the kind or character of the aquifer opening with reference to the screen peculiarities, the adjacent drilling area, and the deviation from the linear law of flow. Both these imperfections give rise to an additional seepage resistance (with regard to a perfect well) and therefore to an additional lowering of the level in the well and in the vicinity of it.

The problem of the additional resistance according to the opening degree ξ_1 has an exact theoretical solution since the parameter value is determined as a function of quantities $\frac{l}{M}$, $\frac{M}{r_0}$, and of the screen position in the aquifer. The problem of the additional resistance according to the kind of opening ζ_2 of the aquifer is much more complex, because of the diversity and to the variation with time of its conclusive factors. The first factor is the screen and its constructive peculiarities, especially its active surface. The second factor is the area adjacent to the well which can be essentially different from the rest of the aquifer as concerns its hydraulic properties. The radius of this area can be as much as tens of centimeters. This seepage resistance of the area surrounding the well can be considered as a consequence of the skin-effect.

The skin-effect appears principally in the case of the hydraulic drilling or when the water bearing formation includes fine sand, silty sand and clayish fractions. These factors contribute to the appearance of certain mechanical and physico-chemical processes such as the disturbance of the natural structure of the rock, the piping process, the clogging of the porous space as well as of the gravel filter and even of the orifices of the screen, the chemical processes, etc.

The seepage resistance due to the skin-effect ζ_{2z} can be diminished by using dry drilling and by using the correct size of filter but it will however remain of a noticeable magnitude in wells working at large drawdowns, which leads unavoidably to the carrying away and sedimentation of fine particles around the well. There are cases when the well loss due to the existence of this area with small hydraulic properties is important and its omission liable to produce noticeable errors in the determination of the hydraulic properties and of the pumping well productivity. As a matter of fact, in the case of production wells, the creation by natural development of the surrounding area with a greater permeability is a technical necessity.

The flow régime factor can determine an additional resistance ζ_{2r}, owing to the deviation from the linear law of flow in the neighbourhood of the well as a result of the high values of the hydraulic gradients in this area.

The complexity of the factors producing the additional resistance ζ_2 does not always permit us to separate it into its three components ζ_{2f}, ζ_{2z} and ζ_{2r}. That is why in practice the total seepage resistance $\zeta_0 = \zeta_1 + \zeta_2$ of the well is estimated on the basis of the pumping test data from which is afterwards separated the seepage resistance due to the kind of the aquifer opening $\zeta_2 = \zeta_0 - \zeta_1$.

Theoretically, a perfectly penetrating experimental or production well (with filter) cannot be made according to the kind of opening of the aquifer because of the screen, the seepage resistance ζ_{2f} of which is determined by the sum of the pressure losses occurring at the water passage from the rock through filter and into the well hole. It is however possible to obtain a well with a negligible resistance ζ_{2f} in the case of shallow aquifers, by using filters with a large active surface. At small drawdowns, it is possible to have equally $\zeta_{2z} = 0$; the deviations from the linear flow ($\zeta_{2r} > 0$) appears at large drawdowns and coarse grained aquifers. There are also cases when the parameter of the total seepage resistance appears to be negative ($\zeta_0 < 0$) in the case of a completely penetrating well according to the opening degree ($\zeta_1 = 0$); such a case is due to the formation during the development around the hole of an area with a much higher permeability than that of the aquifer.

2.1.3 Well loss

The hydrodynamic study of an experimental or a production well can equally be made with the help of the well loss since each seepage resistance corresponds to the loss of an additional load.

Experience showed that the values of the hydraulic conductivity determined by pumping water level measured in the wells are much lower than those measured (under the same hydrogeological conditions) in the observation wells. This is due to the fact that the pumping water level in the experimental well, which is used as a rule in the computation formulae, has an elevation lower by Δh than the water level on the outside wall of the screen. In the case of unconfined aquifers Δh is called *well loss* (Figure 50 *a*); the well loss was studied qualitatively and quantitatively by many researchers who proposed several rational and empirical formulae for its estimation [35].

This phenomenon also appears in the case of the wells made in confined aquifers (Figure 50 *b*) and then is called a *piezometric loss* (ΔS); its occurrence is explained by the fact that the water pressure on a horizontal plane depends directly on the permeability of the medium and reaches its maximum when this medium is devoid of pores, therefore in the well hole. When a great difference of permeability exists at the boundary between the two media (the aquifer and the well screen) the pressure shows a jump. The function of

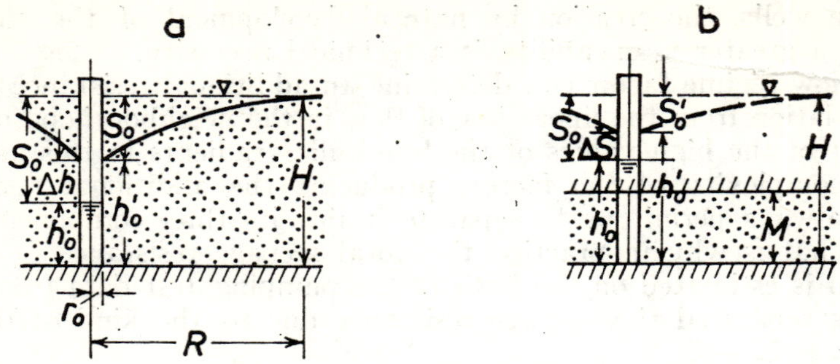

Figure 50. Delimitation of the well loss in a fully penetrating well: *a*, in an unconfined aquifer; *b*, in a confined aquifer.

pressure is only continuous within the limits of a homogeneous permeable medium (or, in other more intricate situations when a mean hydraulic conductivity can be admitted); consequently, its integration can be made up to the external well of the drain. This was Dupuit's solution of the problem, while Kozeny took as a seepage domain limit, the well axis. So as to emphasize the importance of the well loss, Kozeny computed its maximum value which appears in the case of a total lowering of the water level in the well and concluded that $\Delta h_{max} \geqslant H/2$.

In confined aquifers, the problem of the well loss has been studied less. It is note worthy that, after pumpings made in the magnetic anomaly area at Kursk (in USSR) by using in computation the dynamic level within the well, i.e. by disregarding the well loss, the values obtained for the hydraulic conductivity were 2—5 times smaller than those determined by taking the well loss into account. With the same objective in the case of an artesian aquifer, this ratio between the hydraulic conductivities increases up to ten. The author [25] points out that the well loss for an isolated well can reach 50—80% of the drawdown of the well and that it depends linearly on Q and S_0.

Initially, the experimental and analytical study of the well loss started from the assumption that it is determined by the constructive characteristics of the screen and several analytical and empirical formulae were so elaborated. Recent studies tend to explain the well loss by the simultaneous action of several factors: the screen resistance and its variation with time, the characteristic of the well adjacent area with an altered permeability, the boundary conditions, etc. Therefore, the omission of the well loss in determining the parameters of the aquifer is liable to introduce negative or positive gross errors, as a function of the natural development. All analytical or empirical formulae which also contain some constructive characteristics of the screen offer the advantage of requiring no other data but those supplied by the experimental well. But the use of these formulae excludes the presence of an adjacent area with altered permeability and admits an area of homogeneous filtration according to the permeability.

Under these conditions, the well loss cannot be anything but positive The absence of an adjacent area with altered permeability is a very rare occurrence in both the experimental and the production wells. That is why the correct computation of a well loss can only be made with the hydrodynamic data supplied by the central well and by at least one of the observation wells.

2.1.4 Hydraulic conductivity

The hydraulic conductivity is a complex parameter reflecting both the intrinsic rock permeability and the physical properties of the water. It represents as a whole, the facility of the water circulation within an aquifer. Its physical meaning results from Darcy's law which is valid for the laminar flow

$$v = kI$$

where

v is the seepage velocity

I is the hydraulic gradient.

From this relation it follows that the hydraulic conductivity represents the velocity at an unitary hydraulic gradient. It has therefore the dimensions of a velocity. It is considered that for gasless fresh waters at low temperatures hydraulic conductivity is a constant. In actual cases however it was remarked that the value of this parameter depends on the direction of the water flow (horizontal, vertical, ascending or descending) in the rock as well as on the value of the hydraulic gradient. This kinematic peculiarity determine a certain variation of the hydraulic resistance $R = \gamma_a I$, (where γ_a is the water specific weight), which appears in the value of the hydraulic conductivity. As a consequence, the conditions of a determination both in the laboratory or in the field should be much the same as those required for achieving the catchment or dewatering process.

The hydraulic conductivity is moreover known by other names such as *filtering coefficient*, *Darcy's coefficient of permeability*, etc.

In case of the ground waters (mineral waters, thermal waters, etc.) the physical properties of which are variable it is better to use the *coefficient of intrinsic permeability* (k_p) instead of the hydraulic conductivity. This parameter is, in fact, the resistance characteristic of the flow within a porous medium independent of the liquid properties. The two parameters are related to each other by the expression

$$k_p = k \frac{\mu}{\gamma_a} \quad \frac{L}{T} \cdot \frac{FTL^{-2}}{FL^{-3}} = L^2 \qquad (122)$$

where μ and γ_a are the water dynamic viscosity and specific weight. The coefficient k_p has the dimensions of a surface and depends on the geometric properties (structure and texture) of the porous medium.

To characterize the water flow in an unsaturated régime a parameter similar to the hydraulic conductivity can be used, to wit k_n; it is no more a constant of a given area since its value depends mainly on the soil moisture. It is quite a useful parameter for the analysis of the dewatering of the sandy horizons (§ 3.3).

2.1.5 Transmissivity (T)

This parameter, introduced by Theis, represents the flow rate of the aquifer against a 1 m width of it, at a unit hydraulic gradient. It can be expressed in m^3/m.day, by the following formulae:

$T = kM$ — for confined aquifers,

$T = kH_m$ — for unconfined aquifers.

The cartographic representation of the potential of an aquifer by means of this parameter is undoubtedly more satisfactory than the two distinct images viz. of the hydraulic conductivity and of the aquifer thickness, since the catchment or dewatering wells are usually placed in the areas with the highest values of the yielding capacity. Another advantage of using the "transmissivity" notion appears in case of the confined aquifer with large thickness variations, when thickness mean value cannot be calculated on the whole depression area.

In this manner the transmissivity appears as the product of the mean values of parameters k and M:

— high potential	$T > 500$ m^3/m. day,
— moderate potential	$T = 50 - 500$ m^3/m. day,
— low potential	$T = 5 - 50$ m^3/m. day,
— very low potential	$T = 0.5 - 5$ m^3/m. day,
— negligible potential T	$T < 0.5$ m^3/m. day.

2.1.6 Coefficient of storage

This is a parameter used when studying the unsteady flow of the ground water. It represents the volume of water stored or collected by an aquifer against a surface unity, owing to the unitary change of its normal component on the area.

Confined aquifer. In this case, the storage coefficient represents the water volume supplied by a prism the base cross-section of which is equal to unit, when the piezometric surface is reduced by a unit of height (Figure 51). This water volume is due to the water compressibility as well as to the water-bearing formation that undergoes a slight reduction of its porosity. Its values vary from 0.00001 to 0.001. Since the aquifer has a purely elastic behaviour, the respective parameter should also be called a *coefficient of elastic storage* (n_e) the value of which is directly proportional to that of the coefficient of the elastic capacity, according to relation

$$n_e = \gamma_a M \beta \tag{123}$$

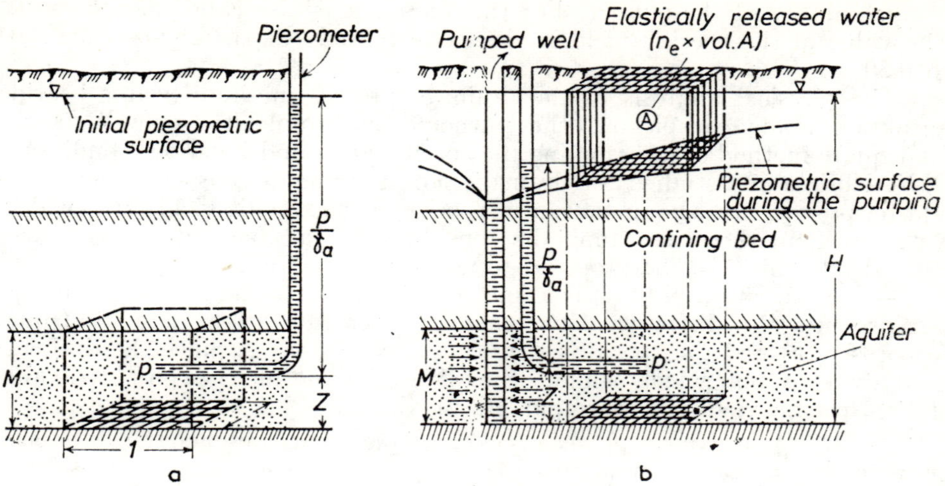

Figure 51. Physical interpretation of the elastic coefficient of n_e of a confined aquifer under natural (*a*) and pumping (*b*) conditions (after De Wiest, 1965).

where γ_a is the specific weight of the water, M the thickness of the aquifer and β the coefficient of elastic storage of the aquifer [56], which represents the variation of the elastic resource of water per unit volume, at a variation of a unit pressure; it is defined by relation

$$\beta = n \, \beta_a + \beta_r \tag{124}$$

where n is the rock porosity, β_a the water coefficient of compressibility (for soft and slightly mineralized waters it varies from $4.5 - 5 \times 10^{-5}$ cm²/kg; β_r the rock coefficient of compressibility (the values of which vary from $0.3 - 10^{-5}$ cm²/kg for limestone to 3×10^{-5} cm²/kg for sands).

The coefficient of elastic storage n_e[1)] must needs be known in the case of strong aquifers with a regional extension and high pressures for which the value of the elastic resource should be computed, as it can be an important component of the total value of the exploitation resources. The elastic resource can be computed by relation

$$W_e = n_e \cdot \Delta S \; \Omega \quad [\mathrm{m}^3] \tag{125}$$

where ΔS is the average drawdown on area Ω assignable to the drainage.

Unconfined aquifers. The applying of the concept of coefficient of storage to this kind of aquifers is more intricate since the volume of water released is particularly due to the gravity drainage and only in a small proportion to the

---

[1)] This symbol was adopted because its relation of definition is similar to that of the porosity (n), while symbol S (used in certain works) can be taken for the drawdown.

compressibility of the aquifer. For this reason it is also called *specific yield*. The value of the specific yield in the case of sandy deposits vary from 0.05 to 0.30.

The physical analysis of the drainage process in sandy water bearing formation made conspicuous the phenomenon complexity as well as the inadequate method of determining the specific yield under various conditions. O. N. Nosova [50] studied the drainage process in aquiferous sands by stages and he proved that the instantaneous value of the specific yield varies widely with time and therefore cannot be considered as a constant of the area studied. On this basis he proposed to use several parameters:

— The specific yield (n_0) is a final characteristic of the gravity drainage; it represents the difference between the water saturation capacity (equal to porosity n) and the specific retention ($n_0 = n - m$) [1]. Since the drainage is an unsteady flow which cannot be achieved in due time the computed value of the specific yield is lower than its maximum value n_0.

— The *specific yield of the capillary fringe* n_c represents the difference between the saturation capacity and the capillary capacity and is considered as a mean value all along the height of the sample ($n_c = n - w$). Under actual conditions, this parameter characterizes the average drainage state of the rock in the area of capillary ascension due to the existence of the depression surface.

— The common specific yield n' characterizes the draining process in time and is given by the ratio between the water volume stored from the beginning of the drainage process up to the considered time and the volume of the drained ground. Consequently, the value of this parameter depends on the drainage time.

— The superficial specific yield n_s is defined as the ratio between the released water volume during a certain time period and the drained rock volume. The coefficient of storage is a very important parameter appearing in the equations of the unsteady flow.

The opinions differ widely as to the ratio between the specific yield and the porosity. The name of *effective* or *active* porosity is given to the portion of the pore section occupied by the moving water, whence

$$n_a = n - w_s \tag{126}$$

where w_s is the humidity in the static state. Of course, this humidity depends on the value of the hydraulic gradient and is not therefore a constant of the rock. In the case of sandy rocks this dependence practically does not appear and therefore we can consider that $w_s = w_m$ which finally leads to the numerical equality between the specific yield n_0 and the active porosity n_a.

Without discussing all the relations between the above coefficients, it is to be remarked that in the hydrodynamic analysis of the draining process, the superficial specific yield is quite important. If the draining is considered as a process which takes place by isolated stages, i.e. as uniform régime, it

---

[1] Humidities are expressed as a ratio between the water volume and the respective rock volume.

appears that coefficient n_s depends on the decreasing velocity of the water current surface, expressed by the hydraulic gradient (I). For $I = 0.001—0.02$ the approximate value of the superficial specific yield is supplied by the following semi-empirical relation [50]

$$n_s = n_0(1 - I) \tag{127}$$

For very small hydraulic gradients ($I < 0.001$) the coefficient n_s can be considered as a constant, practically equal to the specific yield n_0. Nosova [50] offers some data concerning the relation between superficial specific yield n_s and grain-size uniformity of the rock U:

$I = 0.1 - 0.01$ $\quad n_s = (0.7 - 0.8)\ n_0$ for sands with $U < 5$
$\quad n_s = (0.6 - 0.7)\ n_0$ for sands with $U > 5$

$I = 0.01 - 0.001$ $\quad n_s = (0.8 - 0.9)\ n_0$ for sands with $U < 5$
$\quad n_s = (0.7 - 0.8)\ n_0$ for sands with $U > 5$

2.1.7 Hydraulic diffusivity

V. N. Schelkatchev defined this complex parameter of a confined aquifer as a characteristic of the redistribution rhythm of the pressure in case of its natural or forced alteration. Its equation of definition is

$$a = \frac{kM}{n_e} = \frac{k}{\gamma_a \beta}\ [L^2/T^{-1}] \tag{128}$$

An analogous parameter is also defined for the unconfined flow (a_B) called Boussinesq's parameter:

$$a_B = \frac{kH_m}{n_s} \tag{129}$$

where H_m is the mean thickness of the saturated aquifer all during the pumping in the area of the depression cone. Table 48 gives informative values of coefficients a and a_B for a few detrital rocks. This parameter makes the hydrodynamic computations in unsteady flow much easier since it permits us to compute the water levels on the whole surface of the aquifer affected by the pumping.

Table 48

Values of hydraulic diffusivity for some detritical rocks

Type of rocks	Coarse sand	Medium sand	Fine sand	Sandy silt	Sandy clay
a(m²/day)	—	10^5	$5 \times 10^3 \ldots 10^5$	10^3	100 ... 500
a_B(m²/day)	1.000... 3.000	500...2.000	50... 500	20... 100	—

2.1.8 Radius of influence

A correct estimation of the radius of influence of the exploration hydrogeological wells is useful for the determination of the other hydrogeological parameters as well as for the hydrodynamic computation of the catchment and dewatering systems, that are characterized by very high values of the radii of influence. In some formulae for determining the hydraulic conductivity the radius of influence enters by its logarithmic value (meaning that its approximate value does not introduce gross errors), but in most of the drainage schemes its actual value is used.

The radius of influence of the pumpings under unsteady conditions can be determined by taking into account the forces of inertia, elasticity and gravity. In the case of steady flow conditions of the pumping and of an "infinite" aquifer (the most frequent case), the effect of the drainage is felt as far as the boundary of the recharge area, the position of which in the plane cannot however be well defined since, at a certain distance from the experimental well, the hydraulic gradient of the piezometric curve is so small that it can no longer be practically determined, being confused practically with the initial piezometric level. Consequently, most of the relations use the notion of *reduced radius of influence* which corresponds to a simplified hydrodynamic scheme with a circular recharge boundary and an experimental (or production) well placed in the centre of this outline. The passage from the real, more intricate, conditions to these simple schemes is made with the help of many empirical and semi-empirical relations which contain several easily estimated parameters. The number of these relations is high enough since each of them is established for certain hydrogeological conditions.

The real radius of influence can only be computed in experimental groups containing at least two observation wells for each radius.

The determination of the hydrogeological parameters can be made by hydrogeological, geophysical and laboratory methods.

Hereafter, examples of application of the hydrodynamic methods (pumpings under steady and unsteady flow conditions) will be given, laying especial stress on the degree of certainty of the respective computation schemes. From the solved examples, the methods of the parameter computation as well as of the pumping execution will result.

2.2 Aquifer pumping test in steady flow condition

Pumpings in a steady flow constitute the classical hydrodynamic research method of aquifers. They can be accomplished by isolated experimental wells or by experimental groups (stations) that together with the central well, comprise one or more observation wells (piezometers) placed along one or more radii. The number of pumping tests can vary from 1 to 3. The work volume (observations wells and number of pumping stages) will depend, in the case of an experimental station, on the lithological homogeneity and on the aquifer hydraulic state, on the research stage, on its purpose within the overall problem as well as on the depth of the aquifer, "prescriptions" or rules being out of the question. Let us consider for instance the testing of a phreatic unconfined aquifer recommended as a water supply source: if the initial hydraulic gradient is small (of the order of thousandths) and the lithological homogeneity is obvious, it is sufficient to have two piezometers placed along the isopiestic lines direction and three experimental pumpings. Should this phreatic aquifer be tested with a view to drainage by a cut-off trench, the number of pumpings could be reduced to 1—2 since in this case the supplied data are only used for determining the hydraulic properties so that the yield-drawdown diagram is superfluous. In the case of an aquiferous current with a high initial hydraulic gradient and a lithological non-homogeneity, the asymmetry of the depression cone requires the setting of the observation wells along 2—3 radii. When applying the hydrodynamic test to confined aquifers at greater depths, the necessity of using observation wells should also be considered from the economic standpoint. When the purpose of testing is water supply, for instance, such an aquifer may be tested by a single experimental well (without piezometers); the data supplied by the three experimental pumpings would only serve to ascertain the yield-drawdown diagram, the hydraulic properties being then only informative.

When choosing the distance of the piezometers to the discharging well, the type of the water bearing rock and the hydraulic characteristics of the aquifer (Table 49) should be taken into consideration. Shorter distances correspond to the detrital rocks and the unconfined aquifers, while the longer distances are recommended in the case of fissured-karstified rocks; these different distances are related to the growth rate of the cone of depression during the pumping. On the other hand, within the same experimental group, in the case of a supposedly asymmetric cone of depression, the longest distances will be up-stream and the shortest ones down-stream from experimental well, while along the isopiestic lines the values will be intermediate.

Table 49

Distances (in m) from the experimental well to the piezometers

Number of piezometers along a radius	High permeability $k > 1$ cm/s	Medium permeability $1 > k > 10^{-2}$	Low permeability $k < 10^{-2}$ cm/s
1	1—3	3—5	5—10
2	3—10	10—15	15—30
3	10—30	30—50	50—100
4	30—100	100—150	150—300

The *length* and the *position* of the screen in the observation wells exert a great influence on the accuracy of the level measurements because the level is not uniform along the vertical.

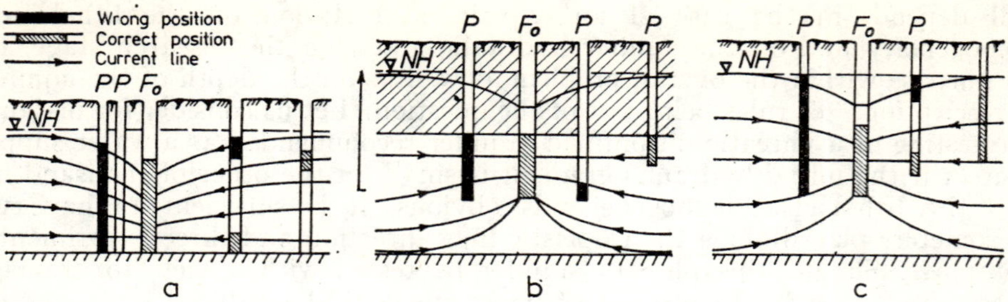

Figure 52. Positioning of the screens in the observation wells:
a, unconfiend aquifer and pumped fully penetrating well; *b*, confined aquifer and pumped partially penetrating well; *c*, unconfined aquifer and partially penetrated well (after [80], 1967).

— In the case of pumpings from unconfined aquifers, even if the discharging well is fully penetrating and the aquifer is homogeneous, a noticeable deformation (chiefly at large drawdowns) of the stream lines and equipotential lines occurs on a radius of about 1.5—2 H. So as to avoid measurements of erroneous pumping water levels, the short screen will be placed at the level where the flow line can be considered as horizontal, i.e. practically near the impermeable bed; the recommended length of the screen is 1/5—1/10 of the aquifer thickness (Figure 52 *a*). Outside the zone with a strong deformation of the stream lines, the position of the screen is less important; in this case, short screens 2—3 m long are recommended, placed in the upper part of the aquifer:

— In the case of pumpings from fully penetrating wells in homogeneous confined aquifers on the vertical, where the stream lines are horizontal along the whole thickness of the aquifer, the screen position is not important.

— When the aquifers are formed by a sequence of strata with various permeabilities, the screen should be placed in the most permeable stratum.

— In the case of partially penetrating wells, the screens of the piezometers should be equally placed at the level where the stream lines are practically horizontal (Figure 52 *b, c*).

The piezometers may be considered as punctiform if the screens are short and also have small diameters (75—85 mm).

In the case of 2—3 pumping steps, the test can be achieved in two ways:

— each pumping should be followed by a recovery period of the initial water level;

— the pumpings are made in stages, passing from one rate of discharge to the following one without waiting for the water level recovery.

In this problem no rules can be established, the principal purpose of an experimental pumping in a steady flow being the stabilization of the yield and of the pumping water level. On the other hand, it is not advisable to forego altogether the curves of level recovering, which supply quite useful field data for estimating some characteristics of the tested aquifer.

Usually, the processing of the experimental data is made on the assumption that the upper and lower formations are practically impermeable. Under these admitted conditions of hydraulic isolation along the vertical, some errors appear within the limits of the cone of depression which are liable to affect the computed hydrogeological parameters. To this effect, A. Schmieder's conclusions [59] are interesting as regards the errors assignable to the hydraulic conductivity and the specific character of these errors in regard to the accuracy required by this parameter. His conclusions are based on the measurements supplied by two experimental groups provided with many observation wells, the processing and interpretation of the data being made from the standpoint of the dewatering process requirements.

In any dewatering scheme for a laminar régime, the function of the pressure reduction may be expressed in a generalized form

$$\Phi = f(x_i)\, k \tag{130}$$

where $f(x_i)$ denotes the technical and geometrical conditions of the aquifer and of the drainage system. The parameters involved in term $f(x_i)$ can be determined within the limits of the admissible errors while coefficient k is liable of gross errors. Such situations lead to erroneous estimations of the dewatering possibilities and then to non-concordances between the designed parameters and the results of the dewatering. If $\pm U_m$ denotes the relative error of the measurements, $\pm U_c$ the relative error of the investigation method and $\pm U_i$ the relative error resulting from interpolations (since the experimental wells are 1—2 km away) the relative error of k will be

$$\pm r_k = \pm \frac{\Delta k}{k_0} = \sqrt{U_m^2 + U_c^2 + U_i^2} \tag{131}$$

where k_0 represents the mean value, unaffected by errors. The experimental value will be:

$$k_x = k_0(1 \pm r_k) \tag{131 bis}$$

So as to be able to estimate the admissible error, the results supplied by the usual research methods on the assumption of a fully penetrating well were compared with those supplied by a confined aquifer.

1. By using concomitantly all the data of the observation wells in Dupuit-Thiem formula, treated by the method of least squares:

$$k_1' = \frac{Q[(\Sigma \ln r_i)^2 - n\Sigma(\ln r_i)^2]}{2\pi M[n\Sigma S_i \ln r_i - \Sigma S_i\, \Sigma \ln r_i]} \tag{132}$$

2. From the measurements of every group of two observation wells:

$$k = \frac{Q \ln \dfrac{r_n}{r_i}}{2\pi M(S_i - S_n)} \tag{133}$$

3. Exclusively on the basis of the data supplied by the experimental well:

$$k_0' = 0.366 \frac{Q}{S_0 M} \log \frac{R}{r} \tag{134}$$

It follows from the comparative computations that parameter k determined solely on the basis of the observation wells presents errors of about $\pm 7.5\%$, which means that this method offers a satisfactory accuracy. When k was determined by means of the data supplied by the experimental well and by an observation well, large variations of the hydraulic conductivity appeared as well as the non-uniform character of the errors, owing to the seepage resistance of the experimental well. This seepage resistance is materialized by the well loss expressed by relation:

$$S = S_0 - S_0' = S_0 - 0.366 \frac{Q}{kM} \log \frac{R}{r} \tag{135}$$

where S_0 is the drawdown measured in the well, and S_0' is the real drawdown on the external wall of the well. The difference of the k values against the distance to the experimental well is due to its dependence on the seepage velocity. That is why, so as to avoid the gross errors, the observation wells must be placed at short distances from the experimental well. Schmieder points out that with only the help of the observation wells, the hydraulic conductivity can be determined within the admissible errors ($\pm 20 - \pm 30\%$) while the use of only the data supplied by the experimental well, or by the experimental well together with those of an observation well, leads to relative errors going up to $- 300 \ldots + 130\%$.

To exemplify and compare some modern processing techniques of the pumping test data in a steady régime, the 26 H Cuzap experimental group with three observation wells was chosen. Three experimental pumpings were performed from the central well. As shown by the hydrogeological section (Figure 53), the aquifer under study (at the bottom of coal bed II) is clearly individualized being separated from the other lower aquifer by a bank of marly-clays about 8 m-thick. The computation elements supplied by the measurements are recorded in Table 50.

Constructive elements to be mentioned are:

— in the experimental well, the isolation of the higher layers was achieved by a permanent casing of $d = 12\,3/4$ in.

— the aquifer was traversed by a dry, mechanical and semi-mechanical drilling;

— the working casing of the experimental well had $d = 10$ in, where the lost screen was introduced;

— the observation wells were provided with $d = 89$ mm screens watertight at the plain casing of $d = 6\,5/8$ in, by which the upper layers are isolated;

— the filter elements of the experimental well are: a gravel pack with $d = 2$—3 mm gravel size, casings with $d = 205$ mm and a 20% perforation degree, mapped by mesh;

— the natural development (about 100 hours) and the three experimental pumpings were made with the help of the compressed air pump.

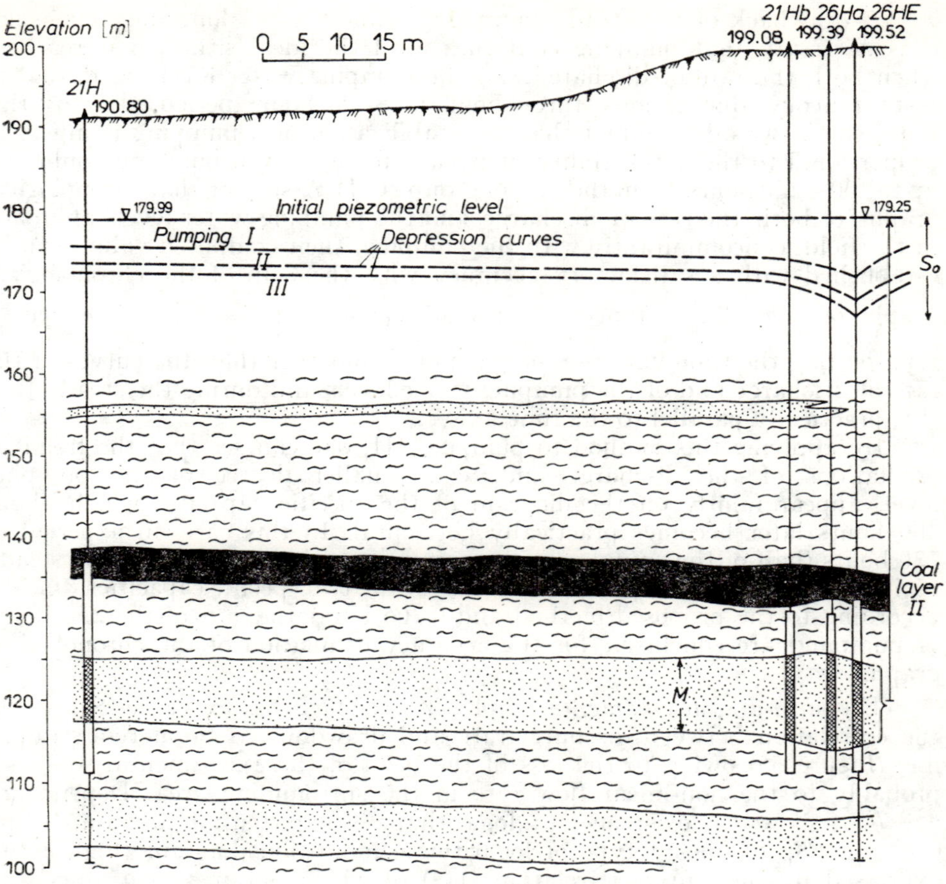

Figure 53. Hydrogeological cross-section through experimental group of wells 26H Cuzap.

Table 50

Experimental pumping data of group 26 H

	Unit of measure	Pumping	Experimental well 26 HE(*0*)	Observation wells		
				26Ha(*1*)	26Hb(*2*)	21H(*3*)
Drawdown (*S*)	m	I	7.70	6.50	5.52	3.50
		II	10.39	8.95	7.40	5.34
		III	11.93	10.82	8.59	6.48
Discharge rate (*Q*)	m^3/day	I	147	—	—	—
		II	192	—	—	—
		III	212	—	—	—
Aquifer thickness (*M*)	m	—	10.92	11.75	10.44	6.62
Screen lenght (*l*)	m	—	9.96	10.87	9.35	0.62
Elevation of initial piezometric level	m	—	179.27	179.25	179.26	178.99
Hydrostatic pressure (*H*)	m col. H_2O		65.08	66.80	67.73	60.04
Distance to 26 HE well-axis	m	—	$r_0=0.103$	$r_1=2.97$	$r_2=8.18$	$r_3=94.30$

A drawback of the steady pumping régime was its long duration, since it is necessary that pumping continues until the ideal situation is reached, when both the rate of discharge and the pumping water level are devoid of instantaneous fluctuations throughout a 6—8 hour period. If to this requirement we add the fact that the stabilization of a pumping is inversely proportional to the permeability, it is easy to see why a pumping could last up to 100—250 hours (as in the case of group 26 H). Assuming that chronological graphs of both the rate of discharge and the pumping water levels are made in the field, concommitantly with the pumping, their computed values can be established with a satisfactory accuracy with the help of the chronological graph of the "instantaneous" specific capacity $\left(q = \frac{Q}{S_0}\right)$. The curve representing the time variation of q is much smoother than the curves of the rate of discharge and of the pumping water levels and during the steady flow it is practically parallel to the time axis.

By applying this method to pumping III, one can see that the frequent oscillations of the discharge and to a smaller degree of the pumping level, appear faintly on the diagram of the specific capacity, which about 100 hours later becomes practically horizontal. In this case, in a period of 120 hours (half of the actual one) the pumping would have reached the steady régime (Figure 54). The quite close values of the two specific capacities (0.21 l/s according to the proposed method and 0.205 l/s at the end of the 250 hours of pumping) are conclusive for the accuracy estimation of the chronological graph.

The *yield-drawdown diagram* emphasizes the fact that for values $S_0 > 5$ m, the correlation between discharge rate and drawdown is no more a straight line (Figure 55) owing to the loss of the head of the experimental well and probably to the non-linear flow rate in the surrounding area. Dependence $q = f(S_0)$ shows the existence of a logarithmic or exponential diagram. Graph $Q = f(\log S_0)$ defines a type of logarithmic indication diagram (Figure 56), expressed by the empirical relation (114) of which parameters a and b are estimated by formulae (115) and (116) with the help of table 51; thus we obtained equation

$$Q = -1.81 + 3.96 \log S_0$$

As shown on § 2.1.1, the logarithmic correlation can be extrapolated up to $S_M = 2-3\ S_{max}$. In the above case the lowest limit should be considered since at about 1 Km south of the experimental group, the surface recharge of the aquifer is found. This means that the latter is of a "semi-infinite" character. Therefore, the highest extrapolated values are:

$$S_M = 2 \times 11.93 = 23.9$$

$$Q = -1.81 + 3.96 \times \log 23.9 = 3.65 \text{ l/s}.$$

In the main, the extrapolation of the drawdown cannot exceed value $H - M$ since in a contrary case the flow in the neighbourhood of the experimental well would become an unconfined flow.

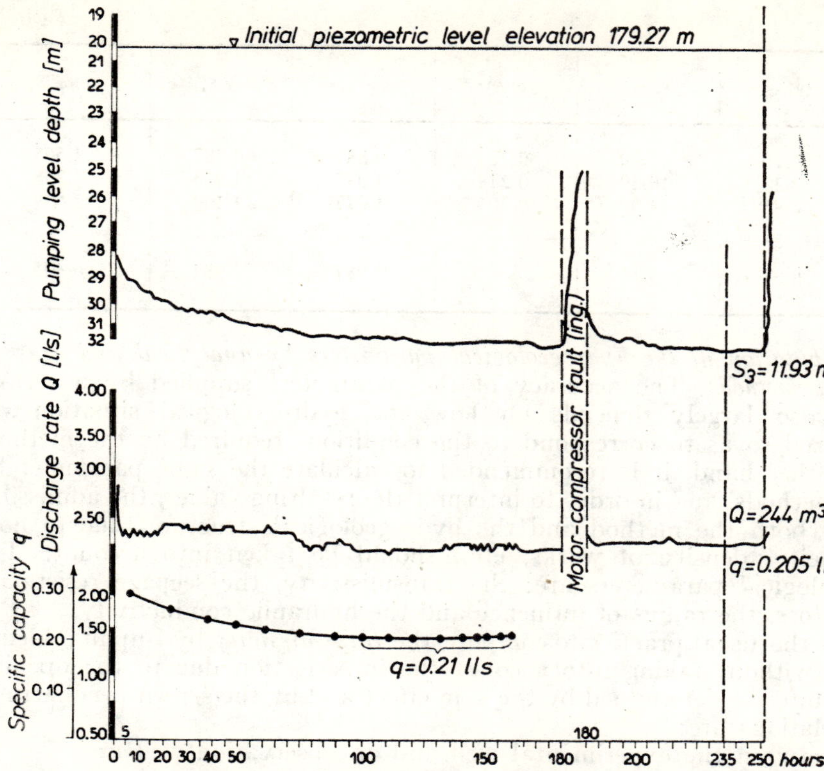

Figure 54. Chronological graphs of pumping III (well 26H Cuzap).

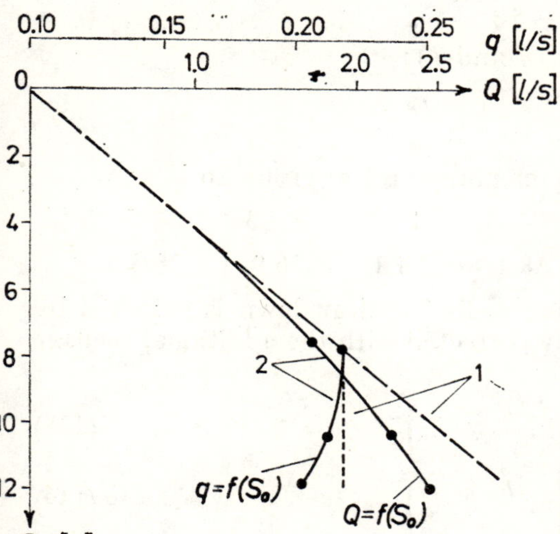

Figure 55. Yield drawdown diagram of well 26 HE: *1*, Theoretical; *2*, Experimental.

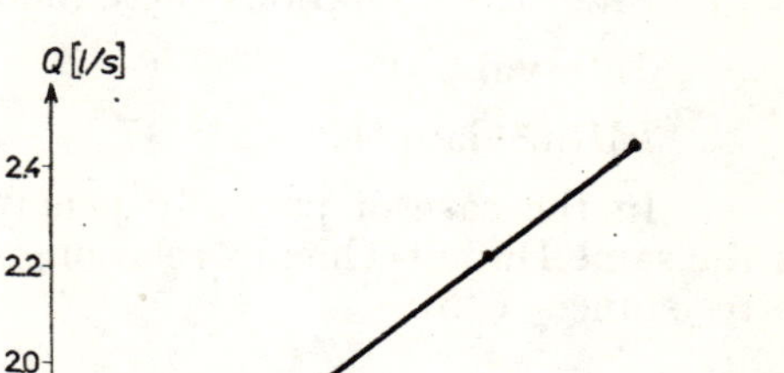

Figure 56. Linearized drawdown diagram of well 26 HE.

	Q(l/s)	S_0(m)	q(l/s)	log S_0	(log S_0)^2	Q log S_0
1	1.70	7.70	0.221	0.887	0.787	1.508
2	2.22	10.39	0.214	1.017	1.034	2.258
3	2.45	11.93	0.205	1.077	1.160	2.639
Σ	6.37			2.981	2.981	6.405

Computation of the hydrogeological parameters by analytical and graphoanalytical methods. The accuracy of the parameters supplied by a certain method also largely depends on how the hydrogeological situation was schematized so as to correspond to the conditions required by the method. On the other hand, it is recommended to calculate the same parameter by several methods, and in order to interpret the resulting values, the admissible errors of both the method and the hydrogeological structure (due to non-homogeneity, density of works, etc.) should be taken into account. The hydrogeological parameters are: the transmissivity, the seepage resistance, the well loss, the radius of influence and the hydraulic conductivity.

It is the usual practice to compute the *transmissivity* by Dupuit-Thiem's relations without taking into account the imperfection due to the opening of the aquifer as determined by the skin effect and by the screen peculiarities. These relations are:

— between the experimental well and an observation well n:

$$kM = \frac{0.366\,Q}{(S_0 - S_n)} \log \frac{r_n}{r_0} \tag{136}$$

— between two observation wells (n and p);

$$kM = \frac{0.366\,Q}{(S_n - S_p)} \log \frac{r_p}{r_n} \tag{137}$$

From these relations there follows for pumping I of group 26 H:

Interval	0—1	0—2	0—3	1—2	1—3	2—3
kM(m^2/day)	66.0	47.2	38.1	24.1	26.9	28.3

In the case of partially penetrating wells the drawdown is supplied by the same Dupuit-Thiem's relations, only corrected with the additional seepage resistance ζ [5]:

$$S_0 = \frac{Q}{2\,\pi kM}\left(\ln\frac{R}{r_0} + \zeta_0\right) \tag{138}$$

$$S_1 = \frac{Q}{2\,\pi kM}\left(\ln\frac{R}{r_1} + \zeta_1\right) \tag{139}$$

$$S_2 = \frac{Q}{2\,\pi kM}\left(\ln\frac{R}{r_2} + \zeta_2\right) \tag{140}$$

By subtraction, we can draw from these relations the formulae for determining the transmissivity along alignments comprised between two wells:

$$(kM)_{0-1} = \frac{Q}{2\pi(S_0 - S_1)}\left(2.3 \log \frac{r_1}{r_0} + \zeta_0 - \zeta_1'\right) \tag{141}$$

$$(kM)_{1-2} = \frac{Q}{2\pi(S_1 - S_2)}\left(2.3 \log \frac{r_2}{r_1} + \zeta_1' - \zeta_2'\right) \tag{142}$$

where

ζ_0 is the total seepage resistance of the experimental well;

ζ_1' and ζ_2' are the seepage resistance due to the opening degree of the observation wells.

In the case of group 26 H (from Cuzap—Romania), the wells can be considered as practically fully penetrating according to the degree of opening (at well 26 H only 0.63 m of the upper part and 0.33 of the lower part of the aquifer were left without screens) so that $\zeta_1' = \zeta_2' = 0$ while the value of ζ_0 will be determined solely by the skin effect and by the filter ($\zeta_0 = \zeta_{2z} + \zeta_{2f}$). By so doing, the values of the transmissivity between the observation wells will be equal to the previously computed ones. If $kM_{(1-2)}$ is known, relation (141) will let us determine the total seepage resistance

$$\zeta_0 = \frac{2\pi(S_0 - S_1)(kM)_{1-2}}{Q} - 2.3 \log \frac{r_1}{r_0} \tag{143}$$

Value $\zeta_0 = -2.15$ represents a peculiar situation in the case of well 26 H, seeing that this parameter is normally positive. This results from the cumulative effect of the gravel pack and of the strong natural development (about ten hours) at a rate of discharge and a drawdown equal to those of the last pumping which by washing away the finer particles led to the increase of the permeability within a certain radius. The seepage resistance being known we can accurately determine the transmissivity between well 26 H and the other observation wells:

$$(kM)_{0-1} = \frac{147}{6.28(7.70 - 6.50)}\left(2.3 \log \frac{2.97}{0.10} - 2.15\right) = 24.2 \text{ m}^2/\text{day}$$

$$(kM)_{0-2} = \frac{147}{6.28(7.70 - 5.52)}\left(2.3 \log \frac{8.18}{0.10} - 2.15\right) = 24.1 \text{ m}^2/\text{day}$$

$$(kM)_{0-3} = \frac{147}{6.28(7.70 - 3.50)}\left(2.3 \log \frac{94.30}{0.10} - 2.15\right) = 26.1 \text{ m}^2/\text{day}$$

If the experimental group consists of at least three observation wells the grapho-analytical process can be applied which makes use of equations (141) and (142). To this end, a graph (Figure 57) is drawn on the ordinate of which the differences of the drawdown between two wells ($\Delta S = S_n - S_p$)

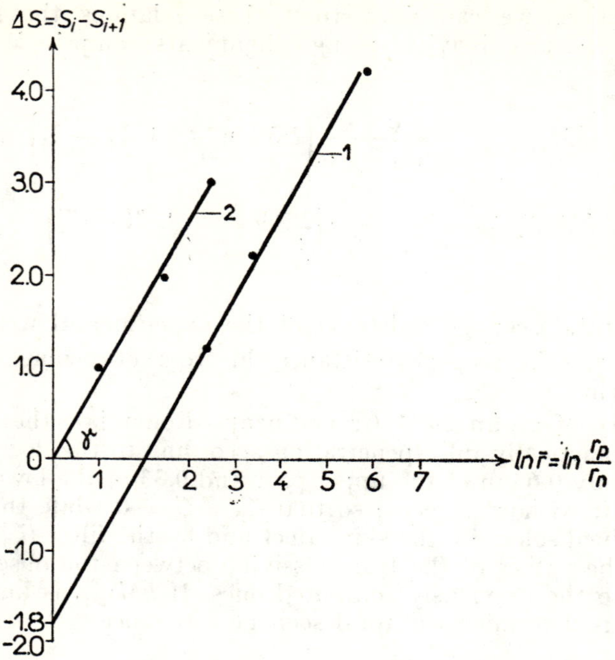

Figure 57. Grapho-analytical determination of kM and ξ_0.

are written and on the abscissa of which the logarithm of the ratio of the distances $\left(\ln \bar{r} = \ln \frac{r_p}{r_n}\right)$ is written. The resulting points are marked on two straight lines: line *1* corresponding to the pairs of wells one of which being the experimental one, and line *2* corresponding to the pairs of observation wells. Line *1* which is the graphical expression of equation (141) has C as initial ordinate. Straight line *2* which results from equation (142) passes therefore through the origin. For geometrical reasons it follows that

$$\zeta_0 = \frac{C}{B} \tag{144}$$

where B is the angular coefficient of the straight line

$$\mathrm{tg}\gamma = \frac{\Delta S_1 - \Delta S_2}{\ln \frac{\bar{r}_2}{\bar{r}_1}}$$

The transmissivity is supplied by relation

$$kM = \frac{Q}{2\pi B} \tag{145}$$

The advantage of the grapho-analytical method over the analytical one consists in the fact that the mean values of kM and ζ_0 are supplied by a greater number of experimental data. In the case of group 26 H line *1* drawn from the data in table 52, has the negative initial ordinate $C = -1.8$.

Table 52

Pairs of wells	ΔS_1	ΔS_2	ΔS_3	$\ln \bar{r}_1$	$\ln \bar{r}_2$	$\ln \bar{r}_3$
Straight line (*1*)	1.20	2.18	4.20	3.39	4.40	6.84
Straight line (*2*)	0.98	2.02	3.00	1.01	2.44	3.45

The angular coefficient of line *2* can be established by relation

$$B = \frac{\sum_{i=1}^{n} \left(\frac{\Delta S}{\ln \bar{r}} \right)_i}{n} = \frac{2.68}{3} = 0.89$$

From the above-mentioned relations it follows that

$$\zeta_0 = \frac{-1.8}{0.89} = -2.02$$

$$kM = \frac{147}{6.28 \times 0.89} = 26.3 \text{ m}^2/\text{day}$$

The difference between the two values of ζ_0 (-2.15 and -2.02) results from the transmissivity being only computed between wells 26 Ha and 26 Hb (which is an approximation) while the graph takes into account all the values and is therefore more accurate. But even if the difference is considered as an error, it is admissible (6 %). The mean transmissivities are however very close to one another: 26.3 and 26.1 m²/day.

For analytical reasons (§ 2.6.2) it follows that in the case of partially penetrating wells the relations valid for fully penetrating wells can be used provided that the reduced radius of the well [4], estimated by relation

$$r_0' = r_0 \, e^{-\zeta_0} \tag{146}$$

is introduced.

In the case of group 26 H, it follows that

$$r_0' = 0.103 \times e^{+2.15} = 0.103 \times 8.58 = 0.88 \text{ m}$$

The value of the reduced radius is greater than the real one (the geometrical one) owing to the greater permeability around the well. Had the skin effect consisted solely in a reduction of the permeability, then the reduced radius [1]) would have been much smaller than the real one. The transmissivity computed by the perfect well formula is

$$(kM)_{0-1} = \frac{0.366\,Q \log \dfrac{r_1}{r_0}}{S_0 - S_1} \tag{147}$$

$$(kM)_{0-1} = \frac{0.366 \times 147 \times \log \dfrac{2.97}{0.88}}{7.70 - 6.50} = 23.6 \ \text{m}^2/\text{day}$$

The practically acquired value is the same as that previously found (24.2 m²/day).

The parameters of the area with a degraded permeability (k_d, r_d) and degradation factor ζ_{2z} are supplied by the following relations deduced for radial-plane flow:

— the mean hydraulic conductivity between the experimental well and the first piezometer

$$k_m = \frac{\ln r_1/r_0}{\dfrac{1}{k_d} \ln \dfrac{r_d}{r_0} + \dfrac{1}{k} \ln \dfrac{r_1}{r_d}} \tag{148}$$

— the degradation factor (the resistance due to the skin-effect)

$$\zeta_{2z} = \left(\frac{k}{k_d} - 1\right) \ln \frac{r_d}{r_0} \tag{149}$$

In the case of group 26 H, since $\dfrac{kM}{k_m M} = \dfrac{24.2}{66.0} = 0.37 < 1$ it follows that around the drilling hole there is an area of radius r_d where $k_d > k$, a situation also emphasized by the negative value of parameter ζ_0.

If the seepage resistance is only due to the degraded area ($\zeta_{2z} = \zeta_0$), then relations (148—149) will only contain two unknowns k_d and r_d. Since the perfection degree of well 26 HE is $\dfrac{l}{M} = \dfrac{9.96}{10.92} = 0.91$ we can estimate that according to the opening degree the well is practically fully penetrating so that $\zeta_{2z} = \zeta_0 = -2.15$. Under these conditions relation (149) takes the form

$$k_d = \frac{k \ln r_d/r_0}{\zeta_0 + \ln r_d/r_0} \tag{150}$$

[1]) r_0' corresponds to the assumption that the rate of discharge of the aquifer is granted under the conditions of a fully penetrating well.

which introduced in (148) supplies an implicit equation with a single unknown r_d

$$k_m = \frac{\ln r_1/r_0}{\frac{\zeta_0 + \ln r_d/r_0}{k \ln r_d/r_0} \ln r_d/r_0 + \frac{1}{k} \ln \frac{r_1}{r_d}} \tag{151}$$

$$5.82 = \frac{\ln 2.97/0.1}{\frac{-2.15 + \ln r_d/0.1}{2.14 \ \ln r_d/0.1} \ln r_d/0.1 + \frac{1}{2.14} \ln \frac{2.97}{r_d}}$$

Quantities $k_m = 5.82$ and $k = 2.14$ m/day result from dividing the transmissivities by the mean thickness between wells 26 HE and 26 Ha ($M_m = 11.34$ m). Solving relation (151) by trials, $r_d = 1$ m is obtained which introduced in (150) result $k_d = 32.8$ m/day. To conclude, an efficient natural development is obtained:

$$\eta = \frac{k_d}{k} = \frac{32.80}{2.14} = 15.3$$

Another analytical relation between the parameters of the degraded area around the experimental wells is

$$r_0' = \left(\frac{r_0}{r_d}\right)^{k/k_d} \cdot r_d = \left(\frac{0.1}{1.0}\right)^{2.14/32.8} \cdot 1 = 0.86 \text{ m} \tag{152}$$

It follows that the reduced radius of the well (r_0) has practically the same value as the previously computed one. Of course, $k_d = 32.8$ m/day represents a mean permeability in the naturally developed area.

The *well loss* represents the additional head loss due to the total seepage resistance of the experimental well. Owing to an insufficient theoretical study of this problem it is more convenient to apply the method of the complex estimation of the total resistance of the well (ζ_0). Nevertheless, the well loss can be determined with a satisfactory accuracy by the theoretical relation

$$\Delta S = S_0 - S_0' = S_0 - S_1 - \frac{Q}{2\pi k M} \ln \frac{r_1}{r_0} \tag{153}$$

Using the data of wells 26 HE and 26 Ha we obtained:

$$\Delta S = 7.70 - 6.50 - \frac{147 \times 2.3}{6.28 \times 24.2} \log \frac{2.97}{0.1} = -2.08 \text{ m}$$

The well loss has been also computed for the second observation well, and the result is the same

$$\Delta S = 7.70 - 5.32 - \frac{147 \times 2.3}{6.28 \times 24.1} \log \frac{8.18}{0.1} = -2.09 \text{ m}$$

The negative sign of the well loss tallies with that of parameter ζ_0 and has a clear physical meaning (Figure 58). Owing to the creation by the strong natural development of an area around well 26 HE with a hydraulic

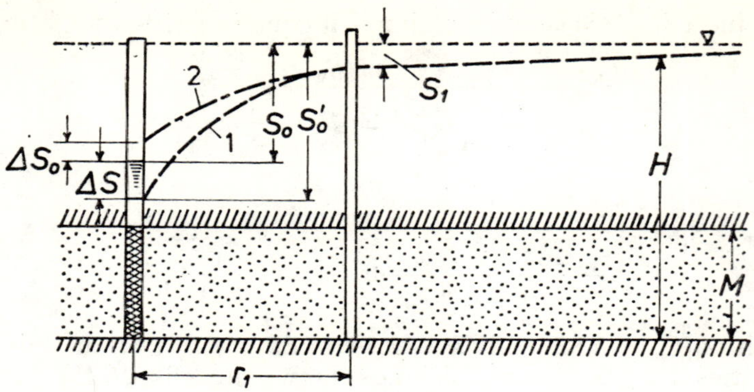

Figure 58. Diagram of the well loss in the case of a strong natural development:

1, Depression curve before development; *2*, Depression curve after development.

conductivity k_0 greater than the natural one (k_i), the drawdown curve is higher and a secondary well loss ΔS_0 appears. The well loss proper (ΔS) appears to be the level difference between the intersection of the initial drawdown curve with the outer wall of the well and the pumping level within the well under the new conditions of permeability; that is why its value is negative. Evidently the secondary well loss can also be computed provided we have an observation well within the boundaries of the high permeability area.

The analytical relation between the well loss and parameter ζ_0 is deduced from relations (143) and (153)

$$\Delta S = \frac{Q}{2\pi kM}\zeta_0 \tag{154}$$

For well 26 HE, if $\zeta_0 = -2.15$ is known, we obtain:

$$\Delta S = \frac{147}{6.28 \times 24.2}(-2.15) = -2.08 \text{ m}$$

which corresponds to the value analytically computed above.

By computing the transmissivity with the drawdown corresponding to k_i, we find:

$$(kM)_{0-1} = \frac{0.366\ Q}{(S_0' - S_1)}\log\frac{r_1}{r_0} = \frac{0.366 \times 147}{(7.70 + 2.08) - 6.50}\log\frac{2.97}{0.1} = 24.2 \text{ m}^2/\text{day}$$

This being equal to the value computed with the help of the total seepage resistance, it follows that the secondary well loss is negligible in the case of group 26 H. In connection with the magnitude of the high permeability area, we can only appreciate that its radius is smaller than r_1 (2.97 m) because the well loss computed with the data of well 26 Ha is equal to that supplied by the data of well 26 Hb.

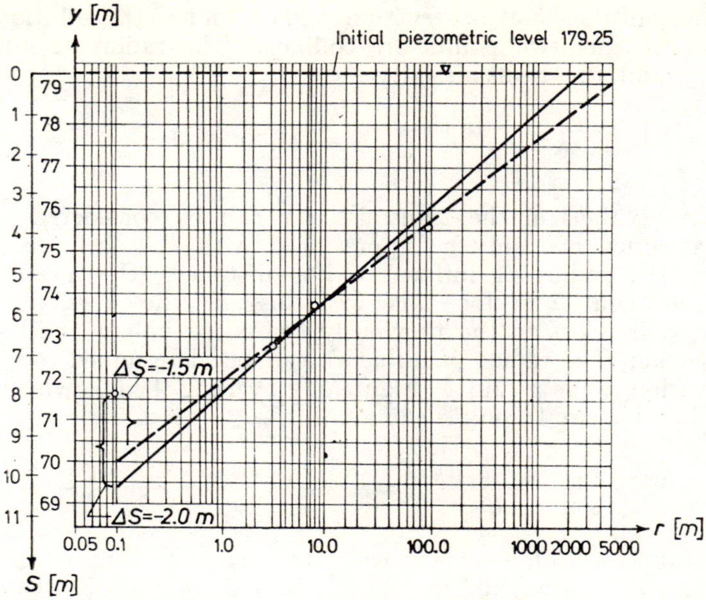

Figure 59. Logarithmic anamorphosis graph of wells 26 H — pumping I.

The *method of the logarithmic anamorphosis* starts from the fundamental equation

$$kM = \frac{0.366\, Q(\log r_2 - \log r_1)}{(y_2 - y_1)} = \frac{0.366\, Q}{\operatorname{tg}\alpha}, \tag{155}$$

where y is the pumping water level.

It is considered that $\operatorname{tg}\alpha = \dfrac{y_2 - y_1}{\log r_2 - \log r_1}$ represents the angular coefficient of the straight line determined on the graph of the logarithmic anamorphosis. On the abscissa of the graph are marked the distances between the observation and experimental wells and on its ordinate the corresponding pumping water levels. The linearization of the drawdown curve is found by applying the principle that the deviations of the points against the straight line should be minimal, while the sums of the deviations on both sides should be equal. Obviously, in the case of group 26 H, the experimental well cannot be taken into account because of its considerable well loss. The other two versions remain.

— Taking into account all the observation wells, factual radius of influence results (the distance to the intersection of the straight line with the non-pumping water level) of over 5 km and the transmissivity

$$kM = \frac{0.366 \times 147}{1.93} = 27.9\ \text{m}^2/\text{day}$$

— Taking into account observation wells *a* and *b*, the solution is more accurate because the two points are collinear. The radius of influence is $R = 2400$ m and the transmissivity

$$kM = \frac{0.366 \times 147}{2.27} = 23.70 \text{ m}^2/\text{day}$$

These two values of the transmissivity can be considered acceptable but their accuracy can only be appreciated with the help of the well loss and of the radius of influence. For instance, the graph supplies the value of about 2 m for this well loss in the version with two observation wells (closs to the previously computed value) and —1.5 m in the other version. The radius of influence can be determined by the data from observation wells *a* and *b* (Figure 53), by using the following analytical formula deduced from Dupuit's equation:

$$\log R\ ^{1)} = \frac{S_1 \log r_2 - S_2 \log r_1}{S_1 - S_2} = \frac{6.50 \times 0.913 - 5.52 \times 0.473}{6.50 - 5.52} = 3.40 \quad (156)$$

The result is a radius of influence of about 2500 m which shows a higher degree of accuracy to the data supplied by the observation wells. From the analysis of the schematic hydrogeological map of the Cuzap area (Romania) it follows that this radius of influence is a physical reality since north-westwards the extension of the aquifer is very large. But southwards the aquifer offers a "semi-infinite" character since at about only 1 km from the pumping group its recharge boundary is to be found at the basis of the crystalline formation (Rez Mountains). Obviously along this direction the radius of influence to be considered in computations will be $R = 1$ km, but it could not be brought into relief because of the lack of observation wells.

The *method of least squares*, elaborated by E. E. Kerkis [33] offers the advantage that even in the case of palpable deviations of the experimental points from the straight line it is possible to establish the most probable piezometric line and to compute the transmissivity and the radius of influence from the data of all the observation wells placed along a radius without using the graphical construction. The following transformations are required for the linearization of the drawdown curve:

$$S = \frac{0.366\,Q}{kM} (\log R - \log r) \quad (157)$$

$$\frac{0.366Q}{kM} \log R = X; \qquad \frac{0.366\,Q}{kM} = Y \quad (158)$$

$$S = X - Y \log r \quad (159)$$

[1]) This is the factual radius of influence and not the reduced one, as it is supplied by the empirical relations.

Starting from the assumption that the deviations of the factual data from the theoretical line are random, Kerkis found, by applying the method of least squares, the following final relations for the confined aquifer:

$$kM = \frac{0.366\, Q[n\, \Sigma(\log r_i)^2 - (\Sigma \log r_i)^2]}{(\Sigma S_i \Sigma \log r_i - n\, \Sigma\, S_i \log r_i)} \tag{160}$$

$$\log R = \frac{\Sigma S_i \Sigma(\log r_i)^2 - \Sigma S_i \log r_i\, \Sigma \log r_i}{\Sigma S_i\, \Sigma \log r_i - n \Sigma S_i \log r_i} \tag{161}$$

$$X = \frac{\log r_i\, \Sigma S_i \log r_i - \Sigma(\log r_i)^2\, \Sigma S_i}{(\Sigma \log r_i)^2 - n\Sigma(\log r_i)^2} \tag{162}$$

$$Y = \frac{n\Sigma S_i \log r_i - \Sigma \log r_i \Sigma\, S_i}{(\Sigma \log r_i)^2 - n\Sigma\, (\log r_i)^2}$$

Expressions (158) allow moreover to obtain:

$$kM = \frac{0.366\, Q}{Y}; \quad \log R = \frac{X}{Y} \tag{163}$$

Independently of the analytical computation, equation (159) permits us to draw the linearization graph of the drawdown curve (Figure 59). It is obvious that owing to the large well loss, the data of experimental well 26 HE will not be taken into account in the computation. Table 53 is used for this calculation.

Table 53

Well	r_i(m)	$S_i(m)$	$\log r_i$	$(\log r_i)^2$	$S_i \log r_i$
26 Ha	2.97	6.50	0.473	0.224	3.074
26 Hb	8.18	5.52	0.913	0.834	5.040
21 H	94.30	3.50	1.974	3.897	6.909
	Σ	15.52	3.360	4.955	15.023
	Σ*	12.02	1.386	1.058	8.114

*) Without the 21 H well data

By entering all the observation wells into the calculation, we obtain:

$$kM = 27.2 \text{ m}^2/\text{day}; \; R = 5400 \text{ m}; \; S = 7.39 - 1.98 \log r$$

If only wells 26 Ha and 26 Hb are considered, we shall find:

$$kM = 24.2 \text{ m}^2/\text{day}; \; R = 2500 \text{ m}; \; S = 7.54 - 2.22 \log r$$

This computing method emphasizes once more the accuracy of the measurements in wells 26 Ha and 26 Hb as well as the existence of a systematic error in the data from well 21 H. This systematic error is chiefly due to the great distance from the experimental well. The practical rules recommend that in the case of a small permeability ($k < 10^{-1}$ cm/s) and of three piezometers along the radius, the last one should be placed at about 15 m from the experimental well. But well 21 H was only accidentally worked as a piezometer, since it was made for other hydrogeological purposes. Of course, at first sight the value of the radius of influence ($R = 2500$ m) seems exaggerated. The certainty of the data from the observation wells (a and b in Figure 52) as well as the coincidence of the results obtained by various methods of data processing attest however the validity of this result. The explanation of the high value of R is to be found in the fact that the respective aquifer has only limited dynamic resources.

The *drawdown graph — logarithm of the distances* may constitute a method for determining the transmissivity and the radius of influence. But even without applying the method of least squares the angular coefficient of the straight line [1] is found to be:

$$C = \mathrm{tg}\alpha = \frac{0.366\,Q}{T} = \frac{\mathrm{d}S}{\mathrm{d}(\log r)} \qquad (164)$$

This method is, as a matter of fact, a version of the logarithmic anamorphosis where the pumping water levels are replaced by the drawdown. Taking into account S and r of the observation wells (Figure 59) we obtain:

$$C = \frac{\mathrm{d}S}{\mathrm{d}(\log r)} = \frac{0.98}{0.913 - 0.473} = 2.23 \text{ m} \qquad (165)$$

$$T = \frac{0.366\,Q}{C} = \frac{0.366 \times 147}{2.23} = 24.1 \text{ m}^2/\text{day} \qquad (166)$$

As it was to be expected (since these two points are situated exactly on the straight line) the transmissivity is practically equal to that supplied by the method of least squares. At the intersection point, $S = 0$, hence $Q = 0$, and therefore:

$$\log \frac{R}{x} = 0; \quad R = x = 2500 \text{ m}$$

In conclusion, the graph drawdown-logarithm of distances, which in fact represents the linearization of the drawdown curve is a quick and satisfactory method that permit to determine both the transmissivity and the radius of influence.

The *hydraulic conductivity* can be deduced from the transmissivity if the aquifer testing was accomplished by means of an experimental wells group. Such a case is likely to occur if the aquifer under study will have to

---

[1] The starting point is equation $S = \frac{0.366\,Q}{T} \log \frac{R}{}$

undergo a process of strong gravitational drainage, the effect of which would depend on the hydraulic conductivity.

From Table 50 and Figure 53 it follows that the aquifer thickness has a small variation between wells 26 HE and 26 Hb and that its estimated mean value is:

$$M_m = \frac{10.92 + 11.75 + 10.44}{3} = 11.04 \text{ m}$$

The mean value of the hydraulic conductivity along this alignment is:

$$k_{0-2} = \frac{kM_{(0-2)}}{M_m} = \frac{24.1}{11.04} = 2.18 \text{ m/day}$$

By the same method we found $k_{0-1} = 2.13$ m/day and $k_{1-2} = 2.17$ m/day which emphasizes a good homogeneity of the permeability on the stretch 26 HE — 26 Hb.

The *processing of the data from pumping III* emphasizes the drawdown influence on the possible errors affecting the values of the hydrogeological parameters. If the skin effect is not taken into account, the transmissivities computed by relations (136, 137) will be:

Interval	0 – 1	0 – 2	0 – 3	1 – 2	1 – 3	2 – 3
kM[m²/day]	102	44.1	42.2	22.0	26.8	39.0

As already has been remarked for pumping I, the highest accuracy belongs to $kM_{(1-2)}$. With the help of this value and making use of relation (138) $\zeta_0 = -2.64$ was determined which emphasizes the fact that in proportion of the drawdowns increase, the importance of the high permeability area around the well increases too (owing to its extension in the plane or to an additional dragging away of the fine particles). Taking into account the seepage resistance, the real transmissivities were computed:

$$kM_{(0-1)} = 22.5 \text{ m}^2\text{/day}; \quad kM_{(0-3)} = 26.0 \text{ m}^2\text{/day}$$

The results point out that for a drawdown increase of about 50% (from 7.70 to 11.93 m) the error affecting the transmissivity is still greater if the seepage resistance ζ_0 is not taken into account; if this parameter is considered, the transmissivities for both the pumping tests have practically the same value (the diminution of kM by the drawdown increase is negligible).

By comparing the computed values of the transmissivity of group 26 H Cuzap, it is to be remarked that when the data are supplied by the experimental well and an observation well, this parameter is affected by gross errors: in this case the error reaches 300% for pumping I and 500% for pumping III. The appearance of these errors results from the fact that the seepage resistance of the experimental well is not taken into account; let us remark that the importance of this seepage resistance increases alongside with the drawdown.

The *testing of the aquifer by an experimental well* (without observation wells) introduces great difficulties in the processing of the field data, owing to the impossibility of estimating the hydraulic resistance of the well and of the degraded area around it. When this is the case, the radius of influence in Dupuit's relation may be expressed by the empirical relation of Sichardt:

$$R = 10\, S_0 \sqrt{k} \tag{167}$$

$$k = 0.366 \frac{Q}{S_0 M} [\log (10\, S_0) + \log \sqrt{k} - \log r_0] \tag{168}$$

Since it is an implicit relation, its solution can be obtained by a grapho-analytical way, as follows:

$$0.366 \frac{Q}{S_0 M} = C$$

$$k - C \log \sqrt{k} = C\, [\log(10\, S_0) - \log r_0] \tag{169}$$

The argument value may be computed so as to obtain

$$k - C \log \sqrt{k} = A \tag{170}$$

The hydraulic conductivity k satisfying this equality can be determined by the graphical way. A difficult problem arises from the computation value of thickness M. If the network of the exploration wells allows us to establish the variation law of the thickness, a corresponding quantity will be computed according to the indications of § 2.2.4. In the contrary case the thickness in the experimental well will be considered.

For experimental well 26 HE, pumping I, we obtained:

$$C = 0.366 \frac{147}{10.92} = 4.93$$

$$k - 4.93 \log \sqrt{k} = 4.93(\log\ 77 - \log 0.103)$$

$$k - 4.93 \log \sqrt{k} = 14.16$$

From the graph in Figure 60, we found $k = 17.2$ m/day; by a similar computation $k = 27$ m/day were obtained for pumping III. As compared

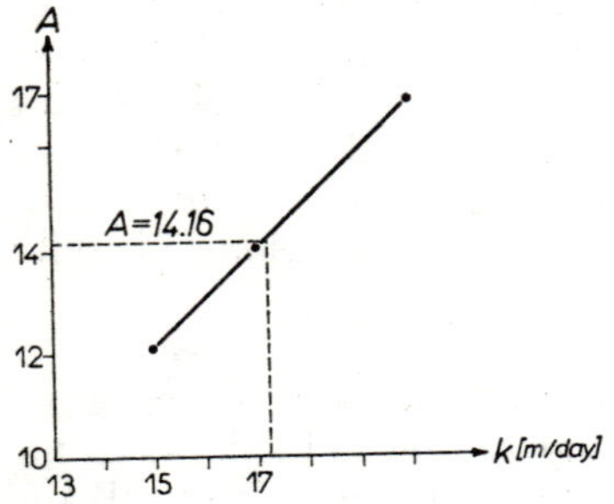

Figure 60. Graph for estimation of the hydraulic conductivity.

with the real value ($k = 2.2$ m/day) these values are 8 and 12 times respectively greater, which emphasizes the inaccuracy of the method. This is due to the presence of the natural development area around the well. If the deteriorated area were filled in so that its permeability was smaller that than of the aquifer (if parameter ζ_0 and the well loss had positive values) the errors would be reversed, i.e. the values obtained for the hydraulic conductivity by this method would be much smaller than the actual ones.

2.2.2 Unconfined aquifers

In the case of unconfined aquifers, phreatic as a rule, the processing of the field experimental data is made upon a larger volume of informations than in the case of confined aquifers, because the shallow depths permit to achieve experimental groups with several observation wells. To this advantageous situation the small values must be added of the systematic and accidental errors assignable to the static and pumping water levels as well as of the possibility of executing wells without drilling mud (dry or water drilling system), which excludes the clogging of the area around the well.

The computation method of the hydrogeological parameters applied under these conditions is presented in the form of a concrete example, to wit the experimental Plopu station in the Titu area from Romania (Figure 61) where the phreatic tested aquifer occurs in a coarse sand and gravel complex. The complex pertaining to the late Pleistocene has a regional development, while its lower part consists of a complex formed by clays and clayey sands. The geometric and hydrodynamic elements of the Plopu experimental group are recorded in Table 54.

For simplification's sake, the wells were conventionally denoted by 0. 1. 2. ..., 5. Among the other technical elements we mention here:

— the wells were executed in a manual and dry system;

— the observation wells were tubed with 7 inches-screens;

— the experimental pumpings were made with centrifugal pumps.

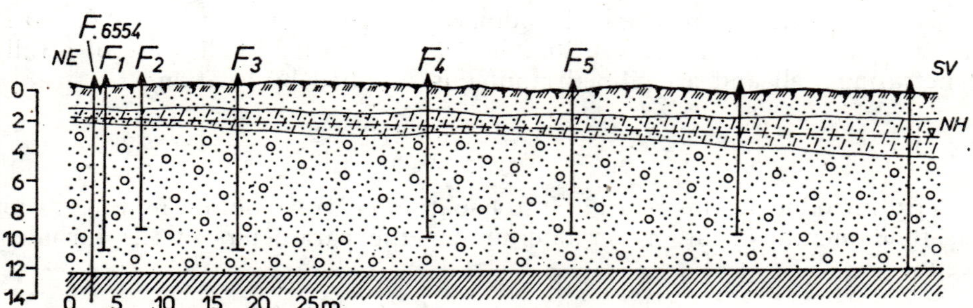

Figure 61. Hydrogeological cross-section through experimental group of wells Plopu.

In the case of phreatic aquifer located in rough alluvial deposits, deviations could appear from the laminar flow within a certain radius of the experimental well. The ignorance of this area with turbulent flow leads to erroneous values of the hydrogeological parameters. When at least two observation wells exist along a radius and when two steady flow state pumpings in a completelly penetrating well are performed, then the parameters of the turbulent flow can be determined starting from the general equation of the hydraulic gradient [70]:

$$I = \frac{\mathrm{d}h}{\mathrm{d}r} = \frac{v}{k} + \beta\left(\frac{v}{k}\right)^2 = \frac{v}{k}\left(1 + \beta\frac{v}{k}\right) \tag{171}$$

where v is the Darcy (seepage) velocity and β the flow parameter (dimensionless) that characterizes the deviation from the laminar régime.

In the case of a turbulent régime, provided that in the respective formulae a smaller hydraulic conductivity ($\bar{k}$) is used (smaller than the real one) the resulting expression derives from formula (171)

$$\bar{k} = \frac{v}{I} = \frac{k}{1 + \beta\dfrac{v}{k}} \tag{172}$$

Since the deviations from the laminar régime are often very small it is recommended that the critical velocity (v_{cr}) should be determined with a given maximum error ε. When this is the case, the critical Darcy velocity corresponds to deviation k from $\bar{k}$, equal to ε, i.e. $\varepsilon = \dfrac{k - \bar{k}}{k}$.

From equation (172) it follows that

$$\beta\frac{v_{cr}}{k} = \frac{\varepsilon}{1-\varepsilon}; \quad v_{cr} = \frac{k}{\beta}\frac{\varepsilon}{1-\varepsilon} \tag{173}$$

$$\bar{k} = \frac{k}{1 + \dfrac{v}{v_{cr}}\cdot\dfrac{\varepsilon}{1-\varepsilon}} \tag{174}$$

For engineering problems, the admissible error (ε) varies from 0.05—0.20 the most frequent being $\varepsilon = 0.10$. The Darcy velocity in the case of a fully penetrating well and a confined aquifer is supplied by relation:

$$v = \frac{Q}{2\pi kM} \tag{175}$$

By introducing this relation into equation (171) we obtain by integration the fundamental equation for determining the parameters of the turbulent régime

$$S_1 - S_2 = \Delta = \frac{Q}{2\pi kM}\ln\frac{r_2}{r_1} + \beta\left(\frac{Q}{2\pi kM}\right)^2\left(\frac{1}{r_1} - \frac{1}{r_2}\right) \tag{176}$$

Using the data supplied by two pumpings (Q_I, Q_{II} and Δ_I, Δ_{II}) this equation will supply:

$$k = \frac{Q_I \varphi}{2\pi k M \, \Delta_I} \ln \frac{r_2}{r_1} = k_D \varphi \tag{177}$$

$$\beta = \frac{r_2 \varphi(\varphi - 1)}{\left(\frac{r_2}{r_1} - 1\right) \Delta_I} \ln^2 \frac{r_2}{r_1} \tag{178}$$

$$\varphi = \frac{1 - \frac{Q_I}{Q_{II}}}{1 - \left(\frac{Q_I}{Q_{II}}\right)^2 \frac{\Delta_{II}}{\Delta_I}}; \quad (\varphi \geqslant 1) \tag{179}$$

where φ is a correction factor for linearizing the filtration and k_D is the hydraulic conductivity in laminar régime (Darcy).

From equation (179) it follows that for $\frac{\Delta_{II}}{\Delta_I} = \frac{Q_{II}}{Q_I}$, $\varphi = 1$ while Dupuit classic relation supplies the hydraulic conductivity. When φ fulfils condition

$$1 \leqslant \varphi \leqslant \varphi_{cr}; \quad \varphi_{cr} = \frac{1}{1 - \varepsilon} \tag{180}$$

then the laminar Darcy's law of filtration is true in the case under consideration.

This computation method elaborated for the confined aquifers can also be applied to the unconfined aquifers provided that an approximate mean value of the flow is introduced into the formulae according to relation

$$M = \frac{1}{2}\left[\left(H - \frac{\Delta_I}{2}\right) + \left(H - \frac{\Delta_{II}}{2}\right)\right] = H - \frac{\Delta_I + \Delta_{II}}{4} \tag{181}$$

where H is the thickness of the unconfined aquifer.

Component ζ_{2r} of the total seepage resistance of aquifer ζ_0 (§ II. 1.2) will be estimated by relation

$$\zeta_{2r} = \frac{Q_I(\varphi - 1)}{2\pi k M \, \Delta_I} \ln^2 \frac{r_2}{r_1}$$

By applying this method to observation wells F_1 and F_2 and to pumpings II and III (Table 54), we obtain

$$\Delta_{II} = 0.39 - 0.31 = 0.08 \text{ m}$$

$$\Delta_{III} = 0.61 - 0.48 = 0.13 \text{ m}$$

$$\frac{\Delta_{III}}{\Delta_{II}} = 1.625; \quad \frac{Q_{II}}{Q_{III}} = \frac{604.8}{993.6} = 0.608$$

$$\varphi = \frac{1 - 0.608}{1 - (0.608)^2 \cdot 1.625} = 0.98 \simeq 1$$

In conclusion, the flow régime is laminar even for the highest values of the hydraulic gradient within the experimental group $\left(I = \frac{0.61 - 0.48}{4}\right) = 0.032$. According to relation (178), $\beta = 0$, therefore $v = kI$.

Table 54

Pumping test results of Plopu wells

Pumping	Discharge rate		Drawdown S_0 in the wells F_0	Specific yield (l/s)	Water level lowering S_i in the observation wells					Distances (x_i)
	l/s	m³/day			F1	F2	F3	F4	F5	
I	4	345.6	0.50 m	8.00	0.24	0.20	0.16	0.14		$r_0 = 0.11$ m
II	7	604.8	1.20 m	5.83	0.39	0.31	0.25	0.23	0.08	$r_1 = 1$ m
III	11.5	993.6	3.50 m	3.50	0.61	0.48	0.40	0.36	0.19	$r_2 = 5$ m
										$r_3 = 15$ m
										$r_4 = 35$ m
										$r_5 = 50$ m

Usually it is considered that $\varepsilon = 0.10$ so that $\varphi_{cr} = \frac{1}{1 - 0.10} = 1.11$; the flow régime results from the comparison between experimental quantities φ and φ_{cr}.

2.2.2.2 Yield-drawdown diagram

If after the pumpings, no important change appears in the water yielding properties in the vicinity of the experimental well, the theoretical equation of the yield-drawdown diagram is supplied by formula (106) where parameter Q_{max} is estimated by relation (107); to facilitate the computation Table 55 was made.

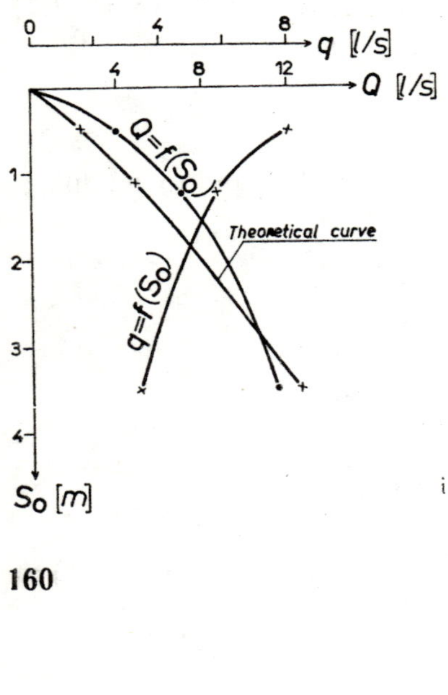

Fig. 62. Yield-drawdown diagram of group of wells Plopu.

According to relation (107)

$$Q_{max} = \frac{172699.0 - 102940.5}{3362.5} = 20.75 \text{ l/s}$$

$$\alpha = \frac{H^2}{Q_{max}} = \frac{9.36^2}{20.75} = 4.22$$

For determining the contingent deviations of the experimental curve from the theoretical one, we shall compute the rates of discharge corresponding to the experimental drawdowns with the help of theoretical equation

$$(9.36 - S_0)^2 = 4.22(Q_{max} - Q) \tag{182}$$

Table 55

Pum-ping	Q, l/s	S_0, m	H^4Q	$H - S_0$	$(H-S_0)^2$	$H^2Q(H - S_0)^2$	$[H^2 - (H - S_0)^2]$	
I	4.0	0.50	30 702.0	8.86	78.50	27 509.5	82.99	$H=9.36$ m
II	7.0	1.20	53 728.6	8.16	66.59	40 833.4	441.84	$H^2=87.61$
III	11.5	3.50	88 268.4	5.86	34.34	34 597.6	2 837.69	$H^4=7\,675.51$
			172 699			102 940.5	3 362.52	

The acquired values ($Q_I = 2.2$, $Q_{II} = 5.0$, $Q_{III} = 12.6$ l/s) point out important deviations. Let us remark that up to $S_0 = 3$ m (Figure 62), the experimental rates of discharge are larger than the theoretical ones. As a matter of fact, the non-integration of the experimental curve within the parabolic (theoretical) correlation is also explained by the lack of a linear dependence between specific capacity q and drawdown S_0 (Figure 62). Under these conditions the logarithmic correlation is as suitable as the exponential one. Figure 63 shows the existence of a logarithmic correlation [1]).

$$Q = 6.54 + 8.90 \log S_0$$

The computed values of the rate of discharge ($Q'_I = 3.9$, $Q'_{II} = 7.2$, $Q'_{III} = 11.4$ l/s) which are very close to the experimental ones prove the validity of this type of correlation. The problem of extrapolating the yield drawdown diagram to the unconfined aquifers the thicknesses of which are as a rule small and medium, should be treated with particular attention and by opting for the low limit of the extrapolation. Since for the logarithmic correlation it is admitted that $S_M = 2 - 3\ S_{max}$, there results that $S_M = 2 \times 3.50 = 7.0$ m and $Q_M = 6.54 + 8.90 \log 7 = 14.1$ l/s respectively.

The form of the yield-drawdown diagram can be connected with certain characteristics of the unconfined aquifer by relations of correspondance

[1]) The computing method of parameters $a = 6.54$ and $b = 8.90$ (formula 114) is presented in § 2.2.1.1.

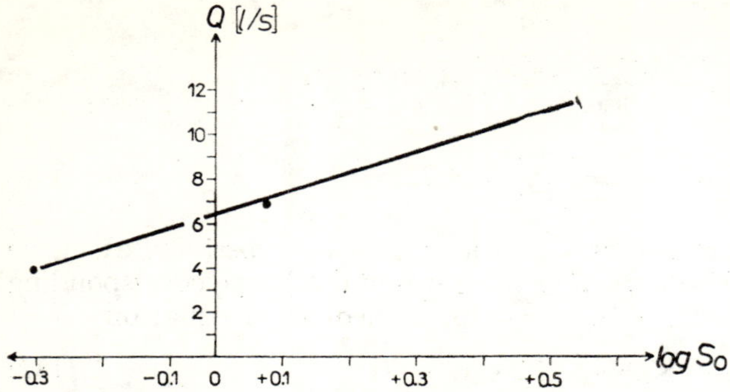

Figure 63. Linearized yield-drawdown diagram.

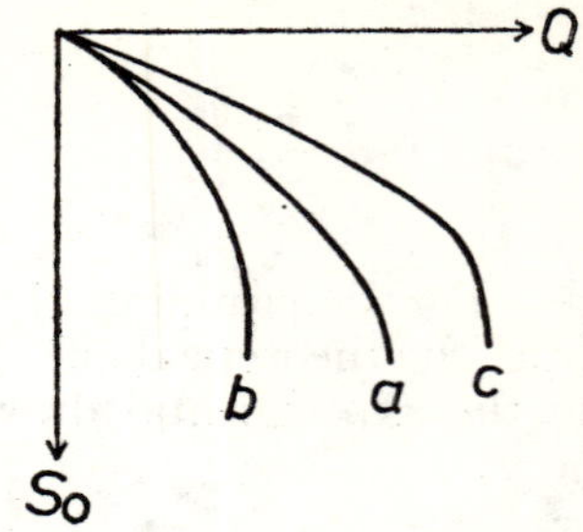

Figure 64. Various yield-draw down diagrams of unconfined aquifers.

materialized by quantitative or qualitative elements. In this case, the parameter of the so-called specific drawdown $\left(s = \frac{S_0}{Q}\right)$, which represents the drawdown determining the rate of discharge unit, can be computed from this curve. This specific drawdown may have a constant value along certain portions of the curve, but it increases usually with the drawdown This increase can be slow (Figure 64, curve *c*) or sudden (curve *b*) as a function of several factors, such as: the permeability and the dynamic resource magnitude, the interception by the cone of depression of a recharge or of a barrier boundary, the non-homogeneity of the permeability in the plane or on the vertical. The contribution of each of these factors to the form of the yield-drawdown diagram can only be established by a complex interpretation of all the hydrogeological data and parameters. Figure 64 presents two cases of the permeability non-homogeneity on the vertical, *b* and *c*, in regard to curve *a* which represents the case of a homogeneous aquifer on the vertical. In case *b*, the specific drawdown presents a sudden increase due to the permeability diminution with the depth, while curve *c* represents a slow increase of the specific drawdown due to the permeability increase with the depth. Figure 62 shows that in the case of experimental group Plopu (Romania), a homogeneous permeability appears on the vertical of the aquifer.

2.2.2.3 Computation of the hydrogeological parameters by analytical and grapho-analytical techniques

The processing of the experimental data in steady flow régime of an unconfined aquifer allows us to compute the following parameters: the hydraulic conductivity, the seepage resistance of an experimental well as well as its well loss and its radius of influence. To this end, all the relations elaborated for the confined aquifer can be used provided that we know the thickness of the aquifer current, which under the new conditions is a space variable. By applying the principle of the arithmetic mean to the passage from a confined aquiferous current to a unconfined one, the following relations can be used:

$$MS_0 = \frac{H+h_0}{2} \cdot (H - h_0) = 0.5(H^2 - h_0^2) = 0.5\, S_0(2H - S_0) \qquad (183)$$

$$MS_1 = \frac{H+h_1}{2} (H - h_1) = 0.5(H^2 - h_1^2) = 0.5\, S_1(2H - S_1) \qquad (184)$$

$$M(S_0 - S_n) = \frac{h_0 + h_n}{2} [(H - h_0) - (H - h_n)] = \frac{h_0 + h_n}{2} (h_n - h_0) =$$

$$= 0.5(h_n^2 - h_0^2) = 0.5(S_0 - S_n)\,(2H - S_0 - S_n) \qquad (185)$$

$$M(S_n - S_p) = \frac{h_n + h_p}{2} [(H - h_n) - (H - h_p)] = \frac{h_n + h_p}{2} (h_p - h_n) =$$

$$= 0.5(h_p^2 - h_n^2) = 0.5(S_n - S_p)\,(2H - S_n - S_p) \qquad (186)$$

In these relations h_0 is the pumping water level of the central well h_n and h_p in the observation wells.

The hydraulic conductivity on various alignments is computed by means of Dupuit-Thiem relations:

— between the experimental well and an observation well

$$k_{0-n} = 0.732 \frac{\log \frac{r_1}{r_0}}{(2H - S_0 - S_n)\,(S_0 - S_n)} \qquad (187)$$

— between two observation wells

$$k_{n-p} = 0.732\, Q \frac{\log \frac{r_1}{r_0}}{(2H - S_n - S_p)\,(S_n - S_p)} \qquad (188)$$

The appplication of these relations to Plopu group, pumping III supplied the values of the hydraulic conductivity on various intervals (Table 56 and Figure 65).

From the comparative analysis of the values computed for k, correlated with certain known hydrogeological conditions of the experimental group, it appears that certain values are erroneous, their deviations from the values considered as normal being too great. For instance, all values computed for the interval between the central well and an observation well are erroneous because the area with a modified permeability around the central well was not taken into account. As to the rest of them the sequence of values emphasizes an increase of the permeability going from the central well towards the boundary of the depression cone with the exception of the intervals that include F_5; the reiteration of these cases justifies the exclusion of this well from the field of investigation due either to the erroneous values of distance r_5 or to the drawdown S_5. In Table 56, the erroneous values of k are between

Hydraulic conductivity of various intervals (Plopu wells)

Interval	0–1	0–2	0–3	0–4	0–5
k(m/day)	(16.9)*	(27.1)	(33.8)	(39.0)	(38.9)
Interval	1–2	1–3	1–4	1–5	
k	222.6	230.2	253.3	(164.4)	
Interval	2–3	2–4	2–5		
k	242.9	286.2	(139.2)		
Interval	3–4	3–5			
k	(371.2)	(99.4)			
Interval	4–5				
k	(36.5)				

* Values in brackets are erroneous

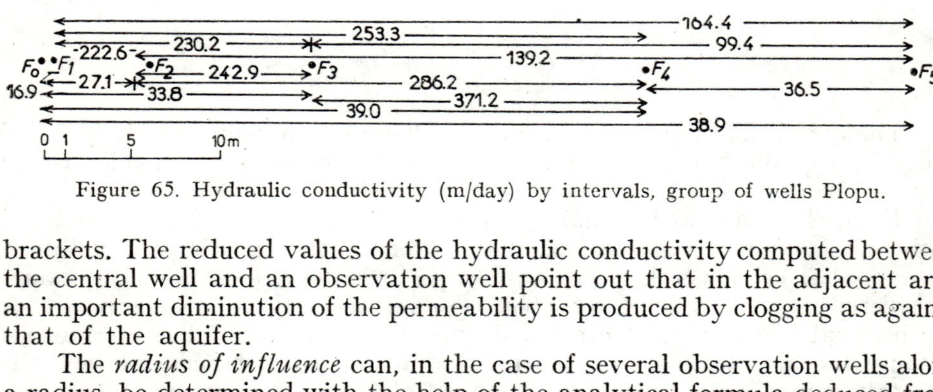

Figure 65. Hydraulic conductivity (m/day) by intervals, group of wells Plopu.

brackets. The reduced values of the hydraulic conductivity computed between the central well and an observation well point out that in the adjacent area an important diminution of the permeability is produced by clogging as against that of the aquifer.

The *radius of influence* can, in the case of several observation wells along a radius, be determined with the help of the analytical formula deduced from the Dupuit equation:

$$\log R^{1)} = \frac{S_n(2H - S_n) \log r_p - S_p(2H - S_p) \log r_n}{(S_n - S_p)(2H - S_n - S_p)} \tag{189}$$

[1]) This is the real radius of influence.

If well F_5 is excluded from the series of accurate data, the radius of influence can be computed by means of the pairs of data 1—2, 1—3, 1—4, 2—3, 2—4 and 3—4. For example, with data from wells 1 and 2 (see also Table 54) we obtain:

$$\log R_{(1-2)} = \frac{0.61(18.72 - 0.61)\log 5 - 0.48(18.72 - 0.48)\log 1}{(0.61 - 0.18)(18.72 - 0.61 - 0.48)} = 3.377$$

$$R_{(1-2)} = 2380 \text{ m}$$

For the other cases we obtained:

$$R_{(1-3)} = 3\,105 \text{ m}; \qquad R_{(1-4)} = 6\,950 \text{ m};$$

$$R_{(2-3)} = 3\,800 \text{ m}; \qquad R_{(2-4)} = 13\,650 \text{ m};$$

$$R_{(3-4)} = 82\,790 \text{ m}.$$

Normally, of course, the radius of influence thus computed should have the same value, whatever pair of observation wells should be taken into account. From the obtained results however, a great variation of the computed values appears, due to some errors of the experimental data (S and r) which differ from one well to another. While when computing the hydraulic conductivity these errors appear as admissible, they cannot be accepted in the case of formula (189) owing to this formula "sensitivity" which is connected with the fact that the radius of influence is deduced by applying the antilogarithms (measurement errors of the pumping water level which though small are liable to affect considerably the radius of influence). In conclusion, the use of formula (189) requires quite accurate measurements of the drawdowns and of the distances between wells.

The *total seepage resistance* (ζ_0) can be estimated by means of relation (143) where thickness M and drawdown S are supplied by relations (183—186) specific to the unconfined aquifer. If observation well F_1 is used, relation (143) will become:

$$\zeta_0 = \frac{\pi(S_0 - S_1)(2H - S_0 - S_1)k_{1-2}}{Q} - 2.3 \log \frac{r_1}{r_0} \qquad (190)$$

From pumping III (Table 54) it was deduced that

$$\zeta_{0(\text{III})} = \frac{3.14(3.50 - 0.61)(18.72 - 3.50 - 0.61)\,222.6}{993.6} - 2.3 \log \frac{1}{0.11} = 27.45$$

Parameter ζ_0 can equally be determined by the grapho-analytic method that uses the data of all the observation wells. Equations (141, 142) modified for an unconfined aquifer (where the hydraulic resistances of the observation wells are negligible) can be represented like those for the confined aquifer except that quantity $\eta = (S_n - S_p)(2H - S_n - S_p)$ appears on the ordinate.

The seepage resistance (ζ_0) is then determined with the help of the parameters of both the straight lines (Figure 66). The angular coefficient of straight line 2 is:

$$B = \frac{\Sigma\left(\frac{\eta}{\ln \bar{r}}\right)_i}{n} = \frac{5.35}{4} = 1.34^* \tag{191}$$

$$\zeta_0 = \frac{C}{B} = \frac{39.0}{1.34} = 29.10$$

The very large and close together values of parameter ζ_0 (computed by two diagrams) point to the existence of a strongly clogged area around the experimental well F-6554 Plopu.

If the total seepage resistance (ζ_0) is known, the accurate values of the hydraulic conductivity between the central well and one of the observation wells can be determined by relation:

$$k_{0-n} = \frac{Q}{\pi(S_0 - S_n)(2H - S_0 - S_n)}\left(2.3 \log \frac{r_n}{r_0} + \zeta_0\right) \tag{192}$$

Starting from $\zeta_0 = 27.45$, we obtained:

$$k_{0-1} = 202.2 \text{ m/day}; \qquad k_{0-2} = 222.4 \text{ m/day}$$

$$k_{0-3} = 223.0 \text{ m/day}; \qquad k_{0-4} = 202.5 \text{ m/day}.$$

The small differences between these values and the computed ones on the only basis of the observation wells (Table 56) confirm the accuracy of quantity ζ_0 as well as the conclusion about the existence of a strongly clogged area.

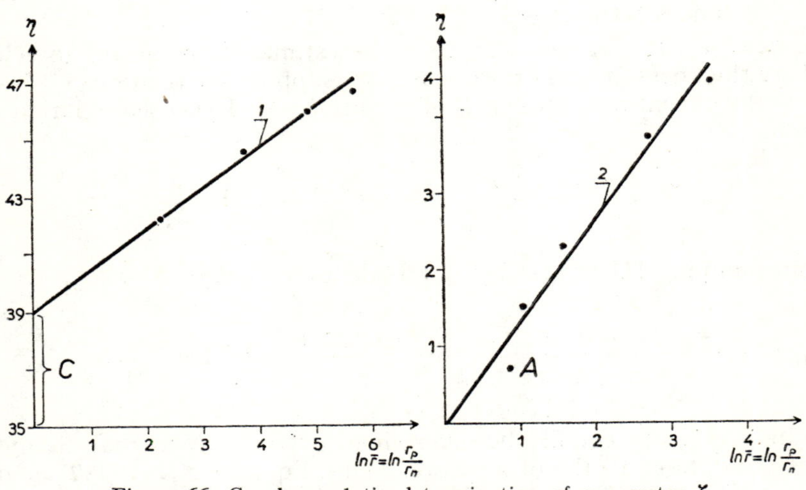

Figure 66. Grapho-analytic determination of parameter ζ_0.

* Point A was discarded because of its great deviation from the straight line.

Formula (190) moreover, allowed us to compute the hydraulic resistances for pumpings II and I, which supplied $\zeta_{0(II)} = 13.25$ and $\zeta_{0(I)} = 6.90$ *).

The *well loss* (Δh) can be determined by the modified Dupuit's equation

$$Q = \frac{\pi k_{0-1}[h_n^2 - (h_0^2 + \Delta h)^2]}{\ln r_n/r_0} \tag{193}$$

$$\Delta h = \sqrt{h_n^2 - \frac{Q}{\pi k_{0-1}} \ln \frac{r_n}{r_0}} - h_0 \tag{194}$$

By using observation well F_1, pumping III will supply (Table 54):

$$\Delta h_{III} = \sqrt{8.75^2 - \frac{993.6}{3.14 \times 202.2} \ln \frac{1}{0.11}} - 5.86 = 2.68 \text{ m}$$

So as to eliminate the inaccuracy of observation well F_1 which results from its situation in the clogged area (owing to the small value $r_1 = 1$ m) the same well loss was equally computed with the data of well F_2, which supplied $\Delta h_{III} = 2.70$ m. The quasi-identity of both these values allows us to conclude that the radius of the clogged area is smaller than 1 m; usually the radius of the modified permeability area (clogged or naturally developed) is of about tens of centimeters.

Analogously, $\Delta h_{II} = 0.70$ m and $\Delta h_I = 0.20$ m were obtained for the other two pumpings. It follows that the well losses are — with regard to the respective drawdown

77% for pumping III

58% for pumping II

40% for pumping I.

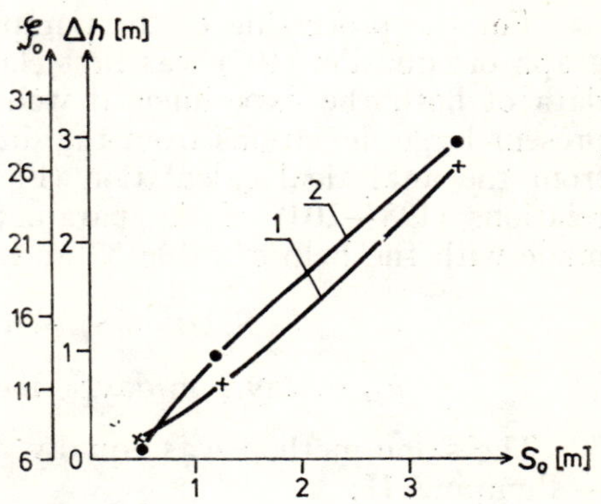

Figure 67. Variation of the well loss (*1*) and of the seepage resistance (*2*).

* In these computations it made use of the hydraulic conductivity $k_{1-2} = 215.2$ m/day supplied by pumping II

Figure 67 allows us to conclude that within the limits of the investigated domain ($S_0 = 0.50 — 3.50$) both the well loss and the seepage resistance are practically linearly dependent on the drawdowns.

The *method of least squares* allows us to determine the most probable line of depression as well as to compute the hydraulic conductivity and the radius of influence by using the data supplied by all the observation wells placed along a radius of the experimental group. The comparison of these data with the theoretical values permits us to estimate the quality of the pumpings as well as the influence of certain factors upon the line of depression. To linearize the curve of depression, the following transformations [36] should be made:

$$S(2H - S) = \frac{0.732Q}{k}(\log R - \log r) \tag{195}$$

$$\frac{0.732Q}{k}\log R = U; \quad \frac{0.732Q}{k} = V \tag{196}$$

$$S(2H - S) = U - V \log r \tag{197}$$

The mathematical processing of these relations by the above mentioned method supplies the following final expressions of the parameters:

$$k = \frac{0.732\, Q[n\Sigma(\log r_i)^2 - (\Sigma \log r_i)^2]}{\Sigma S_i(2H - S_i)\Sigma \log r_i - n\Sigma S_i(2H - S_i) \log r_i} \tag{198}$$

$$\log R = \frac{\Sigma S_i(2H - S_i)\, \Sigma(\log r_i)^2 - \Sigma S_i(2H - S_i) \log r_i\, \Sigma \log r_i}{\Sigma S_i(2H - S_i)\, \Sigma \log r_i - n\Sigma S_i(2H - S_i) \log r_i} \tag{199}$$

$$U = \frac{\Sigma \log r_i\, \Sigma S_i(2H - S_i)\log r_i - \Sigma(\log r_i)^2\, \Sigma(2H - S_i)\, S_i}{(\Sigma \log r_i)^2 - n\Sigma(\log r_i)^2} \tag{200}$$

$$V = \frac{n\Sigma S_i(2H - S_i) \log r_i - \Sigma \log r_i\, \Sigma(2H - S_i)\, S_i}{(\Sigma \log r_i)^2 - n\Sigma(\log r_i)^2} \tag{201}$$

For the processing of the pumping data (Table 54) by this method, the graph of equation (197) was first elaborated. At a glance, it follows that the data of both the experimental well (F-6554) and the observation well (F_5) present large deviations from the line of depression, hence they are removed from the analytical calculation (Figure 68). The computation according to relations (198—201) of the parameters corresponding to pumping III was made with the help of table 57 and supplied the following results:

$$S(2H - S) = 10.94 - 2.92 \log r \tag{202}$$

$$k_M = 249.2 \text{ m/day}; \quad \log R = 3.742; \quad R = 5500 \text{ m}$$

The same method was equally applied to the other pumpings.

— Pumping II:

$$S(2H - S) = 7.09 - 1.94 \log r \tag{203}$$

$$k_m = 228.4 \text{ m/day}; \quad R = 4550 \text{ m}$$

Table 57

Well	r_i m	S_i m	$S_i(2H - S_i)$	$\log r_i$	$(\log r_i)^2$	$S_i(2H - S_i) \log r_i$
Central F_0	0.11	3.50	(53.27)	(−0.959)	(0.920)	(−51.09)
F_1	1.00	0.61	11.05	0	0	0
F_2	5.00	0.48	8.76	0.699	0.489	6.12
F_3	15.00	0.40	7.33	1.176	1.383	8.62
F_4	35.00	0.36	6.61	1.544	2.384	10.21
F_5	50.00	0.19	(3.52)	(1.699)	(2.887)	(5.98)
		Σ	33.75	3.419	4.256	24.95

The values in brackets were not taken into account

— Pumping I:

$$S(2H - S) = 4.47 - 1.22 \log r \tag{204}$$

$$k_m = 208.2 \text{ m/day}; \quad R = 4740 \text{ m}$$

The straight lines (lines of depression) in Figure 68 were drawn on the basis of relations (202—204) so as to make possible their extrapolation up to $r_i = r_0$ and $r_i = R$. The three mean values of the hydraulic conductivity are in normal sequence. The comparison (for pumping III) of k values along various portions (Table 56) with the mean value $k_m = 249.2$ m/day shows the advantage of the method of least squares which supplies the most probable values of the computed parameters. Even in the problem of the radius of influence, this method supplied conclusive results in opposition to the Dupuit relation (189) which, as was proved previously, supplies unlikely results as

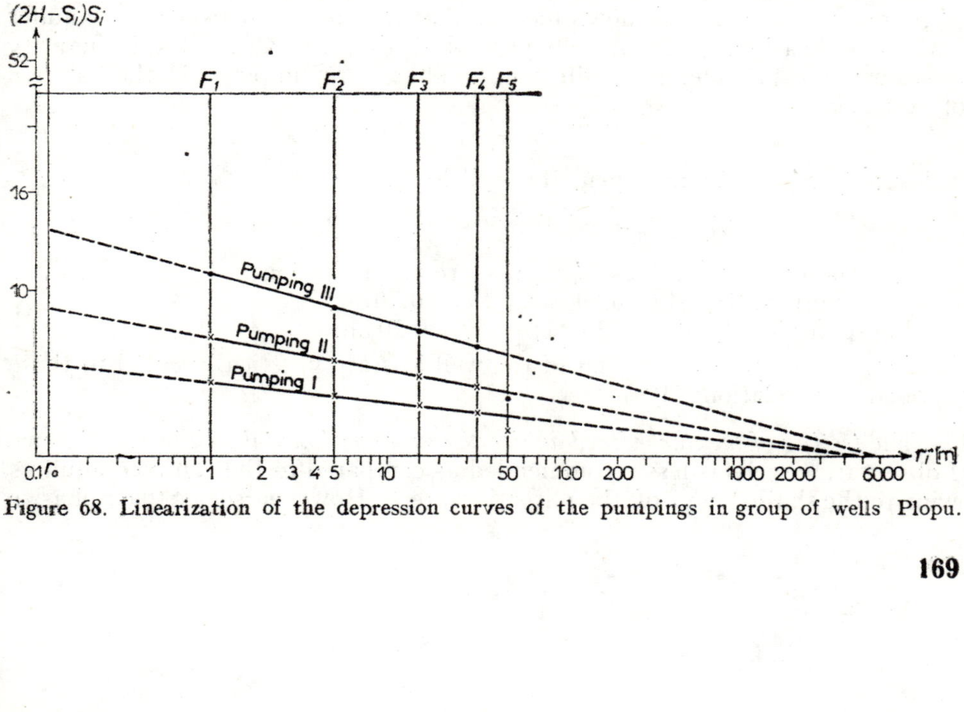

Figure 68. Linearization of the depression curves of the pumpings in group of wells Plopu.

concerns the measurement errors. The three analytically determined values (5 500; 4 500; 4 740 m) do not however present a normal variation which requires a qualitative analysis of the problem. It is admitted that in the case of aquifer currents without a recharge by vertical leakage the radii of influence and the rates of discharge are related by a directly proportional ratio $\left(\frac{R_I}{R_{II}} = \frac{Q_I}{Q_{II}}\right)$ while if there is a vertical capture the dependence between the two parameters changes

$$\frac{R_I}{R_{II}} = \sqrt{\frac{Q_I}{Q_{II}}} \tag{205}$$

The other two values of the radius of influence are obtained by taking $R_{III} = 5500$ m as a basis:

$R_{II} = 3355$ m and $R_I = 1925$ m in the absence of the recharge;

$R_{II} = 4290$ m and $R_I = 3300$ m in the case of a vertical recharge

The investigated aquifer has a natural hydraulic gradient $I = 2\%$ which on the other hand is a consequence of the small value of the recharge originated from atmospheric rainfall. Under these conditions, the differences between the values of the radii of influence should be rather large, which is not the case since the larger difference is of 1000 m only. Consequently, taking into acount the close to one another values of the radii of influence as well as the absence of a dependence (within the two above-mentioned schemes) between these and the corresponding rates of discharge, it is estimated that the considered aquifer is "semi-infinite" having at 5 km distance a recharge boundary consisting of a hydrographic element or of a contact with a much stronger aquifer.

The extrapolation of the lines of depression (Figure 68) up to their intersection with the abscissa, i.e. $(2H—S_i)\, S_i = 0$, supplies graphically the values of the radii of influence. On the other hand, the point of intersection of these lines with vertical $r_i = r_0$ corresponds to quantity $(2H—S_0')\, S_0'$ where S_0' represents the real drawdown measured at the contact between the aquifer and the well casing. If, graphically or analytically, the value of this quantity is known, the parameter S_0' will result. Thus, for pumping III the solution of equation

$$18.72\, S_0' - S_0'^2 = 13.8$$

will supply $S_0' = 0.76$ m. Then, the well loss is

$$\Delta h = S_0 - S_0' = 3.50 - 0.76 = 2.74 \text{ m}$$

For the other pumpings S_0' values are the following:

For pumping II, $S_0' = 0.50$ m; $\Delta h = 0.70$ m;

For pumping I, $S_0' = 0.30$ m; $\Delta h = 0.20$ m.

Let us note that the values of the well loss are practically equal to those supplied by relation (194).

The *testing of the unconfined aquifer by an experimental well* in the absence of observation wells is less recommended as compared to the confined aquifers owing to the shallowness of the phreatic waters. However, so as to emphasize

Figure 69. Determination of the hydraulic conductivity, group of wells Plopu, pumping III.

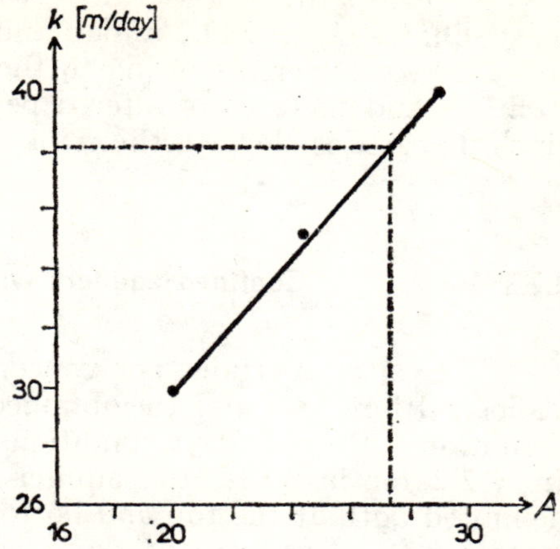

the importance of the systematic errors ascribable to this method, parameters k and R will be computed for all three pumpings of Plopu group. In this case, the radius of influence will be estimated by the empirical relation of Kusakin; introduced into the equation of the water inflow to a well, this will supply an implicit relation, the solving of which can be made grapho-analytically as follows:

$$R = 2S_0\sqrt{kH}$$

$$k = \frac{0.732\,Q}{(2H - S_0)\,S_0}\left[\log(2S_0\sqrt{H}) + \log\sqrt{k} - \log r_0\right] \tag{206}$$

$$\frac{0.732\,Q}{(2H - S_0)\,S_0} = C$$

$$k - C\log\sqrt{k} = C\left[\log(2S_0\sqrt{H} - \log r_0\right] \tag{207}$$

The argument can be computed, so that

$$k - C\log\sqrt{k} = A \tag{208}$$

By means of a graph, the value of the hydraulic conductivity (k) can be determined. By so doing, there resulted for pumping III (Table 54):

$$C = 0.732\,\frac{993.6}{(2\times 9.36 - 3.50)\,3.50} = 13.65$$

$$k - 13.65\log\sqrt{k} = 13.65\left[\log(2\times 3.50\sqrt{9.36} - \log 0.11\right]$$

$$k - 13.65\log\sqrt{k} = 27.52$$

From the graph (Figure 69), we deduced that $k = 38.2$ m/day, which introduced into relation (206) supplies the reduced value of the radius of influence

$$R = 2\times 3.50\sqrt{38.2\times 9.36} = 132 \text{ m}$$

By comparing these values with the ones supplied by the method of least squares ($k_m = 249.2$ m/day and $R = 5500$ m) it follows that this method introduces gross errors owing to the impossibility of taking into account the well loss and the area of altered permeability around the experimental well; since this area is clogged, the error is negative, $e = -85\%$.

2.2.3 Confined aquifers with recharge by vertical leakage

Certain non-concordances appearing between the hydrogeological forecasts and the obtained results could also be due to the incorrect definition of the boundary conditions in the vertical cross section of the aquifer. In § 2.2 for instance, the aquifer was considered as a permeable medium delimited both at the top and bottom by practically impermeable beds. As a matter of fact however, the clay-marly deposits situated at shallow depths, in structures of rather late geological age by gravitational consolidation could not be transformed into "absolutely" impermeable beds; under certain hydrogeological conditions the latter present a certain permeability which determines their behaviour as semi-pervious layers or aquifers. Within a hydrogeological structure, owing to the very great difference between the permeabilities of the water bearing formations and of the semi-pervious layers, vertical currents of interaction appear between aquifers, this phenomenon being also called leakage. This leakage can appear both in natural conditions (though with a lower intensity because of the small pressure differences between the aquifers) and, chiefly, in disturbed conditions, such as experimental pumpings, catchments, drainages.

In the case of catchments for water supply or for thermomineral waters, the leakage can constitute a negative phenomenon both from the hydrodynamic standpoint, because it determines the reducing of the water resources in the hydrogeological structure, and from the hydro-chemical standpoint should the aquifers be affected by the leakage, alter water qualities. The thorough knowledge of the phenomenon may lead to its directing, through the elimination or the reduction to a minimum of the prejudicial aspects. In the case of dewatering of ore deposits, the leakage could have a high intensity owing to the very great drawdowns; in this case, the dewatering rate is small because of the large flow due to the vertical leakage from the neighbouring aquifers. Generally, a groundwater balance (for a certain structure) cannot be conceived in the absence of a quantitative estimation of the leakage.

The theoretical bases of the leakage have been elaborated by M. S. Hantush [29] and by A. M. Miatiev and P. Ya. Polubarinova-Kocina [52] who gave solutions for particular cases. The following hypotheses are admitted in these solutions:

— All the aquifers and semi-pervious layers occurring in the interaction are homogeneous from the standpoint of the water yielding properties.

— The vertical seepage through the semi-pervious layers occurs in the condition of non yielding the water contained by the rocks, a reason for which it is neglected in calculations.

— Aquifers $n - 1$ and $n + 1$, subordinate from the hydraulic standpoint, ensure a constant pressure on their contact surface with the semi-pervious layers.

— From the theoretical standpoint the aquiferous currents of these complexes are three-dimensional. In computations they are considered two-dimensional; this approximation is warranted by the very large difference of permeability between the aquifer (k) and the semi-pervious layers (λ). In such cases, it is considered that the leakage is vertical through the semi-pervious layers, while the flow in the aquifer is horizontal.

Of course, these simplifying hypotheses cannot be admitted unconditionally in every case. Thus, the constant pressure can only be warranted by the subordinate aquifers if their water yielding properties are higher than that of the principal aquifer. In this respect, the acceptance of this hypothesis in the case of depth aquifers in interference is liable to introduce gross errors. On the other hand, direct observations brought into relief micro- and macro-nonhomogeneities of the semi-pervious deposits. Another drawback in solving leakage schemes lies in the fact that up to now there exists no clear representation of the flow process through semi-pervious layers (the relation between Darcy's law and the "initial gradient" of the flow, the manifestation of the elastic régime, etc.).

Figure 70 represents schematically the dynamics of the ground waters during a pumping in aquifer n. In a natural régime, owing to the increase with the depth of the piezometric level of the three aquifers, aquifer n recharged aquifer $n - 1$ while being recharged from the bottom layer. The pumping from aquifer n, besides intensifying the supply from the bottom, changes the

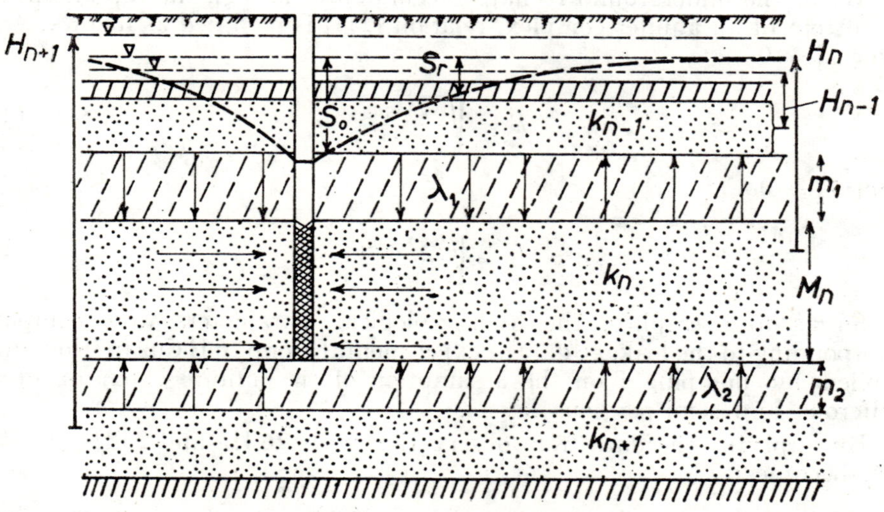

Figure 70. Dynamics of the ground-waters during pumpings in the case of a leakage from top and bottom.

hydraulic ratios to aquifer $n - 1$ and consequently in the area where the pumping levels are lower than H_{n-1} the vertical percolation becomes descendent. In the rest of this area, the ascendent vertical percolation is maintained through the upper layer with hydraulic conductivity (λ) as in the non-influenced régime.

In the case of a pumping in steady flow and under the conditions of Figure 70, the drawdown at a point at a distance r from the experimental well is supplied by relation:

$$S = \frac{Q}{2\pi k_n M_n} K_0\left(\frac{r}{B}\right) \tag{209}$$

where K_0 is the modified Bessel function of II-nd kind of zero degree [75]. For very small values of the argument ($r/B < 0.02$) the Bessel function becomes $K_0\left(\frac{r}{B}\right) \approx \ln\frac{1.12\ B}{r}$ and the drawdown in the experimental well is then supplied by relation

$$S_0 = \frac{Q}{2\pi K_n M_n} \ln \frac{1.12B}{r_0} \tag{210}$$

Parameter B which can be called the "leakage factor" is defined by relation [4]

$$B = \sqrt{\frac{k_n M_n}{\frac{\lambda_1}{m_1} + \frac{\lambda_2}{m_2}}} \tag{211}$$

When the underground recharge occurs only through the top or through the bottom of an aquifer complex, relation (211) is modified as follows:
— for $\lambda_1 = 0$

$$B = \sqrt{\frac{k_n M_n m_2}{\lambda_2}} \tag{212}$$

— for $\lambda_2 = 0$

$$B = \sqrt{\frac{k_n M_n m_1}{\lambda_1}} \tag{213}$$

So as to exemplify, the hydrogeological parameters were computed corresponding to the leakage data of experimental group 26 H Cuzap (Romania), supplied by pumping I on the assumption of the infinite extension of the aquiferous and semi-pervious layers.

By applying the data from observation wells 26 Ha and 26 Hb, relation (209) becomes:

$$S_1 = \frac{Q}{2\pi k_n M_n} K_0\left(\frac{r_1}{B}\right) \tag{214}$$

$$S_2 = \frac{Q}{2\pi k_n M_n} K_0\left(\frac{r_2}{B}\right) \tag{215}$$

$$\frac{S_1}{S_2} = \frac{K_0\left(\frac{r_1}{B}\right)}{K_0\left(\frac{r_2}{B}\right)} \tag{216}$$

When drawdowns S_2 and S_1 and distances r_2 and r_1 are known, factor B can be determined by trials. The graphical solving of the implicit equation (Figure 71)

$$1.178 = \frac{K_0\left(\frac{2.97}{B}\right)}{K_0\left(\frac{8.18}{B}\right)} \tag{217}$$

allowed us to determine the value of $B = 1950$ m. Hence, the value of the transmissivity of the tested aquifer between the two observation wells will be

$$S_1 - S_2 = \frac{Q}{2\pi k_n M_n}\left[K_0\left(\frac{r_1}{B}\right) - K_0\left(\frac{r_2}{B}\right)\right] \tag{218}$$

$$k_n M_{n(1-2)} = \frac{Q}{2\pi(S_1 - S_2)}\left[K_0\left(\frac{r_1}{B}\right) - K_0\left(\frac{r_2}{B}\right)\right] \tag{219}$$

$$k_n M_{n(1-2)} = \frac{147}{6.28(6.50 - 5.52)} K_0\left(\frac{2.97}{1950}\right) - K_0\left(\frac{8.18}{1950}\right) = 24.2 \text{ m}^2/\text{day} \tag{220}$$

The transmissivity thus obtained has practically the same value as in the case of the hydraulic isolation hypothesis of the aquifer both in the upper bed and in the lower bed. Since B is a global parameter it should be approximated whether the recharge originates in both the upper bed and the lower bed or only in the lower bed. Considering that the permeability of the clays is practically the same in the upper bed and in the lower bed

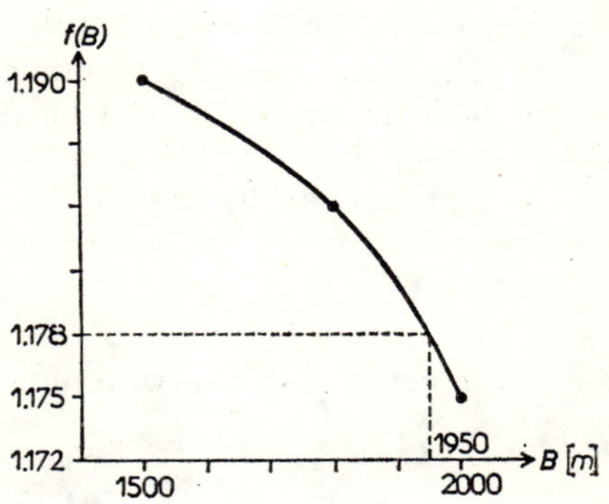

Figure 71. Determination of the leakage factor.

($\lambda_1 = \lambda_2 = \lambda$), relation (211) permits us to determine the value of the hydraulic conductivity

$$\lambda = \frac{k_n M_n \cdot m_1 \times m_2}{B^2(m_1 + m_2)} = \frac{24.2 \times 18.0 \times 8.0}{1950^2(18.0 + 8.0)} = 3.5 \times 10^{-5} \text{ m/day} \qquad (221)$$

On the assumption that the underground recharge originates in the lower bed only, the hydraulic conductivity of the respective clays is

$$\lambda_2 = \frac{k_n M_n m_2}{B^2} = \frac{24.2 \times 8.0}{1950^2} = 5.1 \times 10^{-5} \text{ m/day} \qquad (222)$$

Bocever and Verighin [5] emphasized that the underground recharge of a drained aquifer becomes important when $\frac{k_n}{\lambda} \geqslant 100$ and $\frac{M_n}{m} \geqslant 3$. Taking these criteria into account it can be considered that in the case of group 26 H the vertical recharge could only originate in the lower bed because $\frac{k_n}{\lambda_2} = \frac{2.18}{5.1 \times 10^{-5}} = 4.3 \times 10^4$ and $\frac{M_n}{m_2} = \frac{11.1}{8.0} = 1.4$. The smaller than 1 value of the ratio of the thickness for the upper bed $\left(\frac{11.1}{18} = 0.6\right)$ only determines a trivial leakage from a quantitative standpoint. According to relation (209) the equation of the pumping piezometric surface of experimental group 26 H Cuzap, pumping I, is

$$S = 1.04\ K_0(5.13 \times 10^{-4}\, r) \qquad (223)$$

If this equation is particularized for central well 26 HE ($r = r_0$), there results $S_0' = 9.93$ m. Under these conditions, the computation will supply for the well loss

$$\Delta S = S_0 - S_0' = 7.70 - 9.93 = -2.23 \text{ m}$$

In a leakage working aquifer the absolute value of the well loss will evidently be higher than in an aquifer with upper and lower confining (impervious) beds (on § 2.2.1.2, $\Delta S = -2.08$ m was computed). If the well loss is positive, the leakage will lead to a reduction of its value.

N. N. Bindeman [3] recommends that for the estimation of the reduced radius of influence we start from the equations of the experimental well for the following two cases:

— recharge by leakage

$$S_0 = \frac{Q}{2\pi k_n M_n} \ln \frac{1.12B}{r_0} \qquad (224)$$

— upper and lower confining bed

$$S_0 = \frac{Q}{2\pi k_n M_n} \ln \frac{R}{r_0} \qquad 225)$$

From these relations it follows that $R = 1.12\ B$. For experimental group 26 Cuzap, we obtain

$$R = 1.12 \times 1\,950 = 2\,184 \text{ m}$$

Another parameter characterizing the hydraulic conductivity of the semi-pervious layers is the leakage coefficient (leakance) defined by relation (222)

$$\frac{\lambda_2}{m_2} = \frac{T_n}{B^2} = K_a \tag{226}$$

where T_n is the transmissivity ($T_n = k_n M_n$).

The leakance (K_d) can be defined as the recharge rate passing through the unit area of the interface between the aquifer and the semi-pervious layer on condition that the head loss is equal to one. In this case

$$K_d = 5.1 \times 10^{-5}/8.0 = 24.2/1\,950^2 = 0.6 \times 10^{-5} \text{ day}^{-1}$$

Therefore, unlike factor B which was only introduced in order to facilitate the computation, the leakage coefficient (K_d) has a well defined physical meaning.

To conclude, the value of the aquifer transmissivity is equal to the computed one on the assumption that the aquifer is delimited by upper and lower confining beds. This does not mean however that the additionally computed parameters (B, λ and K_d) are less important. Thus, in the case of complex aquifers and of great drawdown, the inflow due to vertical leakage must be taken into account. Bindeman [3] points out that factor B can have values of 1000—10 000 m, the larger values being characteristic of cases with insignificant leakage. In our case, value $B = 1\,950$ m points out a strong recharge by leakage from the lower bed of the aquifer under test. For a complete characterization of the leakage it is also necessary to compute the flow rate of this phenomenon within the boundaries of the depression cone of the experimental group. In paper [5], the fact is emphasized that during the stabilization of the hydraulic régime only the water due to the leakage, is pumped from the well i.e. $Q = Q_d$ and therefore the resources of the tested aquifer do not any more participate in the water inflow to the well.

So as to make the importance of the additional resource of water (by leakage) more intelligible the variation of the flow rate Q_r should be analysed at various distances from the central well. This parameter can be computed by one of the following relations

$$Q_r = Q \frac{r}{r_0} \frac{K_1(r/B)}{K_1(r_0/B)} \tag{227}$$

or

$$Q_r = Q \frac{r}{B} K_1(r/B) \tag{228}$$

where K_1 is the modified Bessel function of the second kind and of the first degree [75], which for value $x < 0.002$ of the argument is approximated by $K_1(x) \approx 1/x^2$.

In connection with the participation of the leakage flow rate in the aquifers with an infinite extension, paper [5] introduces the notion of recharge radius (R_a) representing the distance (from the experimental well) within which, under the influence of the leakage, the prevailing part of the flow rate is formed. This parameter is defined as follows:

$$\varepsilon = \frac{Q - Q_r}{Q} = 1 - \frac{Q_r}{Q} = \left[1 - \frac{r}{B} K_1(r/B)\right] 100\% \tag{229}$$

Function $\varepsilon = f(r/B)$ points out that $\varepsilon = 90\%$ corresponds to $r/B = 3.2$. It follows that, for experimental group Cuzap, we shall have:

$$R_a = 3.2 \ B = 3.2 \times 1\,950 = 6\,240 \ \text{m}$$

Therefore an area of 6 240 m-radius would be necessary, so that the pumped rate of discharge be ensured by the sole leakage. The drawdown for $r = 6\,240$ m is supplied by relation (223):

$$S = 1.04 \ K_0(5.13 \times 10^{-4} \times 6\,240) = 0.03 \ \text{cm}$$

This result seems to justify the above-mentioned notion of recharge radius. But papers [5] and [52] emphasize the fact that the values of this radius, owing to certain deficiencies of the premises of the leakage theory are overrated and it is preferable to adopt the reduced radius of Darcy's theory. According to this theory, the pumping piezometric curve varies according to a logarithmic law all along the cone of depression, whereas, according to the leakage theory, this law is only valid in the neighbourhood of the experimental well after which the variation slows down. In the case of a strong leakage, with small values of parameter B (of about hundreds of meters) and implicitely of the recharge radius, the pumped rate of discharge can only be furnished by the leakage phenomenon. But in the case of a moderate leakage the hypothesis should be admitted that the pumped rate of discharge in the steady régime is ensured by both the leakage and the dynamic resource of this tested aquifer. This is the situation of experimental group Cuzap.

In Figure 72 *b*, curve $S = f(r)$ was drawn with the help of relation (223). For $R = 2\,184$ m, $S = 0.36$ m was obtained and therefore, on alignment $R_a - R = 6\,240 - 2\,184 = 4\,056$ m, the resulting hydraulic gradient is extremely small, $I = 0.08\%$. On the same figure we have drawn the curve supplied by the approximate relation of Dupuit

$$S = \frac{Q}{2\pi k_n M_n} \ln \frac{1.12 \, B}{r} = 0.97 \ln \frac{2184}{r} \tag{230}$$

The difference of the drawdown between the values supplied by the exact and approximate formulae are quite small; as a matter of fact N. N. Bindeman [3] demonstrated by computations that the error of the approximate formula is only 3—4% of the drawdown in the experimental well.

The computation of the leakage flow rate can be made with the help of modulus of underground recharge (W_s) which represents the water flow rate infiltrated along the vertical by unit area of the upper or of the lower bed. The value of this parameter is equal to that of the leakage velocity due

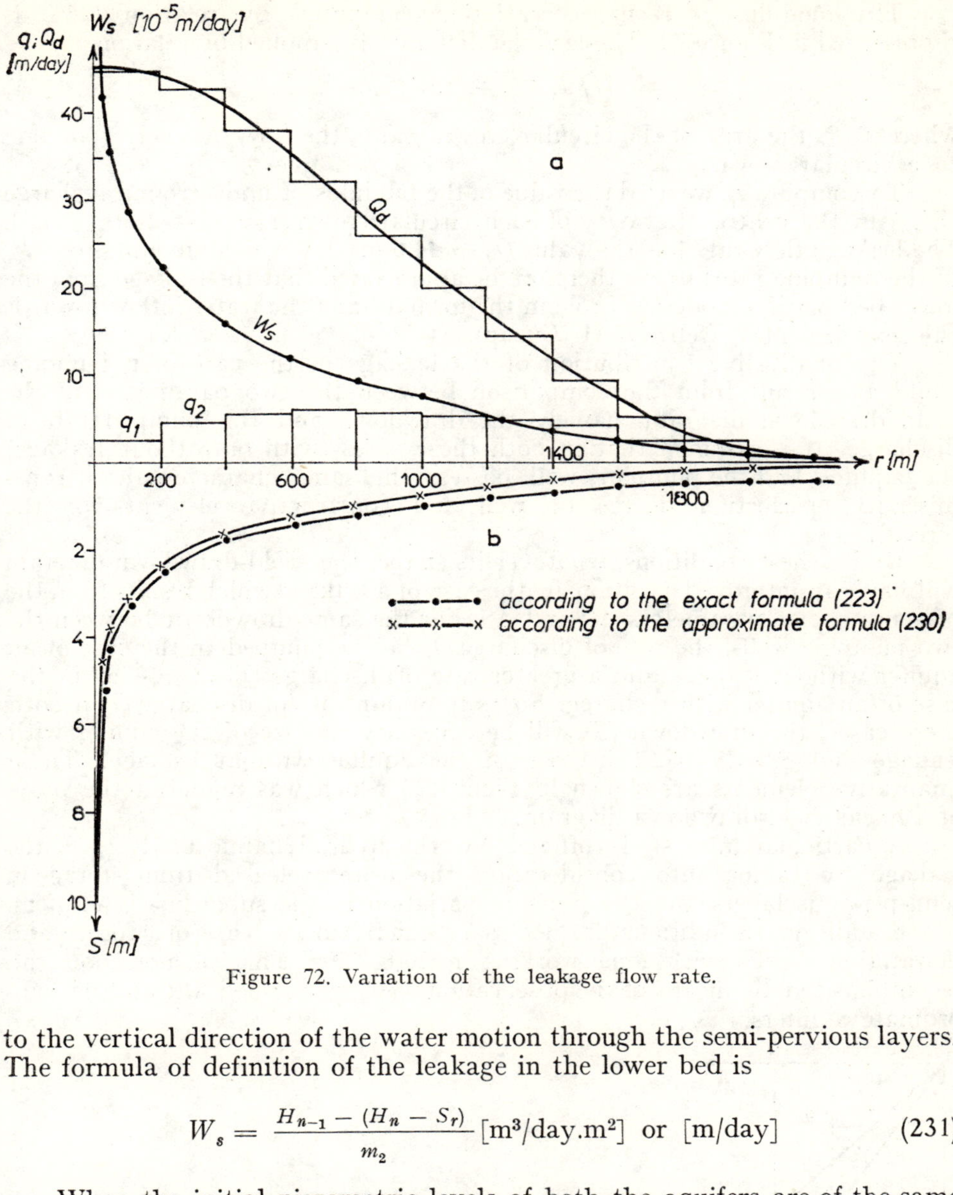

Figure 72. Variation of the leakage flow rate.

to the vertical direction of the water motion through the semi-pervious layers. The formula of definition of the leakage in the lower bed is

$$W_s = \frac{H_{n-1} - (H_n - S_r)}{m_2} \, [\mathrm{m^3/day.m^2}] \text{ or } [\mathrm{m/day}] \tag{231}$$

When the initial piezometric levels of both the aquifers are of the same elevation ($H_{n-1} = H_n$), this relation becomes

$$W_s = \lambda_2 \frac{S_r}{m_2} \tag{232}$$

where S_r is the drawdown at distance r.

The modulus of recharge variation computed by relation (231) is represented in Figure 72. The leakage flow rate is supplied by relation

$$Q_d = \Sigma q_i = \Sigma W_{sm} \Omega_i \tag{233}$$

where Ω_i is the area of the circular crown and q_i the flow rate corresponding to a circular crown.

To compute q_i, we used the value of the modulus of underground recharge (W_{sm}) in the centre of gravity of each circular crown [$r \approx r_n + 2/3(r_p - r_n)$]. The leakage flow rate has the value $Q_d = 44.6$ m^3/day which amounts to 34% of the pumping rate; it can therefore be appreciated that the leakage from the lower bed plays a moderated role in the formation of the water inflow towards the experimental well (26 H Cuzap).

The qualitative contribution of the leakage in the case of a pumping could also result from the comparison between the two parameters of the yield drawdown diagram, namely the drawdown and the pumped rate of discharge. It is assumed that in both these cases (with or without leakage) the aquifer and the pumping well possess the same characteristics (transmissivity, perfection degree of well and constructive elements of the screen).

Under these conditions, whatever its shape, the yield-drawdown diagram will have an improved position in the case of a leakage which results from the presence of two parameters ΔQ and ΔS. For the same drawdown between the two pumping wells, the rate of discharge Q_i can be pumped in the case of an aquifer without leakage and a greater rate of discharge $Q = Q_i + \Delta Q$ in the case of an aquifer with recharge. So as to obtain rate of discharge Q in both these cases, the drawdown (S) will be necessary in case of the aquifer with leakage and $S_i = S + \Delta S$ in case of the aquifer without leakage. These qualitative elements are shown in Figure 73 which was made on the basis of a linear yield-drawdown diagram.

A particular interest is offered by the hydrodynamic analyses of the leakage by taking into consideration the water released from storage in semi-pervious layers and the pressure variation in the subordinate aquifers. These additional conditions are necessary chiefly in the case of capture and dewatering works with long working periods [30]. This phenomenon can be emphasized by means of the observation wells performed within the subordinate aquifers.

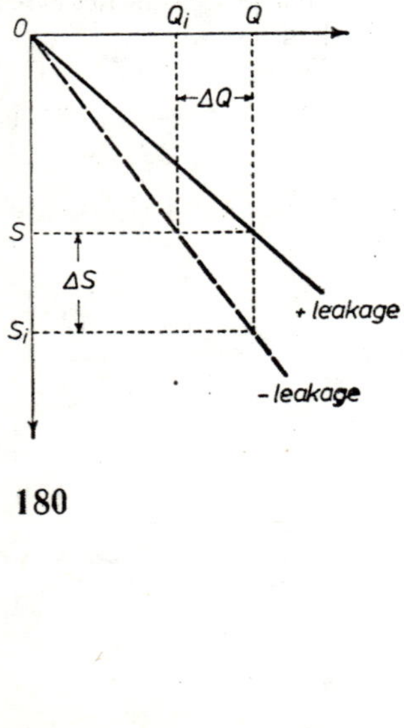

Figure 73. Leakage influence upon the yield-drawdown diagram.

The processing methods of the field data in view of determining the hydrogeological parameters, are based on the differential equation of the radial-plane flow in which aquifer thickness M is uniform all over the area of depression. In most cases, this parameter is however a variable, and this leads to quite another image of the distribution of the drawdown than in the case of the radial-plane flow. The asymmetry, as a matter of fact, the lengthening of the equipotential curves towards the thickness reduction (Figure 74) grows at the same time as its variation gradient $\left(\frac{\Delta M}{\Delta x}\right)$.

In § 2.2.1.2, it was pointed out that the data supplied by well 21 H Cuzap are affected by a systematic error, i.e. the transmissivities deduced from these data are greater than the normal ones. This systematic error is due to the very thickness reduction on the 21 H direction. The alignment 26 HE — 26 Ha — 26 Hb (Figure 53) equally presents a thickness variation but this being rather small (0.83 and 1.31 m) the error results also quite small. On alignment 26 Hb — 21 H, the thickness reduction is however important (10.44 — 6.62 = 3.82) which means that the results supplied by applying the differential equation of the radial -plane flow are not satisfactory.

In the structures with very often lithofacies variations a particular attention should be granted to the processing of the observation data. Each computation scheme depends on the law of the thickness variation which can be considered as either exponential or linear. Thickness variation can be established with the help of the hydrogeological cross-sections or of the mean gradient of variation. Thus, M. S. Hantush gave an approximate solution of the scheme with an

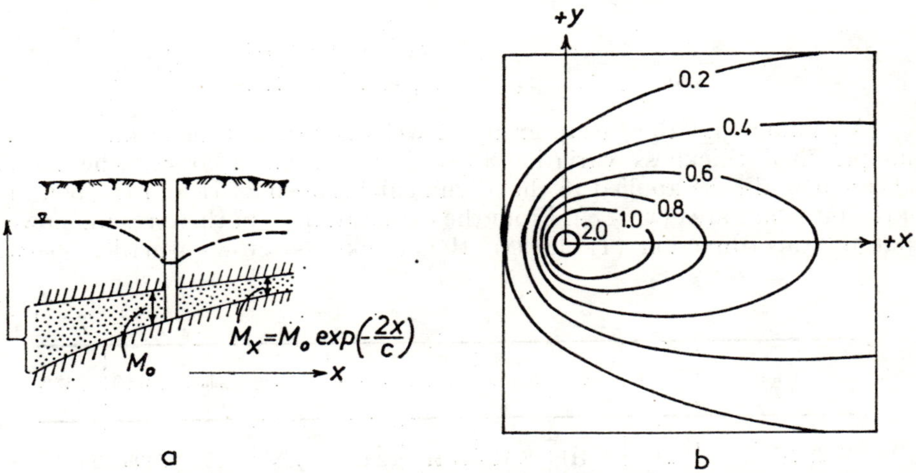

Figure 74. Cross-section (a) and distribution (b) of the drawdowns in the case of a confined aquifer with a variable thickness (after Hantush, 1962).

exponential variation of the thickness in case of the variation gradient $\left|\frac{\Delta M}{\Delta x}\right| < 0.02$. Evidently in the case of small values of the gradient we can also consider a linear variation.

2.2.4.1 Exponential variation of the thickness

For a steady pumping régime, M. S. Hantush established the following equation of the drawdown [31]:

$$S = \frac{Q}{2\pi k M_0} \exp\left(\frac{r}{c}\cos\theta\right) K_0\left(\frac{r}{c}\right) \tag{234}$$

where S is the drawdown at distance r from the experimental well, M_0 the aquifer thickness in the experimental well, θ the polar angle of the computation point and c a constant depending on the variation gradient of the thickness along direction x, the value of which is deduced from equation

$$M_x = M_0 e^{-\frac{2x}{c}} \tag{235}$$

From Figure 74, it follows that the lines of equal drawdown present a more or less oval form, which emphasizes the asymmetric character of the radial current.

In the case of group 26 H Cuzap (Table 58), taking into account that the initial thickness $M = \frac{10.92 + 11.75}{2} = 11.34$ m, there results a mean gradient of variation of 5% between 26 HE and 21 H.

$$\log \frac{M_x}{M_0} = -\frac{2x}{c} \log e$$

$$c = -\frac{2 \log e}{\log M_x - \log M_0} = -\frac{2 \times 94.30 \times 0.434}{\log 6.62 - \log 11.34} = 350 \text{ m}$$

Assuming that the Cuzap group of wells is situated along direction x of the principal thickness variation ($\theta = 0$) the problem arises whether this scheme can also be applied to the alignment between 21 H and the boundary of the influence area ($x > r_3$). From the exploration data there results however that on this direction (Table 58) the thickness grows smaller between

Table 58

x[m]	−220	0	+8.18	+94.30	+250.0	+550.0	+800.0	+950.0	+2200
Well No	3246	26 HE	26 Hb	21 H	22 H	3198	8	3196	54 H
M (m)	6.53	11.43	10.44	6.62	1.50	8.12	10.33	10.43	6.74

$x = 0$ and $x = +250$ m, after which it increases and decreases again up to the limit of the influence area. On direction y (north-eastward) only two thickness measurements can be emphasized: $M = 14.72$ m in well 3175 ($y = +250$ m) and $M = 5.87$ m in well 3085 ($y = +1100$ m). These values of the thickness are only given here to emphasize the fact that in the hydrogeological structures with wide variations of lithofacies it is very difficult to schematize the natural conditions in order to find an accurate calculation scheme. In this case let us remark however that the preferential alignment of maximum diminishing of the aquifer thickness is nevertheless 26 HE — 22 H. Under these conditions, the computation scheme of Hantush can only be applied on this alignment, assumed to be along direction $+x$. By subtracting from one another the equations corresponding to the two observation wells 26 Hb and 26 H, deduced from relation (234), the mean value can be determined of the hydraulic conductivity on the chosen alignment:

$$k_{2-3} = \frac{Q}{2\pi M_0(S_2 - S_3)} \left\{\left[\exp\left(\frac{r_2}{c}\right) K_0\left(\frac{r_2}{c}\right)\right] - \left[\exp\left(\frac{r_3}{c}\right) K_0\left(\frac{r_3}{c}\right)\right]\right\} \quad (236)$$

$$k_{2-3} = \frac{147}{6.28 \times 11.34 \times 2.02} \left\{\left[\exp\left(\frac{8.18}{350}\right) K_0\left(\frac{8.18}{350}\right)\right] - \right.$$

$$\left. - \left[\exp\left(\frac{94.30}{350}\right) K_0\left(\frac{94.30}{350}\right)\right]\right\} = 2.11 \text{ m/day}$$

This is practically equal to the calculated values of hydraulic conductivity (Section § 2.2.1.2) between wells 26 HE — 26 Hb ($k = 2.13 — 2.18$ m/day) which certifies a good homogeneity of the permeability. On the assumption of the uniform thickness and of the radial-plane flow, the computed transmissivity was $(kM)_{2-3} = 28.3$ m²/day; dividing this to $k = 2.11$ m/day we obtain the mean aquifer thickness between these two wells, $M_{m(2-3)} = 13.4$ m which is certainly erroneous. In conclusion, the fact of not taking into account the diminution of the aquifer thickness (according to an exponential law) along alignment 26 Hb — 21 H, resulted in an error of about 50 % of both the transmissivity and the hydraulic conductivity.

2.2.4.2 Linear variation of the thickness

In some cases, it is possible to establish the structure of the aquifer by means of drillings performed with a geological or hydrogeological objective and which are usually located at much smaller distances. Thus, within the boundaries of the cone of depression the thickness variations of the aquifer can be estimated along the four principal directions. If the variation gradient also has small enough values ($\alpha < 5\%$), which is frequently the case, a computation scheme can be adopted corresponding to a linear variation of the thickness. In the hydrogeological literature, no such particular cases are to be found but they are solved hereafter with the help of the differential equation of the radial-plane flow. Taking into account the complexity of the hydrogeological conditions, three cases were analysed:

1. The variation gradient is considered constant on all the radial directions;

2. There is a preferential direction of maximum variation of the thickness;

3. The variation direction and gradient are not the same on the four radial directions.

The three computation schemes cover most of the hydrogeological conditions.

The variation gradient of the thickness is constant on the four radial directions. There are hydrogeological cases where the experimental well shows a maximum or minimum of the thickness on the four principal directions and along each of these directions the gradient keeps up both its sign and its value. If M_0 is the aquifer thickness in the experimental well and M_1 is the aquifer thickness in the auxiliary wells (piezometers or exploration wells) and their relation is $M_0 < M_1$, then the differential equation of the well yield will be

$$Q = 2\pi k M_x x \frac{dz}{dx} \tag{237}$$

$$M_x = M_0 + \frac{M_1 - M_0}{r_1 - r_0} x \tag{238}$$

$$Q = 2\pi k \left(M_0 + \frac{M_1 - M_0}{r_1 - r_0} x\right) x \frac{dz}{dx} \tag{239}$$

The following notations are adopted

$$M_0 = b; \quad \frac{M_1 - M_0}{r_1 - r_0} = \alpha$$

$$Q = 2\pi k(b + ax) \times \frac{dz}{dx} \tag{240}$$

$$\int \frac{dx}{x(\alpha x - b)} = \frac{2\pi k}{Q} \int dz \tag{241}$$

Since the first member of the equation is a rational fraction, the integration can be made by decomposition into simple fractions:

$$\frac{2\pi k}{Q} \int_{h^0}^{h_1} dz = \frac{1}{b} \int_{r_0}^{r_1} \frac{dx}{x} - \frac{1}{b} \int_{r_0}^{r_1} \frac{dx}{\alpha x + b} \tag{242}$$

where the integration limits are furnished by the experimental well and one of the observation wells

$$\frac{2\pi k}{Q} (h_1 - h_0) = \frac{1}{b} \ln \frac{r_1}{r_0} - \frac{1}{b} \left[\ln \frac{(\alpha r_1 + b)}{(\alpha r_0 + b)}\right] \tag{243}$$

By substituting $h_1 = H - S_1$, $h_0 = H - S_0$ and $b = M_0$, we get the computation formula of the hydraulic conductivity

$$k_{0-1} = \frac{0.366\ Q}{(S_0 - S_1)\ M_0} \log \left[\frac{r_1(\alpha r_0 + M_0)}{r_0(\alpha r_1 + M_0)} \right] \tag{244}$$

The classical formula for determining k when $M = \text{const.}$ is:

$$k_{0-1} = \frac{0.366\ Q}{M(S_0 - S_1)} \log \frac{r_1}{r_0} \tag{245}$$

The equalization of the two equations supplies the expression of the thickness mean value in case of the radial-plane flow:

$$M_m = \frac{M_0 \log r_1/r_0}{\log \dfrac{r_1(\alpha r_0 + M_0)}{r_0(\alpha r_1 + M_0)}} \tag{246}$$

When the well loss is large (for great drawdowns), so as to increase the computation accuracy, the pumping water level in the experimental well is no longer used in the integration:

$$\frac{2\,\pi k}{Q} \int_{h_1}^{H} \mathrm{d}z = \frac{1}{b} \int_{r_1}^{R} \frac{\mathrm{d}x}{x} - \frac{1}{b} \int_{r_1}^{R} \frac{\alpha\,\mathrm{d}x}{\alpha x + b} \tag{247}$$

$$\frac{2\,\pi k}{Q} (H - h_1) = \frac{1}{b} \ln \frac{R}{r_1} - \frac{1}{b} \ln \left[\frac{\alpha R + b}{\alpha r_1 + b} \right] \tag{248}$$

$$k_{1-R} = \frac{0.366\,Q}{S_1 M_0} \log \left[\frac{R(\alpha r_1 + M_0)}{r_1(\alpha R + M_0)} \right] \tag{249}$$

If the pumping water level is only measured in the experimental well, the integration limits change

$$\frac{2\,\pi k}{Q} \int_{h_0}^{H} \mathrm{d}z = \frac{1}{b} \int_{r_0}^{R} \frac{\mathrm{d}x}{x} - \frac{1}{b} \int_{r_0}^{R} \frac{\alpha\,\mathrm{d}x}{\alpha x + b} \tag{250}$$

$$k_{0-R} = \frac{0.366\,Q}{S_0 M_0} \log \left[\frac{R(\alpha r_0 + M_0)}{r_0(\alpha R + M_0)} \right] \tag{251}$$

By applying the same method when $M_0 > M_1$, the equations become

$$Q = 2\,\pi k \left(M_0 - \frac{M_0 - M_1}{r_1 - r_0}\, x \right) x\, \frac{\mathrm{d}z}{\mathrm{d}x} \tag{252}$$

$$M_0 = b; \quad \frac{M_0 - M_1}{r_1 - r_0} = \alpha \tag{253}$$

$$Q = 2\pi k(b - \alpha x)\, x\, \frac{\mathrm{d}z}{\mathrm{d}x} \tag{254}$$

$$\int \frac{\mathrm{d}x}{x(b - \alpha x)} = \frac{2\,\pi k}{Q} \int \mathrm{d}z \tag{255}$$

On the alignment: experimental well — observation well we have

$$\frac{2\pi k}{Q}\int_{h_0}^{h_1} dz = \frac{1}{b}\int_{r_0}^{r_1}\frac{dx}{x} - \frac{1}{b}\int_{r_0}^{r_1}\frac{(-\alpha)\,dz}{b-\alpha x} \tag{256}$$

$$\frac{2\pi k}{Q}(h_1 - h_0) = \frac{1}{b}\ln\frac{r_1}{r_0} - \frac{1}{b}\ln\left[\frac{(b-\alpha r_1)}{(b-\alpha r_0)}\right] \tag{257}$$

$$\frac{2\pi k}{Q}(S_0 - S_1) = \frac{1}{b}\ln\left[\frac{r_1(b-\alpha r_0)}{r_0(b-\alpha r_1)}\right] \tag{258}$$

$$k_{0-1} = \frac{0.366\,Q}{(S_0 - S_1)M_0}\ln\left[\frac{r_1(M_0-\alpha r_0)}{r_0(M_0-\alpha r_1)}\right] \tag{259}$$

The computation scheme is obviously valid provided that $M_0 > \alpha r_1$. If equation (259) is equalized to the one that would correspond to $M =$ const. the mean aquifer thickness will be expressed by

$$M_m = \frac{M_0\log r_1/r_0}{\log\left[\dfrac{r_1(M_0-\alpha r_0}{r_0(M_0-\alpha r_1)}\right]} \tag{260}$$

If only the pumping water level in the observation well is taken into account, the integral will be converted into

$$\frac{2\pi k}{Q}\int_{h_1}^{H} dz = \frac{1}{b}\int_{r_1}^{R}\frac{dx}{x} - \frac{1}{b}\int_{r_1}^{R}\frac{(-\alpha)\,dx}{b-\alpha x} \tag{261}$$

$$\frac{2\pi k}{Q}(H - h_1) = \frac{1}{b}\ln\frac{R}{r_1} - \frac{1}{b}\ln\left[\frac{(b-\alpha R)}{(b-\alpha r_1)}\right] \tag{262}$$

$$\frac{2\pi k}{Q}S_1 = \frac{1}{b}\ln\left[\frac{R(b-\alpha r_1)}{r_1(b-\alpha R)}\right] \tag{263}$$

$$k_{1-R} = \frac{0.366Q}{S_1M_0}\log\left[\frac{R(M_0-\alpha r_1)}{r_0(M_0-\alpha R)}\right] \tag{264}$$

If only the experimental well is considered, the integration limits will be

$$\frac{2\pi k}{Q}\int_{h_0}^{H} dz = \frac{1}{b}\int_{r_0}^{R}\frac{dx}{x} - \frac{1}{b}\int_{r_0}^{R}\frac{(-\alpha)\,dx}{b-\alpha x} \tag{265}$$

$$k_{0-R} = \frac{0.366Q}{S_0M_0}\log\left[\frac{R(M_0-\alpha r_0)}{r_0(M_0-\alpha R)}\right] \tag{266}$$

This relation is only valid if $M_0 > \alpha R$.

A preferential direction exists of maximum variation of the thickness. There are cases where the aquifer can be clearly characterized by the two geometric elements: strike and slope. In these cases on direction x the gradient is maximum, while on direction y the thickness is constant (Figure 75).

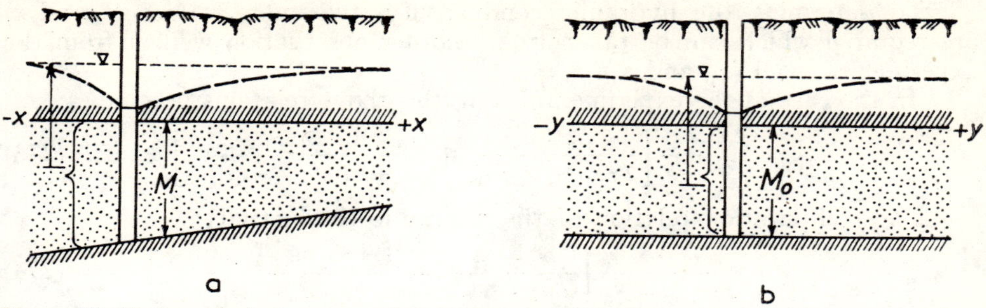

Figure 75. Hydrogeological cross-sections along the direction and the dip of a confined aquifer.

The differential equation of the radial-plane flow, if the position of the computation point is considered, will be

$$Q = 2\,\pi k[M_0 - \alpha x \cos\theta]\, x \frac{dz}{dx} \tag{267}$$

where M_0 is the aquifer thickness at the origin of the system (the experimental well), α is the variation gradient $\left(\frac{\Delta M}{\Delta x}\right)$ and θ the polar angle of the computation point.

The computation is facilitated by introducing a new quantity

$$\frac{Q}{2\,\pi k} = q, \quad q = [M_0 - \alpha x \cos\theta]\, x \frac{dz}{dx} \tag{268}$$

By substituting $\cos\theta = \dfrac{x}{\sqrt{x^2 + y^2}}$ we obtain the differential coefficient of first degree:

$$\frac{\partial z}{\partial x} = \frac{q\sqrt{x^2 + y^2}}{M_0 x\sqrt{x^2 + y^2} - \alpha x^3} \tag{269}$$

The integration of this differential equation can be made with the help of the Taylor series

$$z = z_1 + \frac{x - x_1}{1} z_1' + \frac{(x - x_1)^2}{2!} z_1'' + \frac{(x - x_1)^3}{3!} z_1''' + \ldots \tag{270}$$

The solution being an approximate one, only the first three terms of the series will be taken into account and we shall thus obtain the equation of the piezometric curve resulting from the experimental pumping

$$z = z_1 + (x - x_1) \frac{q}{x_1(M_0 - \alpha x_1)} + \frac{(x - x_1)^2}{2} \frac{q(2\alpha x_1 - M_0)}{x_1^2(M_0 - \alpha x_1)^2} \tag{271}$$

$$z = z_1 + (x - x_1) \frac{Q}{2\pi k x_1(M_0 - \alpha x_1)} + \frac{(x - x_1)^2}{4} \frac{Q(2\,\alpha x_1 - M_0)}{\pi k x_1^2(M_0 - \alpha x_1)^2} \tag{272}$$

To determine the hydraulic conductivity, two more values (z and x) are required which can be taken from another observation well or from the experimental well (x_0 and z_0).

In the case of the experimental well, the above relations become

$$z_0 = z_1 + \frac{Q}{k}\left[\frac{x_0 - x_1}{2\,\pi x_1(M_0 - \alpha x_1)} + \frac{(x_0 - x_1)^2\,(2\,\alpha x_1 - M_0)}{4\pi x_1^2(M_0 - \alpha x_1)^2}\right] \tag{273}$$

The computation relation of the hydraulic conductivity

$$k_{0-1} = \frac{Q}{2\pi x_1(S_1 - S_0)}\left[\frac{x_0 - x_1}{(M_0 - \alpha x_1)} + \frac{(x_0 - x_1)^2(2\,\alpha x_1 - M_0)}{2\,x_1(M_0 - \alpha x_1)^2}\right] \tag{274}$$

will be obtained by transforming the elevation of the pumping level z into drawdown S.

The relation (274) is valid provided that beside the experimental well there is an observation well placed at point $x = x_1$. In the absence of observation wells and if the co-ordinates of the experimental well are $x = x_0$ and $y = 0$ the equation of the piezometric curve will be:

$$z = z_0 + (x - x_0)\frac{q}{x_0(M_0 - \alpha x_0)} + \frac{(x - x_0)^2}{2}\,\frac{q(2\,\alpha x_0 - M_0)}{x_0^2(M_0 - \alpha x_0)^2} \tag{275}$$

At the limit of the cone of depression the piezometric curve has $z = H$ and $x = R$:

$$H = Z_0 + (R - x_0)\frac{Q}{2\pi k x_0(M_0 - \alpha x_0)} + \frac{(R - x_0)^2}{4}\,\frac{Q(2\,\alpha x_0 - M_0)}{\pi k x_0^2(M_0 - \alpha x_0)^2} \tag{276}$$

The well radius (x_0) being very small in regard to the radius of influence it can be neglected:

$$H = Z_0 + \frac{Q}{k}\left[\frac{R}{2\pi x_0(M_0 - \alpha x_0)} + \frac{R^2(2\alpha x_0 - M_0)}{4\pi x_0^2(M_0 - \alpha x_0)^2}\right] \tag{277}$$

$$k_{0-R} = \frac{Q}{2\pi x_0 S_0}\left[\frac{R}{(M_0 - \alpha x_0)} + \frac{R^2(2\alpha x_0 - M_0)}{2x^2(M_0 - \alpha x_0)^2}\right] \tag{278}$$

The solving of computation schemes the observation wells of which are placed at other points than on axis $+x$, entails great mathematical complications and very laborious final computation relations.

The thickness gradient presents variations of both value and sign on the four radial directions. Owing to the frequent lithofacies variations, there are many cases where both the gradient and the variation direction of the thickness appear on the four radial directions. An approximate solution of this scheme is supplied by the method of "radial fragments". In Figure 76 when $M_0 < M_n$, the differential equation of discharge rate can be written as follows:

$$Q = \frac{2\pi}{4}[(M_0 + \alpha_1 x) + (M_0 + \alpha_2 x) + (M_0 + \alpha_3 x) + (M_0 + \alpha_4 x)]\,kx\,\frac{dz}{dx} \tag{279}$$

Figure 76. Thickness variation of a confined aquifer.

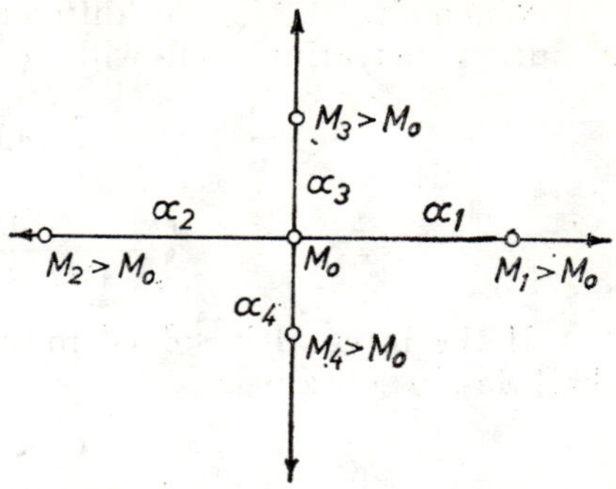

In this equation α is the variation gradient as determined on each direction.

The differential expression of the rate of discharge for the four radial fragments is

$$Q = \frac{\pi}{2}\,[4M_0 x + (\alpha_1 + \alpha_2 + \alpha_3 + \alpha_4)x^2]\,k\,\frac{dz}{dx} \tag{280}$$

$$\int_{r_0}^{R} \frac{dx}{4M_0\,x + \Sigma\alpha_i x^2} = \frac{\pi k}{2Q}\int_{h_0}^{H} dz \tag{281}$$

The integration in x by decomposition into simple fractions supplies the relation for determining the mean hydraulic conductivity on the whole cone of depression

$$k_{0-R} = \frac{0.366\,Q}{S_0 M_0}\log\frac{R(r_0\Sigma\alpha_i + 4\,M_0)}{r_0(R\Sigma\alpha_i + 4\,M_0)} \tag{282}$$

This relation is only valid for the experimental well; the integration between this well and an observation one cannot be made by the above differential equation because the four radial fragments are not in a plane.

Initially, we started on the assumption that all four gradients were positive; in such cases, the results acquired by relation (282) are more accurate. There are however cases where the gradients have different signs along the four directions, and $\Sigma\alpha_i < 0$. In this case, the applicability limit of relation (282) is supplied by inequalities

$$\frac{r_0\Sigma\alpha_i + 4\,M_0}{R\Sigma\alpha_i + 4\,M_0} > 0;\quad R\Sigma\alpha_i + 4M_0 > 0;\quad \Sigma\alpha_i \geqslant \frac{-4\,M_0}{R} \tag{283}$$

In the particular case of compensated gradient values ($\Sigma\alpha_i = 0$), we find the classical Dupuit relation

$$k_{0-R} = \frac{0.366\,Q}{S_0 M_0}\log R/r$$

Evidently, when this is the case the compensation has a negative effect on the hydrodynamic computations.

When $M_0 > M_n$, the differential equation of the radial-plane flow to a fully penetrating well will be

$$Q = \frac{\pi}{2} [4M_0 x - \Sigma\alpha_i x^2]\, k \frac{dy}{dx} \tag{284}$$

$$\int_r^R \frac{dx}{4M_0 x - \Sigma\alpha_i x^2} = \frac{\pi k}{2Q} \int_h^H dy \tag{285}$$

If the integral is solved in x by decomposition into simple fractions we shall lastly obtain

$$k_{0-R} = \frac{0.366\, Q}{S_0 M_0} \log \left[\frac{R(4\,M_0 - R\Sigma\alpha_i)}{r_0(4\,M_0 - r_0\Sigma\alpha_i)}\right] \tag{286}$$

The limit of applicability of this relation is supplied by the following criteria

$$\frac{4\,M_0 - R\Sigma\alpha_i}{4\,M_0 - r_0\Sigma\alpha_i} > 0; \quad 4\,M_0 - R\Sigma\alpha_i > 0; \quad \Sigma\alpha_i < \frac{4\,M_0}{R} \tag{287}$$

By equalizing the two relations (282) and (286) with the one corresponding to $M = \text{const.}$ we obtain the mean values of the thickness in the case of a radial-plane flow.

For $M_0 < M_m$:

$$M_m = \frac{M_0 \log R/r_0}{\log \left[\dfrac{R(r_0\Sigma\alpha_i + 4M_0)}{r_0(R\Sigma\alpha_i + 4M_0)}\right]} \tag{288}$$

For $M_0 > M_m$

$$M_m = \frac{M_0 \log R/r_0}{\log \left[\dfrac{R(4M_0 - R\Sigma\alpha_i)}{r_0(4M_0 - r_0\Sigma\alpha_i)}\right]} \tag{289}$$

All the solutions given in this paragraph take into account the pumping levels in the experimental well because the use of the levels from the observation wells would necessitate the location in the plane of the radial fragments. However, the existence of at least one observation well is required in order to determine the well loss, i.e. the real pumping level in the experimental well.

* * *

To exemplify the computation schemes with a linear variation of the aquifer thickness, experimental group 16 H Șotînga was chosen which consists in central well 16 Ha and observation well 16 Hb situated at distance $x_1 = 97$ m. The pumping III data are:

— specific rate of discharge	Q	$= 27.3$ m/day
— drawdown in well 16 Ha	S_0	$= 15.21$ m
— drawdown in well 16 Hb	S_1	$= 5.73$ m

— radius of the central well $r_0 = 0.073$ m
— thickness in 16 Ha $M_0 = 20.15$ m
— thickness in 16 Hb $M_1 = 17.15$ m

The transmissivity between the two wells can be computed by the classical relation:

$$kM = \frac{0.366\,Q \log \dfrac{x_1}{r_0}}{S_0 - S_1} = \frac{0.366 \times 27.3 \log \dfrac{97.0}{0.073}}{15.21 - 5.73} = 3.22 \text{ m}^2/\text{day}$$

On a first assumption, by taking into account thickness M_0, the value of the filtration coefficient is

$$k = \frac{3.22}{20.15} = 0.16 \text{ m/day}$$

For the investigation of coal deposits, the hydrogeological wells are usually located in the thickest zones of the aquifers. Consequently the values of the hydraulic conductivity are underrated owing to the fact that only this maximum value is taken into consideration instead of the mean value of the whole cone of depression. For a correct treatment of the problem two methods can be applied:

— The direct application of the relations to the computation of k.

— First computing the transmissivity by the classical formulae and then dividing it by the thickness mean value — the very method that will be applied hereafter to group 16 H.

From the analysis of the neighbouring wells it follows that the maximum value of the thickness occurred at well 16 Ha. Between wells 16 Ha and 16 Hb there is the gradient

$$\alpha = \frac{20.15 - 17.15}{97} = 0.03$$

Appreciating that this gradient is the same on all the principal directions, the mean value of the thickness will be given by relation (260)

$$M_m = \frac{20.15 \log \dfrac{97.0}{0.073}}{\log \left[\dfrac{97(20.15 - 0.073 \times 0.03)}{0.073(20.15 - 0.03 \times 97)} \right]} = 19.7 \text{ m}$$

By hypothetically admitting the gradient $\alpha = 0.10$ we obtain:

$$M_m = \frac{20.15 \log \dfrac{97.0}{0.073}}{\log \left[\dfrac{97(20.15 - 0.073 \times 0.10)}{0.073(20.15 - 0.10 \times 97)} \right]}$$

For greater gradients the linear variation is no longer admissible and has to be replaced by the exponential one. The values of the hydraulic conductivity become

$$k = \frac{3.22}{19.7} = 0.16 \text{ m/day}; \quad k = \frac{3.22}{18.3} = 0.18 \text{ m/day}$$

Indeed, the computations show that in the case of great thicknesses the error due to the use of thickness M_0 is admissible so that the application of more intricate relations using the linear variation is no longer compulsory.

Hypothetically, it will be admitted that the maximum thickness in the central well is $M_0 = 5$ m. For gradient $\alpha = 0.03$ therefore

$$M_m = \frac{5.0 \log \dfrac{97.0}{0.073}}{\log \left[\dfrac{97(5.0 - 0.03 \times 0.073)}{0.073(5.0 - 0.03 \times 97)} \right]} = 4.45 \text{ m}$$

For gradient $\alpha = 0.05$ the mean thickness will be

$$M_m = \frac{5.0 \log \dfrac{97.0}{0.073}}{\log \left[\dfrac{97(5.0 - 0.05 \times 0.073}{0.073(5.0 - 0.05 \times 97)} \right]} = 3.35 \text{ m}$$

To these values correspond:

$$k = \frac{3.22}{5.0} = 0.64 \text{ m/day}; \quad k = \frac{3.22}{4.45} = 0.72 \text{ m/day};$$

$$k = \frac{3.22}{3.35} = 0.96 \text{ m/day}$$

These results indicate that in the case of small or moderate thicknesses, the disregarding of a mean value would introduce an important error: e.g. for $M_0 = 5$ m and $\alpha = 0.05$, the resulting values of the hydraulic conductivity would be smaller by 33% than the real ones. In the previously analysed particular case the absolute value of the difference is not too great, owing to the small degree of permeability of the sands.

2.3 Testing the aquifer by pumpings in unsteady state flow

Concomitantly with the development of the theory of the elastic régime the possibility arose of interpreting and processing the data supplied by the nonequilibrium pumpings, namely in unsteady flow. The advantages offered by this method as compared with the steady flow pumpings are: the short pumping time (less than 24 hours), the possibility of estimating more hydrogeological parameters as well as a few computation schemes and formulae encompassing the whole variety of the hydrogeological conditions.

For the hydrodynamic solving of the schemes corresponding to the unsteady pumping régimes, we have to know the initial condition as well as the boundary conditions both in vertical cross-section and in the plane. According to the initial condition, at time $t = 0$ (i.e. before the beginning of the pumping) the drawdown in the whole aquifer must be equal to zero. The boundary conditions in the vertical cross-section refer to the impermeable or semi-pervious character (which determines a ground-water recharge by leakage of the deposits in the upper or in the lower beds. The boundary conditions in the plane are materialized by the variation with time of the hydraulic head or of the flow rate at the limits of the depression domain. If this domain or cone of depression stretches up to the lateral boundaries of the aquifer, the latter should be considered as a limited aquifer (semi-infinite, rectangular, infinite-strip, closed [24]. In most cases however the aquifers may be considered as "infinite" owing to their great extension in regard to the cone of depression. In this case, a certain behaviour condition should be substituted to the boundary conditions at the infinite points, namely that during the pumping, at a very long distance from the experimental well, the pressure could have any finite value whatever.

2.3.1 Confined aquifers with upper and lower impermeable beds

The hydrogeological parameters fulfilling these conditions are: the transmissivity, the hydraulic diffusivity, the coefficient of storage and the seepage resistance of the experimental well. The last one of these parameters can only be determined if there are observation wells which afford moreover a greater accuracy to the other parameters. The literature offers many processing schemes of the measurements. The Soviet literature especially includes the hydraulic diffusivity (or coefficient of piezoconductivity) which is determined at the same time as the transmissivity. The computation schemes in the French, American, etc. literature do not include the hydraulic diffusivity (a), but only its components: transmissivity T and coefficient of storage n_e. This difference is of course only formal, since $a = \frac{T}{n_e}$.

Hereafter, we shall present the computation method of the parameters with the help of the experimental data of pumping group Progresul Bucharest. This experimental group has tested the confined aquifer A of the Frătești Complex, an aquifer system of regional extension. In this group, well S_5-III was the experimental well, while S_5-I and S_5-II were used as observation wells, the pumping working at the constant rate of discharge $Q = 984$ m^3/day. Figure 77 represents a hydrogeological cross-section of the experimental group and Table 59 supplies the hydrogeological and technical characteristics of the wells. The systematic measurements of drawdowns S in central well S_5-III and in observation well S_5-I are recorded in Table 60.

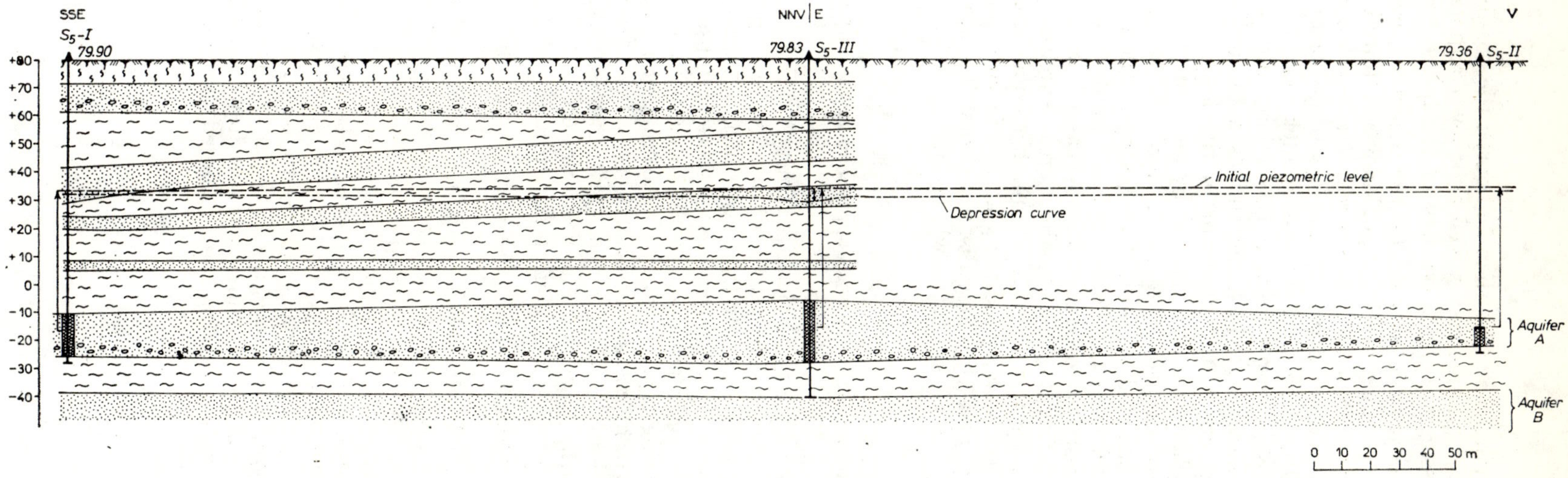

Figure 77. Hydrogeological cross-section through the experimental group of wells Progresul (Frătești complex).

Table 59

Well	Aquifer thickness M(m)	Aquifer static head H (m H_2O)	Piezometric level depth (m)	Screen length (in)	Screen diameter [inches]	Distances r[m]
S5 — III	22.5	61.7	−45.79	21.0	8 5/8"	0.11
S5 — II	10.0	56.0	−45.00	5.8	8 5/8"	236.00
S5 — I	14.5	58.7	−45.84	13.0	8 5/8"	266.00

Table 60

Drawdowns recorded in wells S_5-III and S_5-I

Crt No	Date	Time t (min)	Drawdown (S_0) in well S_5 — III (m)	Drawdown (S_1) in well S_5 — I (m)	*Crt No*	Date	Time t (min)	Drawdown (S_0) in wells S_5-III (m)	Drawdown (S_1) in well S_5-I (m)
1	12 V 1965	0	0	0	26		915	4.08	0.65
2		1	1.50	0	27		975	4.12	0.66
3		3	2.00	0	28		1 035	4.17	0.68
4		5	2.96	0.01	29		1 095	4.25	0.70
5		10	3.35	0.02	30		1 155	4.25	0.70
6		15	3.38	0.05	31		1 395	4.25	0.70
7		20	3.47	0.07	35		1 455	4.25	0.70
8		30	3.51	0.08	36		1 515	4.25	0.72
9		40	3.58	0.14	37		1 575	4.25	0.74
10		50	3.68	0.18	46	14.V.1965	2 115	4.25	0.82
11		60	3.70	0.22	47		2 175	4.25	0.82
12		75	3.72	0.32	49		2 295	4.26	0.82
13		135	3.78	0.41	51		2 415	4.26	0.82
14		195	3.80	0.42	52		2 475	4.26	0.82
15		255	3.84	0.45	53		2 535	4.26	0.82
16		315	3.86	0.48	55		2 655	4.26	0.82
17		375	3.89	0.50	56		2 715	4.26	0.82
18		435	3.90	0.52	70	15.V.1965	3 555	4.27	0.86
19		495	3.91	0.54	71		3 615	4.27	0.86
20		555	3.92	0.56	72		3 675	4.28	0.86
21		615	3.92	0.60	73		3 735	4.29	0.86
22	13.V.1965	675	3.95	0.61	74		3 795	4.36	0.86
23		735	3.98	0.62	75		3 855	4.36	0.86
24		795	4.00	0.63	76		3 915	4.36	0.86
25		855	4.05	0.64	82		4 275	4.36	0.86

So as to use correctly the computation schemes in an elastic régime, graphs S—t and S — ln t should first be drawn by means of the field data. It follows from Figure 78 that the experimental well S — t is characterized by a fast rise up to time $t = 100$ min, after which it very slowly sloped down. The point of inflexion of the observation well (S_5—I) coincides with that of the experimental well after which the two curves become practically parallel to each other. Another point to notice is the prompt reaction recorded by the

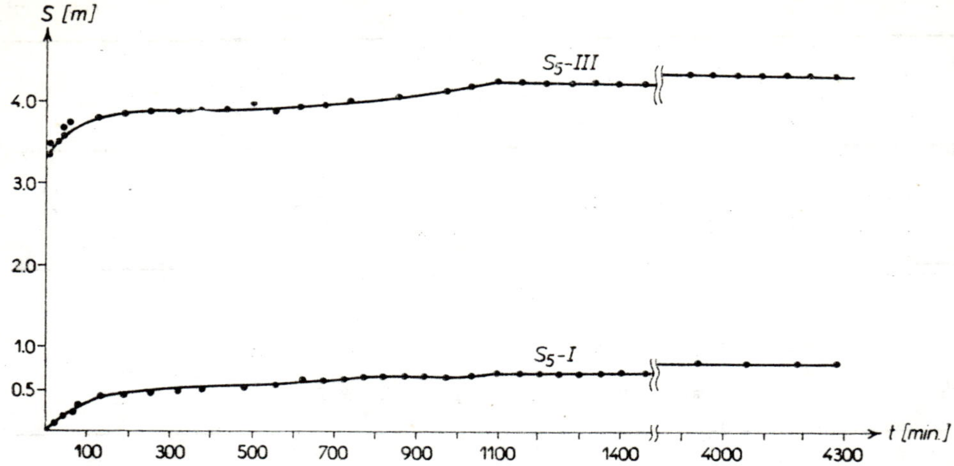

Figure 78. Chronological graph of the drawdown in the experimental group of wells, Progresul Bucharest.

observation well about 3 minutes after the beginning of the pumping. All these events are characteristic of aquifer systems with a high hydraulic diffusivity and important dynamic resources.

The form of graph S—$\ln t$ in its extreme portion gives some indications as to the hydrogeological particularities; when the graph is approximately asymptotic to axis $\ln t$ there is a recharge boundary (H = const) while the opposite tendency of the graph points to the existence of an impermeable boundary with $q = 0$. The use of the extreme portions of graph S—$\ln t$ for estimating certain parameters (the distance to the aquifer limit or to the boundaries between areas of various permeabilities, the geometric form of the boundary, etc.) is difficult owing to the variation of the natural factors and chiefly of the aquifer thickness. The determination of the hydrogeological parameters starts from the equation of the radial-plane flow of a confined aquifer towards a fully penetrating well [24]

$$S = -\frac{Q}{4\pi kM} Ei\left(-\frac{r^2}{4at}\right) \tag{290}$$

where Ei is the integral exponential function (Table 61, Figure 87) that represents the very parameter of the dimensionless seepage resistance of the aquifer current to a fully penetrating well in an "infinite" aquifer with a constant discharge Q. In the case of small values of argument $\frac{r^2}{4\,at}$ from the convergent series of the function it is sufficient to take into account the first two terms; relation (290) then becomes

$$S = \frac{Q}{4\pi kM}\left(\ln\frac{4at}{r^2} - 0.5772\right) \tag{291}$$

$$S = \frac{Q}{4\pi kM}\left(\ln\frac{2.25\,at}{r^2}\right) \tag{292}$$

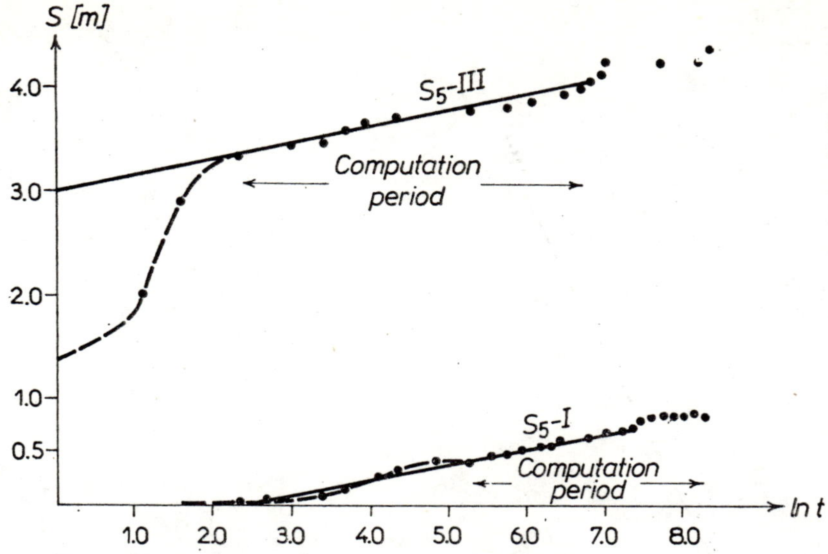

Figure 79. Dependence S — ln t graph of the pumping in the well S_5.

Taking into account the extent of the errors, the use of the logarithmic relation is only admitted if $\frac{r^2}{4\,at} < 0.1$ or $t > 2.5\,\frac{r^2}{a}$ for which the maximum error is 5.7%. As a matter of fact, since the hydraulic diffusivity is not known, these criteria cannot be initially applied. On graph S — ln t it is however possible to establish the time interval within the limits of which the logarithmic approximation is valid. It follows from Figure 79 that in the case of the experimental well this occurs for $\ln t > 2.5$ ($t < 12$ min) and in the case of the observation well for $\ln t > 5.3$ ($t > 170$ min). The deviation of the initial portion of the graph from the linear character is chiefly due to the fact that actually, at the beginning of the pumping, the rate of discharge is not instantaneously stabilized (on the assumption of the theory of the elastic régime) but is variable because of the emptying of the well hole; we could add moreover the influence of the well imperfection as well as the resistances of its adjacent area.

The extreme portions of the two S — ln t graphs present a certain deviation which is probably due to an important variation of both the permeability and the thickness of the aquifer. Even in the linear portion of the graphs there are small deviations (negligible) from the straight line which, can be due to the measurement errors.

Processing of the data from observation well S_5—I, Progresul Bucharest. The equation of logarithmic approximation (291) can be linearized as follows:

$$S = \frac{Q}{4\pi kM} \ln \frac{2.25\,a}{r^2} + \frac{Q}{4\pi kM} \ln t \tag{293}$$

$$S = \frac{0.183\,Q}{kM} \log \frac{2.25\,a}{r^2} + \frac{0.183\,Q}{kM} \log t \tag{294}$$

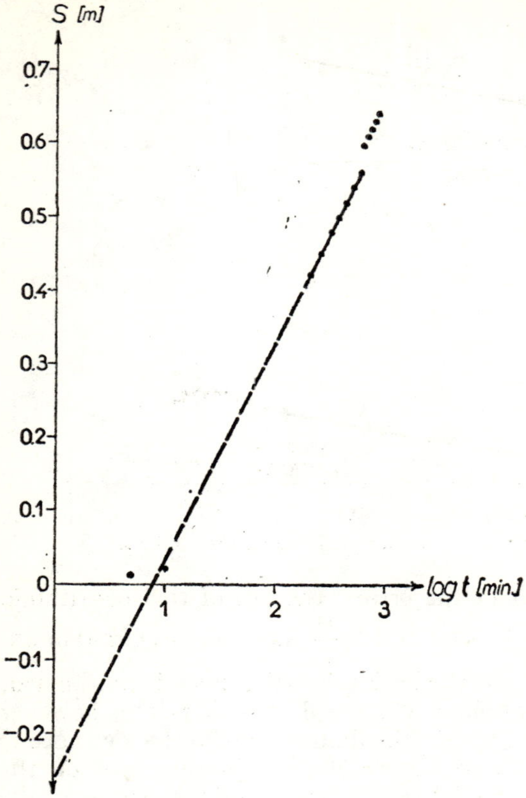

Figure 80. S — log t graph of observation well S_5 — I.

By substituting $\frac{0.183\ Q}{kM} = C$

$$\text{and} \quad C \log \frac{2.25\ a}{r^2} = A, \tag{295}$$

this relation becomes:

$$S = A + C \log t \tag{296}$$

From the graphical representation of the observation data, (Figure 80) there results that the linear domain (S) is comprised between 0.42 and 0.54 m. The graph permits to determine values A and C, the first one being the initial ordinate (= —0.26) and the second one the angular coefficient of the straight line:

$$C = \frac{S_2 - S_1}{\log \frac{t_2}{t_1}} = \frac{0.54 - 0.42}{\log \frac{495}{195}} = 0.296$$

The transmissivity is:

$$kM = \frac{0.183\,Q}{C} = \frac{0.183 \times 0.682}{0.296} = 0.42\ \text{m}^2/\text{min} = 605\ \text{m}^2/\text{day}$$

There are cases where owing to the non-homogeneity of the aquifer and to the measurement errors, the drawing of the straight line "amongst the points" could introduce an important error into the calculation of the hydrogeological parameters. So as to increase the accuracy we can use the analytical method of least squares by applying equation (296). The two normal equations are

$$\Sigma S = A + C\Sigma \log t \tag{297}$$

$$\Sigma S \log t = A\Sigma \log t + C\Sigma(\log t)^2 \tag{298}$$

These equations supply the expressions of parameters C and A

$$C = \frac{\Sigma S \log t - \Sigma S\, \Sigma \log t}{\Sigma(\log t)^2 - (\Sigma \log t)^2} \tag{299}$$

$$A = \Sigma S - C\Sigma \log t \tag{300}$$

To facilitate the computation, a table is used where the sums are made. With these values (the most probable ones) of parameters A and C, it is possible to draw the straight line and then to calculate the hydrogeological parameters.

Relation (295) allows us to determine the hydraulic diffusivity

$$\log \frac{2.25}{r^2} + \log a = \frac{A}{C} \tag{301}$$

$$\log a = \frac{A}{C} - \log \frac{2.25}{r^2} = \frac{A}{C} - 0.35 + 2 \log r \tag{302}$$

$$\log a = \frac{-0.270}{0.296} - 0.35 + 2 \log 266 = 3.59$$

$$a = 3890\ \text{m}^2/\text{min} = 5.6 \times 10^6\ \text{m}^2/\text{day}$$

When equation (290) is used, two relations are first written corresponding to drawdowns S_1 and S_2 at times t_1 and t_2. By dividing them, an implicit equation results which can be solved by the graphoanalytical method, the only unknown being the hydraulic diffusivity (a)

$$\frac{S_2}{S_1} = - \frac{-Ei\left(-\frac{r^2}{4at_2}\right)}{-Ei\left(-\frac{r^2}{4at_1}\right)} \tag{303}$$

This procedure supplied $a = 3750$ m²/min. The hydraulic diffusivity being known, the transmissivity can be determined for any drawdown. Thus, for S_1:

$$kM = -\frac{Q}{4\pi S_1} Ei\left(-\frac{r^2}{4at_1}\right)$$

$$kM = -\frac{0.682}{12.56 \times 0.42} Ei\left(-\frac{70756}{4 \times 3750 \times 195}\right) = 0.41 \text{ m}^2/\text{min}$$

Other graphical methods are proposed moreover [76] for the linearization of the approximately logarithmic formula, namely:

— the drawdown straight line — time logarithm in the case of measurements from a single piezometer and a steady flow;

— the specific drawdown straight line — time logarithm in case the pumping is made by several constant yield steps;

— the drawdown straight line — logarithm t/r^2 when the experimental station contains several piezometers on the same radius;

— the specific drawdown straight line — logarithm t/r^2 when there are several piezometers and a gradual variation of the discharge rate.

In the case of experimental station S_5 Progresul, since there is a single piezometer on a radius (S_5-I), the transmissivity and the coefficient of storage can be determined by means of the drawdown straight line — time logarithm. The straight line equation is obtained as follows:

$$S = \frac{0.183Q}{T} \log \frac{2.25\,T}{r^2 n_e} \log t \qquad (304)$$

or

$$S = \frac{0.183Q}{T}\left[\log\left(\frac{2.25\,T}{r^2 n_e} + \log t\right)\right] \qquad (305)$$

that can be written:

$$S = C(\log b + \log t)$$

Constant $C = \frac{0.183Q}{T}$ represents the very angular coefficient of the line, from which the transmissivity expression results. For the computation of coefficient n_e the straight line will be extended up to the intersection with the abscissa, where $S = 0$ and $\log t = \log t_0$.

By the introduction of these parameters, equation (305) will supply

$$0 = \frac{0.183Q}{T}\left(\log \frac{2.25\,T}{r^2 n_e} + \log t_0\right) \qquad (306)$$

Since $\frac{0.183Q}{T} > 0$, it follows that

$$\log \frac{2.25\,T}{r^2 n_e} + \log t_0 = 0 \qquad (307)$$

$$\log \frac{2.25\,T}{r^2 n_e} t_0 = 0 \tag{308}$$

$$\frac{2.25\,T}{r^2 n_e} t_0 = 1 \tag{309}$$

and lastly

$$n_e = \frac{2.25\,T t_0}{r^2} \tag{310}$$

According to this method, the transmissivity will be computed by the same relation as in the previous case ($T = kM = 0.42$ m²/min).

From the graph in Figure 80, there results that $\log t_0 = 0.90$, hence $t_0 = 7.94$ min.

$$n_e = \frac{2.25 \times 0.42 \times 7.94}{70756} = 1.06 \times 10^{-4}$$

For ascertaining, the hydraulic diffusivity can also be determined by its relation of definition:

$$a = \frac{T}{n_e} = \frac{0.42}{1.06} \times 10^4 = 3962 \text{ m}^2/\text{min} = 5.7 \times 10^6 \text{ m}^2/\text{day}$$

Evidently, the useful domain of the approximate logarithmic relation resulted from graph $S - \ln t$. Since the hydraulic diffusivity is now also known, this domain can also be analytically ascertained, i.e.

$$\frac{r^2}{4at} = \frac{70756}{4 \times 3750 \times 195} = 0.024 < 0.1$$

which means that the above-mentioned condition is fulfilled.

Parameters T and n_e can equally be determined by means of their graphical method of superposition; it requires the drawing of two curves on two log-log-diagrams: the type curve and the experimental curve. The type curve plots values $-Ei(-u)$[1] on the ordinate and argument $u = \frac{r^2 n_e}{4Tt}$ on the abscissa. This curve can be drawn for the desired interval by means of the tables. The graph of the experimental curve carries the drawdowns (S) in the observation well on the ordinate and the r^2/t values on the abscissa. Curve $S = f\left(\frac{r^2}{f}\right)$ shifts parallelly to the co-ordinate axes of the type curve—$Ei(-u) = f(u)$ until a nearly perfect superposition of both the curves is realised. Along this superposition, a point is selected which projected on the axes of co-ordinates supplies the values $-Ei(u)$, u, S and r^2/t. With these values introduced into the formula of the exponential integral the values of T and n_e will be obtained:

$$T = \frac{-Ei(-u)\,Q}{4\pi S} \tag{311}$$

$$n_e = \frac{4Ttu}{r^2} \tag{312}$$

---

[1] In American literature it is also denoted by $W(u)$ and is also called the well function.

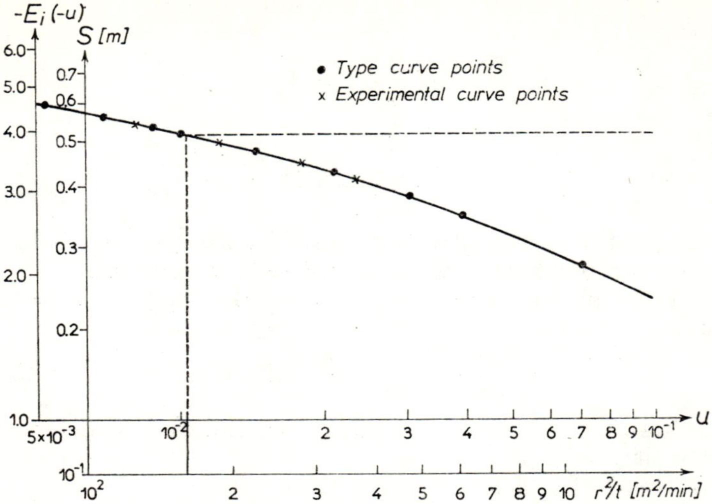

Figure 81. Superimposing experimental curve on the type curve of well $S_5 - I$.

On log-log diagram $S = f(r^2/t)$ of observation well S_5-I the experimental points are collinear on a long enough interval. The superposition of the standard and experimental curves (Figure 81) supplies the parameter values of a common point:

$$u = 1.03 \times 10^{-2} \qquad S = 0.53 \text{ m};$$

$$-Ei(-u) = 4.05; \qquad \frac{r^2}{t} = 160 \text{ m}^2/\text{min}$$

If these values are introduced into the previous relations, $T = 0.41$ m²/min and $n_e = 1.05 \times 10^{-4}$ will result which coincide with the previous results. This method should be used in the case of short pumpings as well as when the piezometer is placed very far from the central well.

In conclusion, the hydrogeological parameters determined by means of the data from observation well S_5-I have practically the same value whether they are computed by the exact method or by the approximate one. The different values of the hydraulic diffusivity (3750, 3890, and 3962 m²/min) should be ascribed to the application of the graphical method that unavoidably introduces a certain error. The maximum error of 5.6% is hower admissible.

Processing of the data from experimental well S_5-III [1] *Progresul București.* Even in the case of unsteady flow pumpings the various resistance form of the well and of its adjacent area can exert an important influence upon the computation of the hydrogeological parameters. Since in practice there is a

[1] The experimental well was considered as perfect owing to the very close values of the screen length and of the aquifer thickness (Table 59).

tendency not to drill observation wells, the hydrogeological parameters will be computed hereafter in both the cases, i.e. with and without taking into account the seepage resistances.

From graph $S = f(\log t)$ it follows that within interval $t = 50 - 555$ min a linear dependence appears between the two parameters so that this interval can be used for computation (Figure 82)

$$S = A + C \log t;$$

$$C = \frac{S_2 - S_1}{\log \frac{t_2}{t_1}} = \frac{3.92 - 3.68}{\log \frac{555}{50}}$$

The transmissivity is

$$kM = \frac{0.183Q}{C} = \frac{0.183 \times 0.682}{0.230} = 0.54 \text{ m}^2/\text{min} = 778 \text{ m}^2/\text{day}$$

The hydraulic diffusivity is supplied by relation

$$\log a_0 = \frac{A}{C} - 0.35 + 2 \log r_0$$

where A is the ordinate at the origin (3.29) and r_0 is the experimental well radius (0.11 m);

$$\log a_0 = \frac{3.29}{0.23} - 0.35 + 2 \log 0.11 = 12.03;$$

$$a_0 = 1.07 \times 10^{12} \text{ m}^2/\text{min}$$

This value does not represent the real hydraulic diffusivity of the aquifer since it also includes the seepage resistance of the well; such being the case, it can be called the *reduced hydraulic diffusivity* (a_0) related to the actual

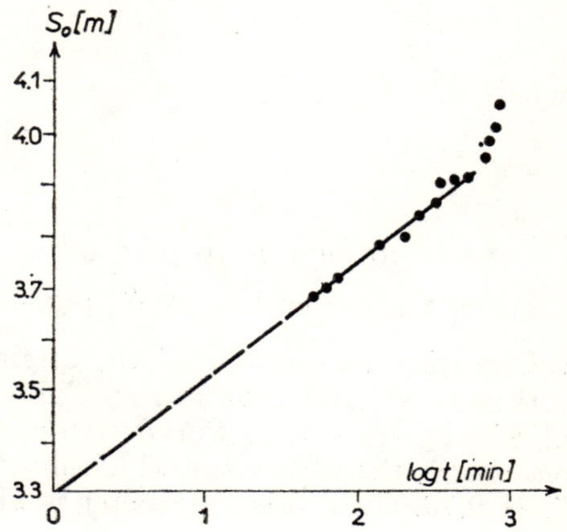

Figure 82. $S_0 - \log t$ graph of experimental well S_5 — III.

parameter (a) which can be deduced from the data of an observation well by relation

$$a_0 = ae^{\zeta_0} \tag{313}$$

Parameter a_0 may be used in the hydrodynamic computations of the capture and drainage systems provided the following two conditions are fulfilled:

— the designed discharges and drawdowns should have values close to the experimental ones;

— the drilling and casing technology of the exploitation wells should be similar to that of the experimental well.

The transmissivity can equally be computed by the following relation

$$T = kM = \frac{Q}{4\pi(S_2 - S_1)} \ln \frac{t_2}{t_1} \tag{314}$$

where two drawdowns are included, measured in the experimental well. This relation (derived from formula (292)) is not affected by errors because it does not contain the well radius. For the interval of logarithmic approximation it will be

$$kM = \frac{984}{4\pi(3.92 - 3.56)} \ln \frac{555}{50} = 785 \text{ m}^2/\text{day}$$

If the values of the hydraulic diffusivity (derived from the observation well data) is known, the parameter of the total seepage resistance (ζ_0) of the well can be estimated by starting from general relation

$$S_0 = \frac{Q}{2\pi kM}\left[-\frac{1}{2} Ei(-u) + \zeta_0\right] \tag{315}$$

$$\frac{2\pi kMS_0}{Q} = -\frac{1}{2} Ei(-u) + \zeta_0 \tag{316}$$

$$\zeta_0 = \frac{2\pi kMS}{Q} + \frac{1}{2} Ei(-u) \tag{317}$$

where

$$u = \frac{r_0^2}{4at}$$

So as to be able to take into account all data of the experimental well a graph with co-ordinates $\bar{S}_0$ and $-\frac{1}{2} Ei\left(-\frac{r_0^2}{4at}\right)$ must be drawn up. On this graph, the ordinate at the origin represents the very value of parameter ζ_0(Figure 83). With data from Table 60 the straight line was drawn the slope of which is $\alpha = 45°$. For determining the value of ζ_0 one more graph was necessary, the abscissa of which must start from zero; it was thus found that $\zeta_0 = 9.70$. Once this parameter is known, it is possible, with the help of the

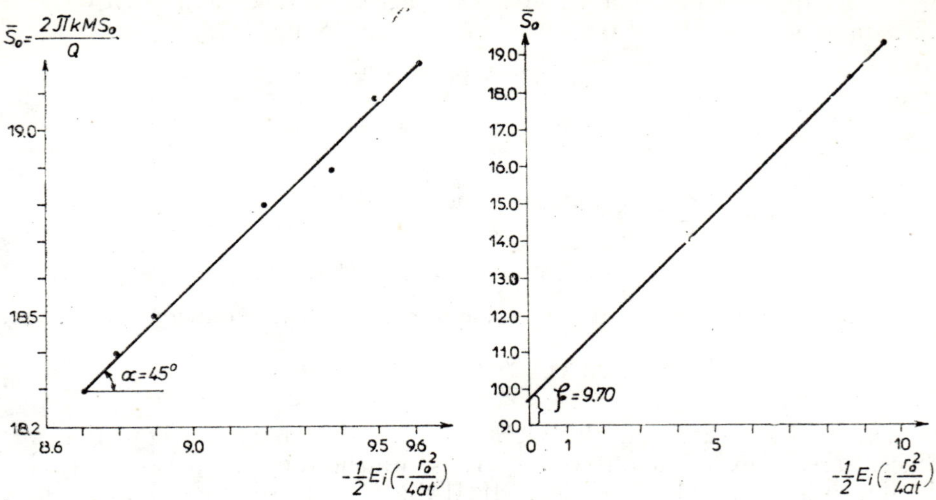

Figure 83. Graphic method for determining ζ_0.

equation of logarithmic approximation, to determine the hydraulic diffusivity by linearizing the equation as follows:

$$S_0 = \frac{Q}{4\pi kM}\left(\ln\frac{2.25\,at}{r_0^2} + 2\zeta_0\right) \tag{318}$$

$$S_0 = \frac{Q}{4\pi kM}\left(\ln\frac{2.25\,at}{r_0^2} + 19.4\right)$$

$$S_0 = \frac{Q}{4\pi kM}\left(\ln\frac{6.64\times 10^8}{r_0^2}\,at\right)$$

$$S_0 = \frac{0.183Q}{kM}\left(\log\frac{6.64\times 10^8}{r_0^2}\,at\right) \tag{319}$$

This equation can be written in the form

$$S_0 = A + C\log t \tag{320}$$

thence

$$S_0 = \frac{0.183Q}{kM}\log\frac{6.64\times 10^8}{r_0^2}\,a + \frac{0.183Q}{kM}\log t \tag{321}$$

$$C = \frac{0.183Q}{kM}; \qquad A = C\log\frac{6.64\times 10^8}{r_0^2}\,a$$

In the graph of Figure 82, the angular coefficient $C = 0.23$ and initial ordinate $A = 3.29$ being known, both the transmissivity and the hydraulic diffusivity can be determined. It is also to be remarked that the kM value is the previously computed one when the seepage resistance of the well was not taken into account, i.e. $kM = 0.54$ m²/min = 778 m²/day.

$$A = C \log 5.49 \times 10^{10} a$$

$$\frac{A}{C} = \log 5.49 \times 10^{10} + \log a$$

$$\log a = \frac{A}{C} - \log 5.49 \times 10^{10} = \frac{3.29}{0.23} - 10.74 = 3.56$$

$$a = 3630 \text{ m}^2/\text{min} = 5.23 \times 10^6 \text{ m}^2/\text{day}$$

If kM and a are known it is possible to estimate the coefficient of storage:

$$n_e = \frac{kM}{a} = \frac{0.54}{3630} = 1.5 \times 10^{-4}$$

The positive sign and the high value of parameter ζ_0 show that the adjacent area of experimental well S_5-III is influenced by a noticeable clogging (the screen is also liable to be clogged). If this is the case, the well loss can be defined by the following relation (Figure 58)

$$\Delta S = S_0 - S_0' = S_0 - S_1 - \frac{Q}{4\pi kM}\left[\ln \frac{2.25\,at}{r_0^2} - \ln \frac{2.25\,at}{r_1^2}\right] =$$

$$= S_0 - S_1 - \frac{Q}{4\pi kM} \ln \frac{r_1^2}{r_0^2} \qquad (322)$$

For $t = 555$ min we obtain (Table 60):

$$\Delta S = 3.92 - 0.56 - \frac{984}{12.56 \times 778} \ln \frac{266^2}{0.71^2} = 1.85 \text{ m}$$

From relation (322) it follows that the value of the well loss is a function of time due to difference $S_0 - S_1$. In the case of well S_5-III a peculiar situation occurs due to the fact that the clogging process took place earlier when this well had been for a long time a catchment work. The actual drawdown measured at the outer wall of the well is therefore $S_0' = 2.12$ m. The introduction into the computation of the drawdown measured within the well ($S_0 = 3.92$) affects the transmissivity by an error of +85%.

* * *

By comparing with each other the parameters obtained for the experimental and observation wells we can make assumptions about the non-homogenity of the aquifer. Thus, the higher transmissivity in S_5-III (0.54 m²/min against 0.42 m²/min in S_5-I) should chiefly be ascribed to the thickness which exceds by 155% the thickness of S_5-I. On the other hand, an increase appears in the coefficient of storage (42%) which means that on the direction of S_5-III the permeability is greater because the gravel fractions prevail here over the sand ones. The rather important increase of the coefficient of storage with regard to the transmissivity led to a small reduction of the hydraulic diffusivity (by 6.2% in S_5-I). In this case we have of course to take into account the systematic error due to the great distance between the two wells and of the thickness variation of the horizontal aquifer, which constitutes a deviation from the conditions required by the computation scheme. For a correct computation of the parameter, the observation well should be placed nearer to the experimental one so that the variation of the thickness and of the water-yielding properties between the two wells should not significantly influence the computation.

For a correct interpretation of the results it should be taken into account that hydrogeological parameters so determined are mean values in the whole cone of depression. Consequently, the advantage of the long pumpings is that the mean values are obtained on a large area with the least random deviations in regard to the means of the actual collectivity.

The reduced radius of influence can be deduced from relation

$$S = \frac{Q}{4\pi kM} \ln \frac{2.25\,at}{r^2} \tag{323}$$

by admitting the boundary conditions of the cone of depression, i.e. $S = 0$ and $r = R$.

$$\ln \frac{2.25\,at}{R_t^2} = 0\,; \quad \frac{2.25\,a}{R_t^2} = 1$$

Thus $R_t = 1.5\sqrt{at}$. Consequently, during the computing interval the radius of influence of the experimental well is

$$R_t = 1.5\sqrt{3630 \times 555} = 2130 \text{ m}$$

And since for the observation well this computation interval was only 495 min, the radius of influence will be

$$R_t = 1.5\sqrt{3870 \times 495} = 2076 \text{ m}$$

The hydrogeological parameters determined by means of the observation well correspond to an annulus, the width of which is $b = 2076 - 266 = 1810$ m since the integration limits are from "infinite" (i.e. the reduced radius of influence) to r_1 (=266 m), The parameters determined by the data of experimental well represent the mean values of the water yielding properties on a 2130 m radius area including also the annulus assignable to observation well. Owing to the complexity of the hydrogeological conditions it was impossible to cover a larger area; the determination of the distance and of the area was made with the help of the linear portion of graph $S = f(\log t)$.

2.3.2 Confined aquifers recharged by leakage

The hydrodynamic analysis of the leakage in unsteady régime presents certain mathematical and methodological difficulties as compared with the steady flow conditions (§ 2.2.3).

Making some assumptions, such as considering the aquifer currents to be two-dimensional, Hantush and Jacob proposed the following equation for the drawdown when testing a confined aquifer by a fully penetrating well [30]:

$$S = \frac{Q}{4\pi T} W\left(u, \frac{r}{B}\right) \tag{324}$$

where

$$W\left(u, \frac{r}{B}\right) = \int_u^{\infty} \exp\left(-y - \frac{r^2}{4B^2 y}\right) \frac{dy}{y} \tag{325}$$

is the well function under recharge conditions (Table 61, Figure 87) and

$$u = \frac{1}{y} = \frac{r^2 n_e}{4tT} \tag{326}$$

From Figure 87 it can be seen that at the limit and in the absence of the leakage, function $W(u, r/B)$ becomes $-Ei(-u)$.

The analytical determination of parameters T, u and B from the above equations is rather difficult and that is why certain graphical and grapho-analytical solutions were sought. One of the methods is e.g. the use of the inflection point, which was elaborated by Hantush. This method consists in drawing curve $S = f(\log t)$ for an observation well and in selecting certain characteristic points on this curve. The curve presents an inflexion point m_i (the angular coefficient) the co-ordinates of which, S_i and t_i are used for computing the hydrogeological parameters. The drawdown (S_i) is equal to half the maximum drawdown (S_M) corresponding to the steady flow ($S_i = 0.5\ S_M$). Figure 84 supplies the computation relations of the hydrogeological parameters when there is an inflexion point. If the pumping is brief and the steady flow was not yet reached, the S_M estimation is made by the extrapolation

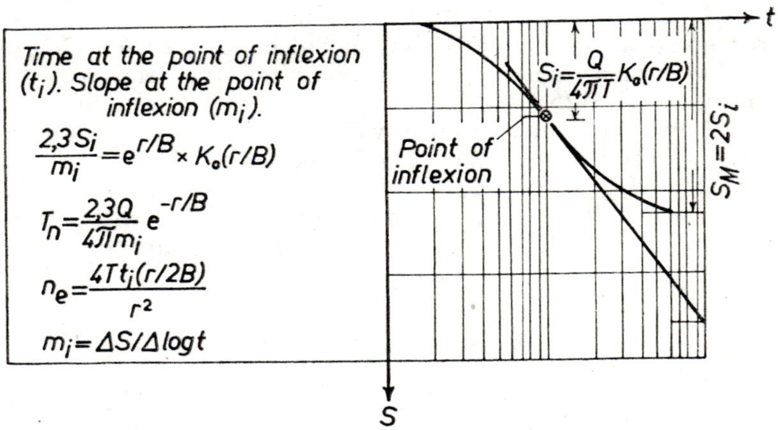

Figure 84. Drawdown variation with time of the observation well (after Hantush 1956).

$u = r^2/4at$	$\times 10^{-15}$	$\times 10^{-14}$	$\times 10^{-13}$	$\times 10^{-12}$	$\times 10^{-11}$	$\times 10^{-10}$	$\times 10^{-9}$	$\times 10^{-8}$
1.0	33.9616	31.6590	29.3564	27.0538	24.7512	22.4486	20.1460	17.8435
1.5	33.5561	31.2535	28.9509	26.6483	24.3458	22.0432	19.7406	17.4380
2.0	33.2684	30.9658	28.6632	26.3607	24.0581	21.7555	19.4529	17.1503
2.5	33.0453	30.7427	28.4401	26.1375	23.8349	21.5323	19.2298	16.9227
3.0	32.8629	30.5604	28.2578	25.9552	23.6526	21.3500	19.0474	16.7449
3.5	32.7088	30.4062	28.1036	25.8010	23.4985	21.1959	18.8933	16.5907
4.0	32.5753	30.2727	27.9701	25.6675	23.3649	21.0623	18.7598	16.4572
4.5	32.4575	30.1549	27.8523	25.5497	23.2471	20.9446	18.6420	16.3394
5.0	32.3521	30.0495	27.7470	25.4444	23.1418	20.8392	18.5366	16.2340
5.5	32.2568	29.9542	27.6516	25.3491	23.0465	20.7439	18.4413	16.1387
6.0	32.1698	29.8672	27.5646	25.2620	22.9595	20.6569	18.3543	16.0517
6.5	32.0898	29.7872	27.4846	25.1820	22.8794	20.5768	18.2742	15.9717
7.0	32.0156	29.7131	27.4105	25.1079	22.8053	20.5027	18.2001	15.8976
7.5	31.9467	29.6441	27.3415	25.0389	22.7383	20.4337	18.1311	15.8280
8.0	31.8821	29.5795	27.2769	24.9744	22.6718	20.3692	18.0666	15.7640
8.5	31.8215	29.5189	27.2163	24.9137	22.6112	20.3086	19.0060	15.7034
9.0	31.7643	29.4618	27.1592	24.8566	22.5540	20.2514	17.9488	15.6462
9.5	31.7103	29.4077	27.1051	24.8025	22.4999	20.1973	17.8948	15.5922

of the curve. In many cases however, the inflexion point cannot be found on the graph owing to the complexity of the leakage system; to solve the problem, the curve properties are used on the branch where it is approximated as the asymptote to the greatest drawdown.

$$m_t \approx \frac{2.3Q}{4\pi T_n} e^{-p} \tag{327}$$

$$S = S_M - \frac{Q}{4\pi T_n}[-Ei(-p)] \tag{328}$$

$$p = \frac{T_n t}{n_e B^2} \tag{329}$$

From relation (327) and (328) we deduce

$$f(p) = e^p[-Ei(-p)] = 2.3\,\frac{S_M - S}{m_t} \tag{330}$$

From the hydrogeological cross-section (Figure 77), we can see that in the Frătești complex the conditions favour the phenomenon of recharge. Thus, aquifer A (tested) is separated from aquifer B by a clayly horizon 13 m-thick ($m_1 = 13$ m). In a non-influenced régime, the recharge has a very small intensity but in the case of pumping from aquifer A (group S_5 Progresul) the difference between the two levels increases which leads to an increased intensity of the phenomenon of ascending vertical filtration.

Table 61

$-Ei(-u)$ (after Wenzel)

$\times 10^{-7}$	$\times 10^{-6}$	$\times 10^{-5}$	$\times 10^{-4}$	$\times 10^{-3}$	$\times 10^{-2}$	$\times 10^{-1}$	$\times 1$
15.5409	13.2383	10.9357	8.6332	6.3315	4.0379	1.8229	0.2194
15.1354	12.8328	10.5303	8.2278	5.9266	3.6374	1.4645	0.1000
14.8477	12.5451	10.2426	7.9402	5.6394	3.3547	1.2227	0.04890
14.6246	12.3220	10.0194	7.7172	5.4167	3.1365	1.0443	0.02491
14.4423	12.1397	9.8371	7.5348	5.2349	2.9591	0.9057	0.01305
14.2881	11.9855	9.6830	7.3807	5.0813	2.8099	0.7942	0.006970
14.1546	11.8520	9.5495	7.2472	4.9482	2.6813	0.7024	0.003779
14.0368	11.7342	9.4317	7.1295	4.8310	2.5684	0.6253	0.002073
13.9314	11.6280	9.3263	7.0242	4.7261	2.4679	0.5598	0.001148
13.8361	11.5330	9.2310	6.9289	4.6313	2.3775	0.5034	0.0006409
13.7491	11.4465	9.1440	6.8420	4.5448	2.2953	0.4544	0.0003601
13.6691	11.3665	9.0640	6.7620	4.4652	2.2201	0.4115	0.0002034
13.5950	11.2924	8.9899	6.6879	4.3916	2.1508	0.3738	0.0001155
13.5260	11.2234	8.9209	6.6190	5.3231	2.0867	0.3403	0.0000658
13.4614	11.1589	8.8563	6.5545	4.2591	2.0269	0.3106	0.0000376
13.4008	11.0982	8.7957	6.4930	4.1990	1.9711	0.2840	0.0000216
13.3437	11.0411	8.7386	6.4368	4.1423	1.9187	0.2602	0.0000124
13.2896	10.9870	8.6845	6.3828	4.0887	1.8695	0.2387	0.0000071

To determine the hydrogeological parameters the experimental graph $S = f(\log t)$ was drawn for observation well S_5-I (Figure 85). The disposition of the points on this graph does not permit however to define an inflexion point, respectively to use the formulae from Figure 84. But in interval $t = 375 - 1000$ min the points are collinear and Table 60 permits us to appreciate

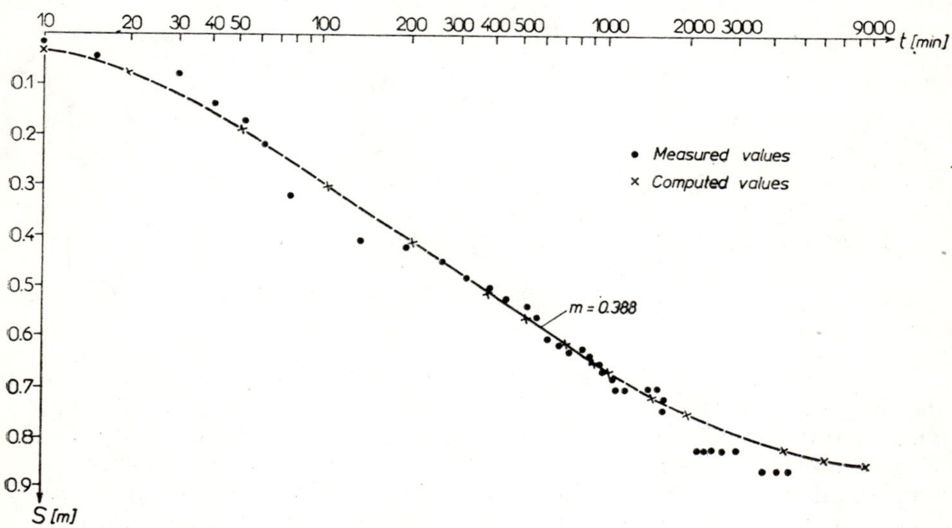

Figure 85. $S - \log t$ graph of observation well $S_5 -$ I.

that $S_M = 0.86$ m. Under these conditions. relation (330) can be used where the co-ordinates of an observation point $S = 0.50$ m at $t = 375$ min (0.26 days) are also included

$$f(p) = e^p[-Ei(-p)] = 2.3 \frac{0.86 - 0.50}{0.388} = 2.13$$

Solving it by the grapho-analytical method, this implicit equation supplies

$$p = 0.087; \quad e^p = 1.091; \quad -Ei(-p) = 1.95$$

$$T_n = \frac{2.3 \times 984}{4\pi \times 0.388 \times 1.091} = 426.2 \text{ m}^2/\text{day}$$

$$S_i = 0.5\, S_M = \frac{Q}{4T_n}\, K_0(r/B)$$

$$K_0(r/B) = \frac{2\pi T_n S_M}{Q} = \frac{6.28 \times 426.2 \times 0.86}{984} = 2.34$$

From the tables we have $r/B = 0.11$, hence $B = \frac{266}{0.11} = 2418$ m

From relation (339) we deduce:

$$n_e = \frac{T_n t}{p \cdot B^2} = \frac{426.2 \times 0.26}{0.087 \times 2418^2} = 2.2 \times 10^{-4}$$

This value permits us to determine the hydraulic diffusivity

$$a_n = \frac{T_n}{n_e} = \frac{426.2}{2.2} \cdot 10^4 = 1.94 \times 10^6 \text{ m}^2/\text{day}$$

In deciding whether the leakage occurs from the upper or lower bed or solely from the lower bed, we have to take into account the hydrogeological conditions in the cross-section. Condition $k_n/\lambda > 100$ is completely fulfilled since the value of k_n is about 20—30 m/day. The other condition (M_n/m > 3) is not fulfilled because thickness M_n of aquifer A is equal to 14.5 — 22.5 m while thickness m of the clayey horizon (semipervious) of both the bottom and the top on alignment S_5-III—S_5-I is about 13 m. According to the horizontal situation established by Emil Liteanu, over the Frătești Complex (early Pleistocene) lies a marly complex (middle Pleistocene) where there are also sandy horizons of small thicknesses. Contrary to the situation at the top, in the bottom of aquifer A, aquifer B develops, the aquiferous potential of which is still stronger than that of aquifer A. To elucidate this problem let us compute the hydraulic conductivity of the semi-pervious grounds (λ) in both these cases:

— In case of the recharge from both the top and the bottom, according to relation (221) we find

$$\lambda_1 = \lambda_2 = \lambda = \frac{426.2 \times 13 \times 13}{2418^2 \times 26} = 4.74 \times 10^{-4} \text{ m/day}$$

where m_1 and m_2 are the thicknesses of the semi-pervious beds in the top and at the bottom, while λ_1 and λ_2 are the corresponding hydraulic conductivities (considered to have the same value).

— In case of the recharge in the bottom (relation 222),

$$\lambda_2 = \frac{426.2 \times 13}{2418^2} = 0.95 \times 10^{-3} \text{ m/day}$$

Taking into account that the semi-pervious formation consists in clays and marls, sometimes sandy, the hypothesis of the vertical percolation in both ways must be admitted by specifying however that the leakage from the lower bed is stronger. This phenomenon is quite important when appreciating the discharge potential of aquifer *A*, much exploited in the Bucharest area.

The leakance (K_d) when this phenomenon occurs both from the upper bed and from the lower bed is defined by the following relation (derived from formula 221):

$$K_d = \frac{\lambda(m_1 + m_2)}{m_1 m_2} = \frac{T_n}{B^2} \tag{331}$$

The resulting value ($K_d = 0.73 \times 10^{-4}$ day^{-1}) emphasizes the larger percolation capacity (on the vertical) of these Pleistocene marl-clays than that estimated for the shales-clays in Cuzap lignite deposits (§ 2.2.3). The difference is due chiefly to the smaller degree of consolidation of the Pleistocene marl-clays with regard to those of Panonian age. But this case does not entail an increase of the leakage intensity as materialized by parameter *B*. Its value ($B = 2418$ m) points out a lower intensity of the leakage than the estimated one for the Cuzap structure ($B = 1950$ m); this can be ascribed to the great thicknesses of the semi-pervious layers in both the upper and the lower bed of aquifer *A*.

In order to verify the adopted computation scheme, the computed values of the drawdowns (324) should be compared with the measured values. From Figure 85 we can see that the two series of values are very close to each other on the linear portion of the graph, which confirms the validity of the computation scheme including the maximum admitted reduction of the drawdown (S_M); in certain cases, this extrapolated quantity is established by successive computations. The noticeable deviation of the other measured values with regard to the computed ones is due chiefly to the variation of the transmissivity. From this standpoint, it is to be remarked that the value $S_M = 0.86$ m was recorded after about 2.5 days of pumping while on the computed curve (on the assumption of a constant transmissivity) the drawdown $S = 0.86$ m corresponds to $t = 8$ days. In the initial portion of the graph it is equally remarkable that the experimental points display deviations from the theoretical curve which should be ascribed to the non-homogeneity of the principal aquifer (*A*).

Besides this procedure, the hydrogeological parameters of the leaky system can also be determined by means of a complex graph by the superposition of the experimental curve over the standard one [14]. This method is simpler than the previously presented one, but as it is graphic it is liable to introduce a certain error.

Confronting the values of parameters T, a and n_e, computed on the assumption of the leakage from both the upper bed and the lower bed, with those determined on the assumption of the impermeable upper and lower bed (T', a' and n_e') the following values will result:

$$\frac{T'}{T} = \frac{605}{426.2} \cdot 100 = 150\,\%; \qquad \frac{a'}{a} = \frac{5.70 \times 10^6}{1.94 \times 10^6} \cdot 100 = 294\,\%$$

$$\frac{n_e'}{n_e} = \frac{1.06 \times 10^{-4}}{2.20 \times 10^{-4}}\, 100 = 48\,\%$$

The differences between the two series of values can be considered as very great in the case of the hydraulic diffusivity and of the transmissivity but merely great in the case of the coefficient of storage yield. The transmissivity increase is somehow considered as equivalent to the rate flow due to the leakage. Of course, these conclusions are valid in the case of the hydrogeological conditions of experimental group Progresul Bucharest. These differences are, however, amplified when these parameters are introduced in the computations schemes of the capture or dewatering systems. That is why a thorough knowledge of the hydrogeological conditions is essential so as to be able to outline the most accurate structure and to choose the most adequate computations schemes. In paper [5], it was shown that sometimes the leakage can be strong enough to account for the whole water supply pumped from the well.

M. S. Hantush elaborated the above-mentioned computation scheme on the assumption that the hydraulic head is constant in the subordinate aquifer and the water within the semi-pervious layers is not released upwards. Later on he also took into account the water released by the semi-pervious formation and showed that if condition $n_e^x/n_e \leqslant 0.1$ (where n_e^x is the coefficient of storage of the semi-pervious deposits) is fulfilled, the water volume from either the upper or the lower bed can be neglected, as in the previous procedure. If $n_e^x/n_e > 0.1$ the computation scheme becomes complicated by the inclusion of the water volume released by the semi-pervious deposits. This case is specific to structures offering large variations of lithofacies where the aquifers present many hydrogeological "windows" in the upper as well as in the lower bed.

Under certain conditions, the computation formulae can be simplified. For instance, at times $t \geqslant (2 - 2.5)\, r^2/a_n$ and $r/B \leqslant 0.2$, function $W(u, r/B)$ can be approximated by relation

$$W(u, r/B) \approx 2\, K_0(r/B) - I_0(r/B)\, [-Ei(-bt)] \qquad (332)$$

where I_0 is the modified Bessel function of the I-st kind and of the zero order while a_n and b are the global parameters in formula (211) of the leakage factor B

$$B = \sqrt{\frac{a_n}{b}} \qquad (333)$$

$$a_n = \frac{k_n M_n}{n_e} \tag{334}$$

$$b = \frac{1}{n_e}\left(\frac{\lambda_1}{m_1} + \frac{\lambda_2}{m_2}\right) = \frac{\lambda(m_1 + m_2)}{m_1 m_2} \tag{335}$$

Between the last parameter (which is used in Soviet papers and can be called the *leakage coefficient*) and parameter K_d from formula (331) (used in American works) the following relation exists:

$$b = K_d / n_e \tag{336}$$

When the values of B and of hydraulic diffusivity a_n are known parameter b can be computed:

$$b = \frac{a_n}{B^2} = \frac{1.94 \times 10^6}{2418^2} = 0.33 \text{ day}^{-1} \tag{337}$$

For observation well S_5-I, condition (332) is fulfilled when

$$t \geqslant 2.5 \frac{266^2}{1.94 \times 10^6} = 0.091 \text{ day} = 132 \text{ min}$$

For the experimental well, the approximate relation can be used in the simplified form from the very beginning of the pumping

$$W(u, r/B) \approx 2 \ln \frac{1.12\,B}{r_0} + Ei(-bt) \tag{338}$$

since $K_0(r_0/B) \approx \ln 1.12\ B/r_0$ and $I_0(r_0/B) \approx 1$.

The determination of the hydrogeological parameters of a leaky system cannot be made by means of experimental well data because for the very small values of ratio r/B (usually $r_0/B < 0.001$), well function $W(u, r_0/B)$ has values practically equal to function $-Ei(-u)$, valid for an aquifer without vertical leakage. Nevertheless, the data of the experimental well correlated with those of the observation wells can supply additional information. Thus, on graph $S_0 = f(\log t)$ of experimental well S_5-III the straight line stretch is comprised in interval 50—555 min and has an angular coefficient $m_e = 0.23$. By the computation within the same interval of drawdowns S_0 with the T_n, B and b values corresponding to the observation well, a straight line with $m_c = 0.41$ was obtained

$$S_0 = \frac{Q}{4\pi T_n}\left[2 \ln \frac{1.12\,B}{r_0} + Ei(-bt)\right] \tag{339}$$

The smaller value of the experimental straight line slope (Figure 86) is due to a higher discharge potential in the area with $r < 266$ m than in cone of depression outside the observation well. This increase of the discharge potential can be due either to the transmissivity increase, or to the leakage (a diminution of parameter B value) or to both these causes. In the case of group S_5 Progresul, the chief cause is the increase of the aquifer thickness i.e. of the transmissivity. When $m_e < m_c$ the explanation is the reverse.

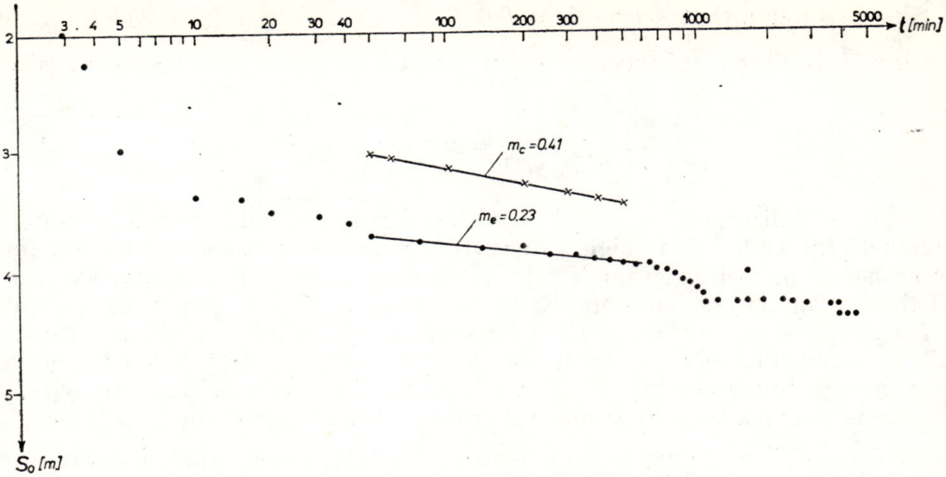

Figure 86. S -- log t graph of experimental well S_5 — III.

If the experimental and the computed lines are parallel, this means that the values of the hydrogeological parameters are constant all along the considered alignment and the difference between the ordinates represents the well loss that can be computed by relation

$$\Delta S = S_0 - S_0' = S_0 - S_1 - \frac{Q}{4\pi T_n}\left[2 \ln \frac{1.12\, B}{r_0} + Ei(-bt) - W(u, r/B)\right] \tag{340}$$

The well loss corresponds evidently to a certain value of the parameter of the total seepage resistance of the well (ζ_0) that can be computed by relation

$$S_0 = \frac{Q}{2\pi T_n}\left[\ln \frac{1.12\, B}{r_0} + 0.5\, Ei(-bt) + \zeta_0\right] \tag{341}$$

This relation can be linearized as follows:

$$\eta = \zeta_0 + 0.5\, Ei(-bt)$$

$$\beta = \frac{2\pi T_n S_0}{Q} - \ln \frac{1.12\, B}{r_0}$$

On a $\eta - \beta$ graph the ordinate at the fixed point is the very value of parameter ζ_0. For determining parameters ΔS and ζ_0 one of the observation wells should be placed close to the experimental well so that the two straight lines would be parallel and consequently the characteristics of the aquifer should have the same values in both the wells.

From the relation (339) the computation period can be established corresponding to the unsteady régime in the neighbourhood of the experimental well; it is determined by the value of exponential function $-Ei(-bt)$. For

$t_n \geqslant \frac{3}{b}$ it is found that $-Ei(-3) = 0.013$, i.e. a very small value which can be neglected. In the case of experimental well S_5-III, this time period is estimated at

$$t_n \geqslant \frac{3}{0.33} = 9.1 \text{ days}$$

The stabilization time of the leakage depends chiefly on the hydraulic conductivity λ: for their high values, the period is relatively small; on the other hand, for their very small values ($\lambda < 10^{-5}$ m/day) the stabilization time of the leakage grows considerably up to tens of days. The appearance of the steady régime does not occur simultaneously on the whole cone of depression; the required time increases with distance r. Since generally the location of the piezometers fulfils condition $r/B \leqslant 0.2$, it can be accepted that the steady régime in the area of an experimental group appears after a time $t \geqslant (2 - 2.5) \frac{r^2}{a_n} + \frac{3 \text{ to } 5}{b}$. The magnitude of this time depends on the value of parameter B. The shorter this period, the quicker the leakage appearance. The horizontal character of curves r/B after a certain period from the recharge beginning indicates the appearance of the steady régime (Figure 87).

As a matter of fact, the unsteady régime of the recharge is particularly important when the pumpings are made in order to estimate the hydrogeological parameters. In the case of isolated works of capture or drainage, owing to their long working periods, the leakage phenomenon should only be analyzed under the steady régime conditions (provided that the rate of discharge is constant); of course when these works are operated in interference, the period of unsteady régime increases to hundreds of days.

From the quantitative analysis of the leakage phenomenon there follows a series of considerations and parameters that help to define the hydraulic régime. Thus, V. M. Shestakov [63] emphasizes the fact that the elastic régime in the semi-pervious deposits can have a considerable influence upon the interaction phenomenon on the vertical of the aquifers because the pressure

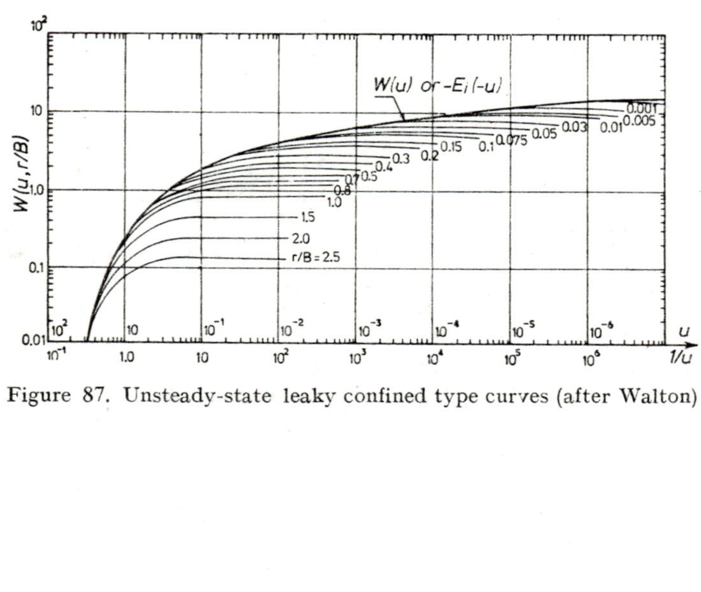

Figure 87. Unsteady-state leaky confined type curves (after Walton)

transmission from an aquifer to another can only be made with a certain time lag (t_0), that can be estimated by relation

$$t_0 = (0.1 \dots 0.15)\,\frac{m_2^2}{a_2} \quad \text{or} \tag{342}$$

$$t_0 = (0.1 \dots 0.15)\,\frac{m_1^2}{a_1} \tag{343}$$

where m and a are the thickness and the hydraulic diffusivity of the lower and upper semi-pervious deposits. In the case of experimental group S_5 with estimation $n_e^x \approx 10^{-5}$, there results that

$$a = a_1 = a_2 = \frac{\lambda m}{n_e^x} - \frac{4.1 \times 10^{-4} \times 13}{10^{-5}} = 533 \text{ m}^2/\text{day}$$

$$t_0 = 0.125\,\frac{13^2}{533} = 0.040 \text{ day} = 57 \text{ minutes}$$

This reduced lag period (which should not be taken into account when computing the leakage) is due to the high permeability of the Pleistocene marly-clays. The marly-clayey deposits with high values of the consolidation degree and of the thickness may determine important lags (t_0) at the input of the subordinate aquifers of the leakage system. The computation period of the leakage corresponding to the straight line on graph $S_0 = f(\log t)$ begins at a time equal to t_0. In the case of well S_5-III, this condition is fulfilled, but in the case of S_5-I, the computation period begins at $t = 375$ m (Figure 85); this should not however be ascribed to the transmission lag of the hydraulic pressure but to the complexity of the principal aquifer that alter the linear dependance $S = f(\log t)$.

The reduction velocity of the pumping level of the experimental well is supplied by equation [5]

$$V = \frac{\partial S}{\partial t} = e^{-bt}\,\frac{Q}{4\pi T_n t} \tag{344}$$

From this relation it follows that the value of this velocity depends chiefly on the value of parameter b; the greater this parameter (i.e. the more intense the leakage), the lower the velocity value, while the stabilization time of the pumping water level grows shorter. Thus, for well S_5-III at the beginning of the leakage phenomenon ($t_0 = 57$ min $= 0.04$ day) velocity v is 2.88 m/day; while at the end of the non-steady régime ($t_0 = 9.1$ days) $v = 1$ mm/day, a value that explains the passage to the steady régime of the pumping level in the experimental well.

No literature offers an accurate solution for estimating the radius of influence under an unsteady règime (R_t). This parameter should nevertheless be known since the leakage computation is only made within the limits of the radius of influence. The following approximate method is therefore proposed which starts from the radius of influence in steady régime $R = 1.12\ B =$

$= 1.12 \times 2418 = 2700$ m (see 2.2.3). The stabilization period of the flow régime for $r = R$ is computed at a first approximation by relation

$$t = 2.5 \frac{R^2}{a_n} + \frac{5}{b} = 2.5 \frac{2700^2}{1.94 \times 10^6} + \frac{5}{0.33} = 25 \text{ days}$$

Quantity $R/B = 1.115$ and argument $u = 0.0393$ (corresponding to $t = 25$ days) permit to determine with the help of Table 62 the value of function $W(u, r/B)$ which for experimental group S is, by interpolation, 0.750. From the function table we can see that for time values $t > 25$ days (and

Values of function $W(u, r/B)$

u \ r/B	0.01	0.015	0.03	0.05	0.075	0.10	0.15	0.2	0.3
0.000001	9.4425	8.6319	7.2471	6.2285	5.4228	4.8541	4.0601	3.5054	2.7449
0.000005	9.4413								
0.00001	9.4176	8.6319	7.2471						
0.00005	8.8827	8.4533	7.2450	6.2285					
0.0001	8.3983	8.1414	7.2122	6.2282	5.4228	4.8541			
0.0005	6.9750	6.9152	6.6219	6.0821	5.4062	4.8530	4.0601		
0.001	6.3069	6.2765	6.1202	5.7965	5.3078	4.8292	4.0595	3.5054	2.7449
0.005	4.7212	4.7152	4.6829	4.6084	4.4713	3.2960	3.8821	3.4567	2.7428
0.01	4.0356	4.0326	4.0167	3.9795	3.9091	3.8150	3.5725	3.2875	2.7104
0.05	2.4675	2.4670	2.4642	2.4576	2.4448	2.4271	2.3776	2.3110	1.9283
0.1	1.8227	1.8225	1.8213	1.8184	1.8128	1.8050	1.7829	1.7527	1.6704
0.5	0.5598	0.5597	0.5596	0.5594	0.5588	0.5581	0.5561	0.5532	5.5453
1.0	0.2194	0.2194	0.2193	0.2193	0.2191	0.2190	0.2186	0.2179	0.2161
5.0	0.0011	0.0011	0.0011	0.0011	0.0011	0.0011	0.0011	0.0011	0.0011

implicitly argument values $u < 0.0393$) the value of function $W(u, r/B)$ remains constant, i.e. the leakage system régime becomes steady. If this condition is not fulfilled the tests will require longer time periods.

Starting from the assumption that the radius of influence in a steady régime corresponds to $t = 25$ days, the following relation is proposed:

$$R_t = \alpha \sqrt{a_n t} \tag{345}$$

where the coefficient α has a distinct value for each of the hydrogeological situations. Thus, for experimental group S_5, the value of this coefficient is

$$\alpha = \frac{R^2}{a_n t} = \frac{2700^2}{1.94 \times 10^6 \times 25} = 0.4$$

For the leaky systems, coefficient α must be lower than 1.5, which is valid for confined aquifers with impermeable top and bottom; this value of coefficient α corresponds to a lower propagation velocity of the pressure wave than in the leakage case.

In the period of unsteady régime the value of modulus W_s of the underground recharge (defined on § 2.2.3) varies with both the space and the time. Thus, in the section corresponding to observation well S_5-I this parameter grows from 0.007 mm/day for $t_0 = 57$ min up to 0.027 mm/day which remains constant during the steady régime.

The relative leakage contribution can also result from the comparative analysis of the drawdowns in the case of the same rate of discharge (Q) from the well:

— with leakage

$$S = \frac{Q}{4\pi T} W(u,\ r/B) \tag{346}$$

Table 62

(after Hantush, 1962)

0.4	0.5	0.6	0.7	0.8	0.9	1.0	1.5	2.0	2.5	5,0
2.2291	1.8488	1.5550	1.3210	1.1307	0.9735	0.8420	0.4276	0.2278	0.1247	0.0074
2.2291										
2.2290	1.8488									
2.2253	1.8486	1.5550	1.3210	1.1307	0.9735	0.8420				
1.7075	1.4927	1.2955	1.2955	1.1210	0.9700	0.8409	0.4276			
1.5644	1.4422	1.3115	1.1791	1.0505	0.9297	0.8190	0.4271	0.2278	0.1247	
0.5344	0.5206	0.5044	0.4860	0.4658	0.4440	0.4210	0.3007	0.1944	0.1174	0.0074
0.2135	0.2103	0.2065	0.2020	0.1970	0.1914	0.1855	0.1509	0.1139	0.0803	0.0073
0.0011	0.0011	0.0011	0.0011	0.0011	0.0011	0.0011	0.0010	0.0010	0.0009	0.0004

— without leakage

$$S_i = S + \Delta S = \frac{Q}{4\pi T} W(u) \text{ [1]} \tag{347}$$

Quantity $\Delta S/S$ results from these two relations and can even be considered as a relative error

$$\Delta S = \frac{Q}{4\pi T}[W(u) - W(u,\ r/B)] \tag{348}$$

$$\eta = \frac{\Delta S}{S_i} = \frac{W(u) - W(u,\ r/B)}{W(u)} = \frac{\Delta W}{W(u)} \tag{349}$$

The leakage contribution can also be deduced from the condition of a constant drawdown and of a variable discharge of the well:

— with leakage

$$Q = 4\pi TS \times W(u,\ r/B) \tag{350}$$

---

[1] The function of the well $W(u) = -Ei(-u)$.

— without leakage

$$Q_i = Q + \Delta Q = 4\pi T.S.W(u) \tag{351}$$

$$\Delta Q = 4\pi TS[W(u) - W(u, r/B)] \tag{352}$$

$$\eta = \frac{\Delta Q}{Q_i} = \frac{W(u) - W(u, r/B)}{W(u)} = \frac{\Delta W}{W(u)} \tag{353}$$

Consequently, the relative error due to the omission of the leakage phenomenon has the same expression in both these cases. Quantities ΔS and ΔQ also have physical meanings: ΔS represents the additional drawdown of the well on the assumption of an aquifer without leakage, while ΔQ represents the additional discharge of the well corresponding to an aquifer with leakage. To conclude, the omission of the leakage phenomenon when designing a capture or drainage system can lead to:

— designed drawdown which are larger (by ΔS) than the real ones, in order to obtain the necessary yield;

— designed yields which are smaller (by ΔQ) than the real ones in order to achieve a certain drawdown.

So as to emphasize the error due to the omission of the leakage, quantity η was computed for various values of time and radii (Table 63, Figure 88). From the table, for the same section, quantity η increases with time. This is due to the evolution of function $W(u)$, the value of which increases in the same proportion as its argument diminishes (Figure 87). By contrast, function $W(u, r/B)$ tends to become stable after a restricted time at a certain value of ratio r/B. In this case, a continuous increase of quantity ΔW and implicitly of relative error η occurs. On the other hand the error value increases in the same proportion as the distance from the experimental well.

So as to have a clearer image of the error importance let us consider the section situated at $r = 560$ m and $t = 100$ days; the relative error being $\eta = 0.55$, it follows that the omission of the leakage phenomenon will lead

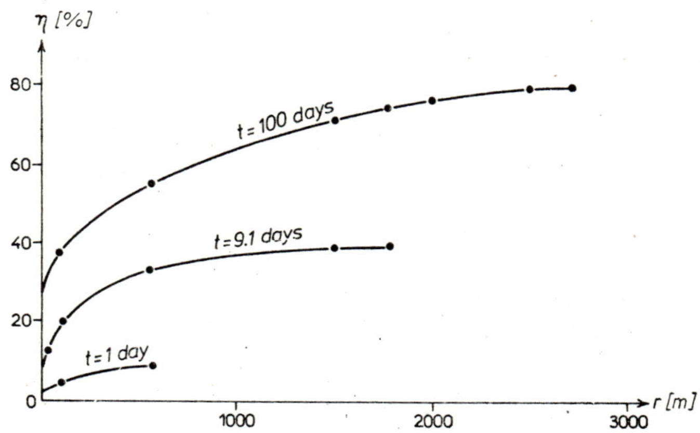

Figure 88. Evolution in time and space of the leakage relative contribution.

Table 63

r [m]	r/B	$t = 1$ day					$t = 9.1$ days					$t = 100$ days				
		u	$W(u)$	$W(u, r/B)$	ΔW	$\frac{\Delta W}{(Wu)} \cdot 100$	u	$W(u)$	$W(u, r/B)$	ΔW	$\frac{\Delta W}{W(u)} \cdot 100$	u	$W(u)$	$W(u, r/B)$	ΔW	$\frac{\Delta W}{W(u)} \cdot 100$
$r_0 = 0.11$	4×10^{-5}	$\rightarrow 0$				$\rightarrow 0$	$\rightarrow 0$				$\rightarrow 0$	$\rightarrow 0$				$\rightarrow 0$
10	0.004	1.29×10^{-5}	10.68	10.40	0.28	2.62	1.42×10^{-6}	12.89	11.23	1.66	12.87	1.29×10^{-7}	15.29	11.275	4.015	26.26
100	0.041	1.29×10^{-3}	6.08	5.81	0.27	4.24	1.42×10^{-4}	8.28	6.62	1.66	20.05	1.29×10^{-5}	10.68	6.620	4.060	38.01
$R_t = 560$	0.232	4.05×10^{-2}	2.66	2.41	0.25	9.40	4.45×10^{-3}	4.84	3.22	1.62	33.47	4.05×10^{-4}	7.25	3.235	4.015	55.38
1500	0.620						5.66×10^{-2}	2.35	1.430	0.920	39.15	2.90×10^{-3}	5.27	1.500	3.770	71.72
$R_t = 1\,780$	0.736						8.0×10^{-2}	2.027	1.210	0.817	40.30	4.09×10^{-3}	4.92	1.250	3.670	74.60
2000	0.827											5.16×10^{-3}	4.69	1.090	3.600	76.75
2500	1.034											8.05×10^{-3}	4.25	0.842	3.408	80.18
$R = 2700$	1.117											9.45×10^{-3}	4.09	0.800	3.290	80.43

to an over-estimation of the drawdown ($\Delta S = 0.55\ S_i$) or to an underrating of the discharge rate ($\Delta Q = 0.55\ Q_i$) with regard to the computed values of S_i and Q_i.

During the period of unsteady régime, the flow rate in any section and at any time can be computed by relation

$$Q(r, t) = Q_0\left[r/BK_1(r/B) - e^{-u-\frac{r^2}{4B^2u}}\right] \quad (354)$$

where Q_0 is the rate of discharge pumped from the well.

The difference between the flow rate of two sections (at the same time) is the contribution of the leakage plus the elastic resources of the tested aquifer. When the pumping period increases, the contribution of the elastic resource decreases while the leakage contribution increases. In steady régime, the rate of discharge pumped from the well can be constituted of:

— the leakage contribution; in this case relation (354) is simplified and becomes (224);

— the leakage contribution and the dynamic resource of the tested aquifer; this represents a hydrodynamic occurrence which from the theoretical standpoint has not yet a satisfying solution.

There exist also other types of structures where the aquifers should interact owing to the pumpings from the wells. Figure 89 and Table 64 offer the possible schemes as well as the relations permitting to compute the leakage parameters.

Table 64

Leakage typical schemes

Scheme	a_n	b	$B = \sqrt{\frac{a_n}{b}}$
a. Tested unconfined aquifer (n) is recharged from the bottom by a confined aquifer $n+1$ (Fig. 94 *a*)	$\frac{k_n H_m}{n_s}$ [1]	$\frac{\lambda_2}{n_s m_2}$	$\sqrt{\frac{k_n H_m m_2}{\lambda_2}}$
b. Tested confined aquifer n is recharged from the bottom by a confined aquifer $n+1$ (Fig. 94 *b*)	$\frac{k_n M_n}{n_e}$	$\frac{\lambda_2}{n_e m_2}$	$\sqrt{\frac{k_n M_n m_2}{\lambda_2}}$
c. Tested confined aquifer n is recharged from the top by an unconfined aquifer $n-1$ (Fig. 94 *c*)	$\frac{k_n M_n}{n_e}$	$\frac{\lambda_1}{n_e m_1}$	$\sqrt{\frac{k_n M_n m_1}{\lambda_1}}$
d. Tested confined aquifer n is recharged by an upper confined aquifer $n-1$ (Fig. 94 *d*)	$\frac{k_n M_n}{n_e}$	$\frac{\lambda_1}{n_e m_1}$	$\sqrt{\frac{k_n M_n m_1}{\lambda_1}}$
e. Tested confined aquifer n is recharged from the top by an unconfined aquifer $n-1$, and from the bottom by a confined aquifer $n+1$, (Fig. 94*e*)	$\frac{k_n M_n}{n_e}$	$\frac{1}{n_e}\left(\frac{\lambda_1}{m_1}+\frac{\lambda_2}{m_2}\right)$	$\sqrt{\frac{k_n M_n}{\frac{\lambda_1}{m_1}+\frac{\lambda_2}{m_2}}}$
f. Tested confined aquifer n is recharged from both top and bottom by confined aquifers $n-1$ and $n+1$ (Fig. 94*f*)	$\frac{k_n M_n}{n_e}$	$\frac{1}{n_e}\left(\frac{\lambda_1}{m_1}+\frac{\lambda_2}{m_2}\right)$	$\sqrt{\frac{k_n M_n}{\frac{\lambda_1}{m_1}+\frac{\lambda_2}{m_2}}}$

[1] H_m is mean thickness of the flow within the whole influence area

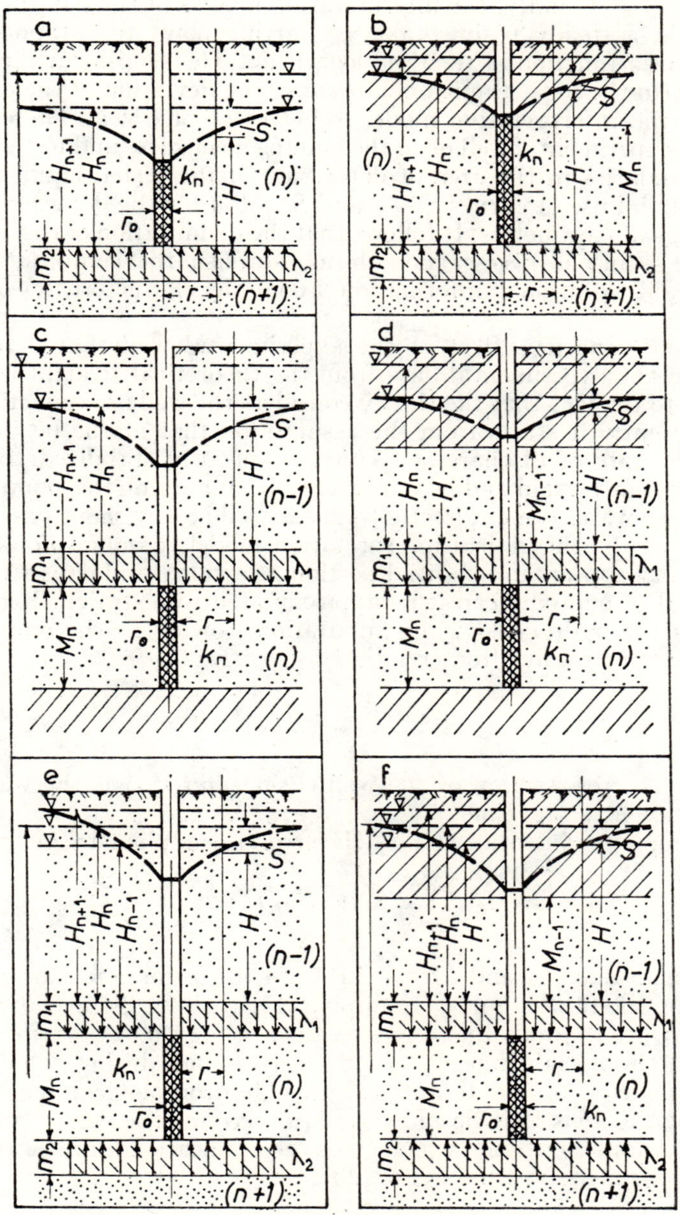

Figure 89. Leakage systems.

2.3.3 Deviation of the experimental data from the interpretation models

Usually, the interpretation models of the data resulting from pumpings in a steady régime take as starting point the graph of the level drawdown as a function of the time logarithm. The semi-logarithmic diagram is prefered (to the logarithmic one) because the deviation of the experimental data from the interpretative model is better emphasized. These deviations, due to the experimental well or to the heterogeneousness and to the presence of certain boundaries in the area of depression of the tested aquifer, decide for the only partial use of curves $S - t$ and $S - \ln t$ for the determination of the hydrogeological parameters. It follows that the delimitation of the deformations on curve $S - \ln t$ is necessary both to establish the computation interval and to obtain additional information about the characteristics of the tested aquifer.

It is the capacity effect of the experimental well that determines the deviation of the experimental data from the theoretical straight line $S - \ln t$. This is due to the fact that the theory considers the well as a punctiform source and consequently it starts from the assumption that at the beginning of the pumping the rate of discharge Q is instantaneously stabilized. But actually the wells have a certain radius and at the beginning of the pumping, a part of the pumped water comes directly from the well bore without passing through the aquifer as the theoretical interpretation models assume. For this reason, graph $S - \ln t$ appears in its initial portion as a curve and only after a certain time interval of pumping can it be approximated to a straight line. The error that affects the hydrogeological computations, owing to the omission of the "capacity" effect results from relation

$$\varepsilon = \frac{V(t)}{Q(t)} = \frac{\omega S_0}{Qt} \tag{355}$$

where ω is the cross-section of the well casing and $V(t)$ is the water volume from the borehole at time t.

Thus, for experimental well S_5-III error is supplied by relation

$$\varepsilon = \frac{0.11 \times S_0}{0.685\, t} = 0.161 \frac{S_0}{t}$$

By using values S_0 and t from Table 64, the error ε variation curve was drawn during the first 30 min of pumping (Figure 90). Some authors [4] recommend for the admissible error values of $\varepsilon \leqslant 0.05$. From Figure 90, for $\varepsilon = 5\%$, the initial time ($t_c = 2$ min) is found during which it is not possible to use, for determining the hydrogeological parameter, the interpretation models that do not take into account the water volume in the well casing. Another criterion which gives the application limits of these models include the aquifer transmissivity [20]:

$$t_0 = \frac{25 r_0^2}{T} \tag{356}$$

Figure 90. Error variation for experimental well $S_5 - III$.

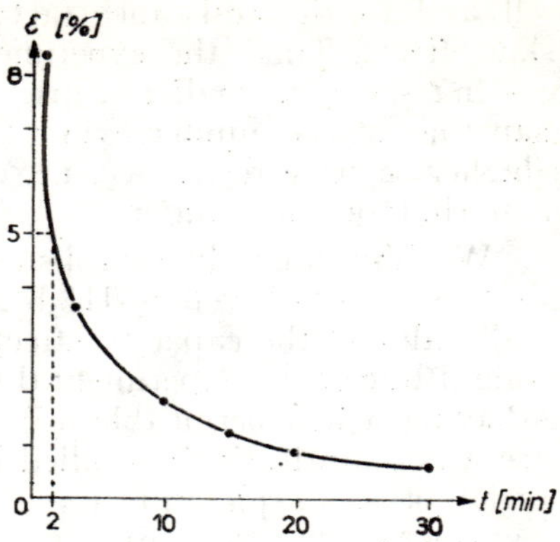

By applying this relation (established for the error $\frac{\Delta S_0}{S_0} = 0.02$) to well S_5-III, we find:

$$t_0 = \frac{25 \times 0.038}{0.30} = 3 \text{ min}; \quad t_0 = \frac{25 \times 0.038}{0.54} = 1.7 \text{ min}$$

In these relations, the transmissivity values 0.30 and 0.54 m^2/min correspond to the schemes with, as well as without leakage. From the acquired results we can conclude that in the case of well S_5-III the capacity effect is of very small importance so that its omission does not introduce gross errors into the hydrogeological computations. This effect is, however, becoming important (characterized by a high enough value of period t_0) in the case of large drawdowns of about tens of meters as well as in the case of less permeable aquifers; relation (356) shows the inversely proportional dependence between t_0 and T. In such cases, the problem can be solved in two ways:

— A long enough period of pumping in unsteady régime so that, after the initial period (t_0), the remaining period would be sufficient to permit the drawing of the rectilinear portion of graph $S - \ln t$; in this connection, it is recommended that the pumping period should be equal to at least 10 t_0.

— The use of integral methods which, by taking into account the part played by the volume of water in the borehole to the balance of the discharged rate, admit that the computation period starts from the very beginning of the pumping.

The capacity effect is equally manifested in the piezometers placed in the immediate neighbourhood of the experimental well; when this is the case the initial period (t_0) is computed by relation

$$t_0 = \frac{12.5 r_0^2}{T} \frac{S_0 + S'}{S'} \tag{357}$$

where S' is the drawdown in the piezometer. Additional information results from the nature of graph $S - \ln t$ deviations on its initial portion. These deviations from the straight line are due to the imperfection of the experimental well and to the resistance factors acting in the well and in its adjacent area (skin effect). Thus, the experimental points situated above the straight line $S — \ln t$ shows a smaller seepage resistance, therefore a greater permeability than that of the aquifer, while the experimental points below this line show a higher seepage resistance, i.e. a smaller permeability (due to the clogging) than that of the aquifer.

We have already established (see 2.3.1) that the rectilinear portion of graph $S — \ln t$ of well S_5-III begins at $t = 50$ min. Taking into account the small value of the capacity effect ($t_0 = 2—3$ min) and the location below the straight line of the experimental points (Figure 82) this well can be considered as having a less permeable adjacent area than the rest of the aquifer. This case was quantitatively studied in § 2.3.1.2., with the help of the rectilinear portion of the graph, and supplied $\zeta_0 = 9.70$. In conclusion, the initial portion as graph $S — \ln t$ can only supply qualitative information concerning the skin effect.

The deviations appearing in the extreme portion of graph $S — \ln t$ are due to some hydrogeological peculiarities of the aquifer within the limits of the influence area, to wit: presence of a recharge or impermeable boundary, variations of transmissivity, recharge by leakage etc. In the case of recharge boundary ($H =$ const.) the graph becomes approximately asymptotic to axis $\ln t$. If within the limits of the influence area there is an impermeable boundary, the graph will deviate upwards, the value of this deviation depending upon the number of impermeable limits. Likewise, the deviations of graph $S — \ln t$ will appear, if at a certain distance of the experimental group, an area is met with different hydraulic characteristics (transmissivity, piezo-conductivity) such situations can occur both in the farthest part of the graph and within it, interrupting then its linear character.

Figure 91 presents a set of theoretical curves ($S — \ln t$) corresponding each to a certain case. In the case of curve *1*, the aquifer behaves as an inifinite, homogeneous medium. In the other cases, however, variations of the angular coefficient appear as against to the corresponding value of the rectilinear portion (m_1). Thus, curve *2* corresponds to an increase of transmissivity T at a certain distance from the experimental well. The horizontal stretch in curve *3* is due to the appearance of the steady régime as a consequence of either an important dynamic resource or the existence of a leakage or of recharge boundary. If the horizontal stretch is followed by a straight line where $m_2 = m_1$ (curve *3'*), then the hydrodynamic situation will be determined by the existence of leakage from the upper or from the lower confining beds with a variable with time flow rate while the subordinate aquifer has weaker hydraulic properties than the principal aquifer. Curve *4* corresponds to a decrease of transmissivity T at a certain distance. If, at a certain distance from the experimental well, there is an impermeable boundary, the angular coefficient of the straight line increases, $m_2 = 2m_1$ (curve *4*). In curve *6* are portrayed the conditions of a difficult circulation of the ground water, generally due to the existence of several impermeable boundary.

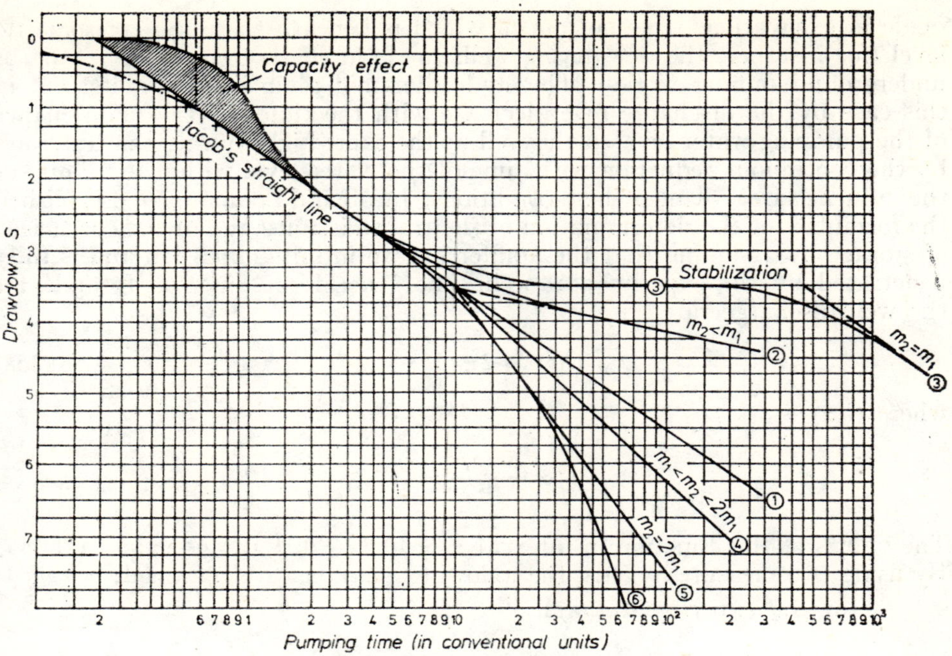

Figure 91. $S - f(\log t)$ graph. Pumping at a constant rate

Figure 91 emphasizes moreover the influence of the capacity effect as well as of the use of the logarithmic approximation — Jacob's straight line (with regard to the exact solution, i.e. Theis' curve) on the form of graph $S - \ln t$.

To conclude, the comparison of the experimental graph with these theoretical graphs presents a particular importance since it offers qualitative elements for the choice of the interpretation model.

2.4 Determination of the hydrogeological parameters in the case of free-flowing (artesian) wells

The hydrodynamic research in an unsteady régime by means of the wells can be achieved in three ways: a) with constant discharge and variable drawdown (by means of pumping); b) constant head pressure in the experimental well ($S_0 =$ const.) and variable discharge; c) by recovering of the water head after pumpings. The testing with constant yield ($Q =$ const.) is applied to the unconfined aquifers as well as to the confined aquifers with rising (but not overflowing) level. In the case of free-flowing wells, the experiment is made at a constant drawdown, the rate of discharge decreasing with time. In this category we find both the artesian flowing wells and the wells made in excavations the bottom of which is beneath the initial piezometric level (for instance, the dewatering wells made in the bottom of open cast or underground mining works). The hydrodynamic régime is characterized in this case by the discharge rate decrease with the time, by the maintenance of the pumping water level in the well at the same hight during the test, and by the continuous reduction with time of the water level within the limits of the influence area. Under these conditions, Jacob and Lohman [79] established the formulae for the determination of the transmissivity and of the coefficient of storage by assuming that the aquifer has an infinite extension and is fully penetrated by the experimental well. The general computation formula for the well discharge rate is:

$$Q(t) = 2\pi kMSG(\alpha) \tag{358}$$

where $G(\alpha)$ is

$$\alpha = \frac{kMt}{r^2 n_e} = \frac{at}{r^2}$$

The values of this function in interval $\alpha = 10^{-4} — 10^{11}$ are given in Table 65. By using two measured yields, the following relations are obtained:

— for the experimental well

$$\frac{Q(t_1)}{Q(t_2)} = \frac{G(\alpha_{01})}{G(\alpha_{01})} \tag{359}$$

— for the observation well situated at distance r from the experimental well

$$\frac{Q(t_1)}{Q(t_2)} = \frac{G(\alpha_{r1})}{G(\alpha_{r2})} \tag{360}$$

By using these two implicit relations it is possible to determine by the graphic method the value of the hydraulic diffusivity (a). The co-ordinates of this graph are $Q(t_1)/Q(t_2)$ and parameter a.

Function $G(\alpha)$ corresponding to the two times t_1 and t_2 is computed for several arbitrary values of the hydraulic diffusivity. The ratio of the two values of the function permit to locate each point on the graph. If the function value is known (from the yields ratio) the actual value of the hydraulic diffusivity can be determined on the graph. Then with the help of one of the yields the transmissivity can be computed

$$T = kM = \frac{Q(t)}{2\pi SG(\alpha)} \tag{361}$$

and afterwards the coefficient of storage

$$n_e = \frac{a}{kM} \tag{362}$$

Table 65

Values of function $G|\alpha|$

α	10^{-4}	10^{-3}	10^{-2}	10^{-1}	1	10	10^{2}	10^{3}	10^{4}	10^{5}	10^{6}	10^{7}	10^{8}	10^{9}	10^{10}	10^{11}
1	56.9	18.34	6.13	2.249	0.985	0.534	0.346	0.251	0.1964	0.1608	0.1360	0.1177	0.1037	0.0927	0.0838	0.0764
2	40.4	13.11	4.47	1.716	0.83	0.461	0.311	0.232	0.1841	0.1524	0.1299	0.1131	0.1002	0.0899	0.0814	0.0744
3	33.1	10.79	3.74	1.477	0.719	0.427	0.294	0.222	0.1777	0.1479	0.1266	0.1106	0.0982	0.0883	0.0801	0.0733
4	28.7	9.41	3.30	1.333	0.667	0.405	0.283	0.215	0.1733	0.1479	0.1244	0.1089	0.0968	0.0872	0.0792	0.0726
5	25.7	8.47	3.00	1.234	0.630	0.389	0.274	0.210	0.1701	0.1449	0.1227	0.1076	0.0958	0.0864	0.0785	0.0720
6	23.5	7.77	2.78	1.160	0.602	0.377	0.268	0.206	0.1675	0.1426	0.1213	0.1066	0.0950	0.0857	0.0779	0.0716
7	21.8	7.23	2.60	1.103	0.580	0.367	0.263	0.203	0.1654	0.1393	0.1202	0.1057	0.0943	0.0851	0.0774	0.0712
8	20.4	6.79	2.46	1.057	0.562	0.359	0.258	0.200	0.1636	0.1380	0.1192	0.1049	0.0937	0.0846	0.0770	0.0709
9	19.3	6.43	2.35	1.018	0.547	0.352	0.254	0.198	0.1621	0.1369	0.1184	0.1043	0.0932	0.0842	0.0767	0.0706
10	18.3	6.13	2.25	0.985	0.534	0.346	0.251	0.196	0.1608	0.1360	0.1177	0.1037	0.0927	0.0838	0.0764	0.0704

As in case of the pumpings with $Q=$ const. the hydraulic diffusivity computed by means of the data supplied by the experimental well includes its seepage resistance too and is defined by relation (313).

After a long working period of the experimental well $\left(\alpha \geqslant 10 \text{ i.e. } t \geqslant 10 \frac{r_0^2}{a}\right)$, function $G(\alpha)$ can be approximated by means of expression

$$G(\alpha) \approx \frac{2}{\ln(2.25\alpha)} \tag{363}$$

which, introduced in relation (356), supplies

$$Q(t) \approx \frac{4\pi T S_0}{2.3 \log(2.25 T t / r_0^2 n_e)} \tag{364}$$

By setting the terms in order we obtain relation

$$\frac{S_0}{Q(t)} = \frac{2.3}{4\pi T} \log \frac{t}{r_0^2} + \frac{2.3}{4\pi T} \log \frac{2.25\, T}{n_e} \tag{365}$$

that defines a straight line on a graph on which the arithmetic values of variable $S_0/Q(t)$ are marked on the ordinate and the logarithmic values of variable t/r_0^2 are marked on the abscissa. Transmissivity T is computed with the help of the expression of the straight line angular coefficient:

$$\frac{\Delta[S_0/Q(t)]}{(\log t/r_0^2)} = \frac{2.3}{4\pi T} \tag{366}$$

$$T = \frac{2.3\Delta(\log t/r_0^2)}{4\pi\Delta[S_0/Q(t)]} \tag{367}$$

The computation of the coefficient of storage can be simplified by taking into account the point of intersection of the straight line with the abscissa (where $S_0/Q(t) = 0$); in this case relation (364) becomes

$$n_e = 2.25\, T(t/r_0^2)_0 \tag{368}$$

Of course, when drawing the graph, the points from the first period of the test will not be taken into consideration since they are not rectilinear; their position with regard to the straight line gives only a qualitative estimation of the permeability in the vicinity of the experimental well. This grapho-analytic method can equally be used in case of the observation well.

For determining the drawdown (S) in the aquifer (except in the experimental well where $S = S_0 =$ const.) in the case of important testing periods $\left(\alpha \geqslant 500 \text{ i.e. } t \geqslant 500 \frac{r^2}{a}\right)$ the approximate formula may be used

$$S \approx \frac{Q(t)}{4\pi T}\left[-Ei\left(-\frac{r^2}{at}\right)\right] \tag{369}$$

which is identical with that used when testing is made at $Q=$ const.

The law of the discharge rate variation with time offers certain qualitative information about the homogeneity and extension of the aquifer. For instance, a linear dependence ($Q = f(t)$) shows the existence of a homogeneous aquifer of very great extension. If within the limits of the influence zone there are impermeable boundaries and considerable variations of the aquifer thickness, the linear dependence is supplied by correlation $\log Q = f(\log t)$.

2.5 Determination of the hydrogeological parameters by means of the water level recovery curve

After an interruption of the pumping or of the free discharge from a well, the aquifer tends to recover its initial equilibrium. The water level both in the experimental well and in the cone of depression is recovered so as to reach the initial piezometric level. The interpretation of the recovery curve of the water pressure both in the experimental well and in the observation wells allows us to obtain particularly useful qualitative and quantitative informations about the storage capacity of the aquifer as well as about the hydraulic state of the experimental (or observation) well, including its adjacent area.

The quantitative interpretation of the water level recovery curves offers a particular interest for the hydraulics of the oil and gas deposits; this problem has recently started to be dealt with in the hydrogeological literature. Usually the hydrogeologic documentation gives these recovery curves without however interpreting them. But this graph processing offers important data concerning the increase of the technical efficiency of the hydrogeological experiments.

The accuracy of the water level recovery curve is determined by the frequency of the observations; taking into account that the recovery rate decreases with time, the following recommendations are made concerning the time intervals between measurements:

— one-minute intervals during the first ten minutes;
— five-minute intervals during the following 15 min;
— ten-minute intervals during the following 30 min;
— 30-minute intervals during a period determined by the recovery rate;
— hourly intervals afterwards.

Marking — on the graph — drawdown during recovery on the ordinate and the time (in min) on the abscissa, the primary diagram results, i.e. the water level recovery curve.

Of late, the hydrodynamic methods of research of the oil-bearing formation on the basis of the field pressure recovery curve were amply developed and improved. Their application in an adapted form to the study of the aquifers by means of wells requires some additional investigations for verifying the accuracy and the theoretical premises.

.1 Criterion for choosing the method of data interpretation

When a qualitative analysis is performed, two segments can be distinguished on a recovery curve; the initial segment which is very inclined, almost subvertical; a slightly inclined segment having its extremity asymptotic to the time axis (Figure 92).

The strongly inclined segment is due to a quick recovery of the water level in the well (that does not entail an instantaneous closing) as well as to its adjacent area the permeability of which can be modified. Owing to these causes the effects of which are cumulative, this segment does not supply significant informations. The segment with a slight slope corresponds to the recovery of the initial equilibrium in the whole depression area created by pumping and especially at the remote points from the experimental well. When this segment is short, the aquifer possesses a good permeability as well as a noticeable recharge (Figure 92, curve *a*). When the inflexion point is low, the slightly inclined segment is very long, the permeability or the recharge of the aquifer is small (curve *c*).

The data processing and the computation of the parameters using the level recovery curve are made by the formulae of the elastic régime and with the additional help of the superposition principle. In this case, it is assumed that during the pumping period (t_0) the actual source (well) worked with $+Q$ discharge rate. It is supposed that this actual source supply the same yield after the pumping is stopped (in the level recovery period t'), so that it becomes necessary to introduce an absorption well with $-Q$ discharge rate which is working during time t' (after the pumping is stopped). This avsorption well has to be introduced into the formula owing to the fact that during period t' the actual source (the experimental well) does not supply water as was assumed, but gets an inflow of intensity $+Q$. By applying the superposition

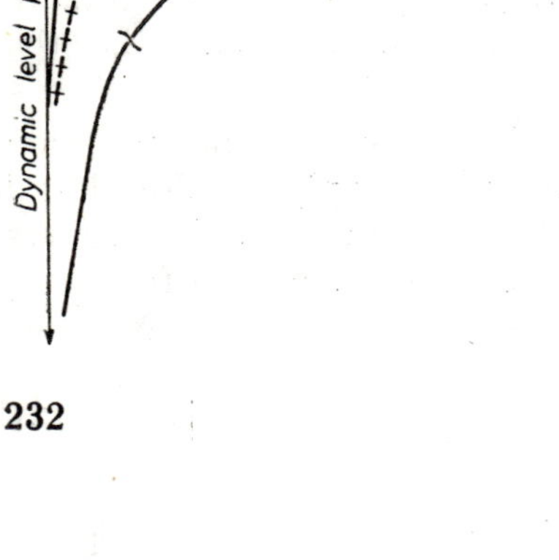

Figure 92. Curves of the water level recovery in the well (after Castany).

principle to both wells, we achieve the following fundamental relation for a fully penetrating perfect well in a confined aquifer with an infinite extension:

$$S_{t_0+t'} = S_{t_0} - \Delta H = \frac{Q}{4\pi kM}\left[-Ei\left(-\frac{r^2}{4a(t_0+t')}\right) + Ei\left(-\frac{r^2}{4at'}\right)\right] \quad (370)$$

where S_{t_0+t} is the level reduction or the residual drawdown at time t' after the pumping is stopped; t_0 is the pumping period; t the level recovery time counted from the pumping halt; S_{t_0} the drawdown at the pumping halt; ΔH the increase of the water level during time t'; r the distance of the computation point from the experimental well (in case of the experimental well $r = r_0$).

For values $\frac{r^2}{4at} < 0.1$ of the argument, the above relation can be approximated by the logarithmic relation (see 2.3.1):

$$S_{t_0+t'} = \frac{Q}{4\pi kM}\ln\left(1+\frac{t_0}{t'}\right) = \frac{0.183\,Q}{kM}\log\left(1+\frac{t_0}{t'}\right) \quad (371)$$

Since the radius (effective or reduced) of the experimental well is not included in this relation, it can be applied to both the fully and partially penetrating wells.

For determining the transmissivity, the graphical representation of relation (371) is used with co-ordinates $S_{t_0+t'}$ and $\log\left(1+\frac{t_0}{t'}\right)$ or $H_{t_0+t'}$ and $\log\left(1+\frac{t_0}{t'}\right)$, since $S_{t_0+t'} = H - H_{t_0+t'}$. After a certain time interval, which is a function of the capacity effect and of the zone with altered permeability, the observed drawdowns are situated along a straight line. Since the value of angular coefficient m is given by the graph, the transmissivity will be

$$T = kM = \frac{0.183\,Q}{m} \quad (372)$$

In connection with the deviation of the experimental points from the straight line during the first period of the level recovery, the time in which the capacity effect occurs can be computed by relation (356). On the basis of the research by means of the electrical models, Miller C. C. [9] draws the following conclusions about the estimation of the altered permeability area (skin effect) around the experimental well:

1. The greater the difference of permeability between adjacent area k_m and aquifer k, the larger is the deviation of the experimental points from the straight line;
2. When $k_m > k$, the initial portion of the graph is above the straight line;
3. When $k_m < k$, this initial portion is below the straight line.

As to the accuracy of the parameter determination with the help of the graph, V. N. Schelkatchev [57] showed that the presence of the zone with an altered permeability (therefore a non-homogeneous aquifer) complicates the deduction of the computation formulae characterizing the variation process of the pumping water level. It can, however, be considered the results so obtained are best when ratios k_m/k and r_m/r_0 tend to the value. These remarks are only valid for the experimental well; on the other hand, it is considered that the transmissivity computed by means of the head recovery straight line is accurate enough since its angular coefficient is not influenced by the altered permeability zone around the experimental well. For a high value of time t', the water level in the well is restored up to its initial position. On the graph, for $t' \to \infty$, quantity $\log\left(1+\frac{t_0}{t'}\right)$ tends to zero. For this reason, the straight line can be extrapolated until its intersection with the ordinate, where $\log\left(1+\frac{t_0}{t'}\right) = 0$, which supplies the initial head (H) of the aquifer.

The shape and the heterogeneity of the aquifer influence the graph form of the level recovery in its final portion (Figure 93). The observed curve will deviate from the theoretical straight line of the "infinite" and homogeneous aquifer for the high values of quantity $\log\left(1+\frac{t_0}{t'}\right)$, as a function of the distance from a boundary or of a change of transmissivity. An increase of the observed curve slope shows the influence of an impermeable boundary or a diminution of transmissivity, while a decrease of the slope shows an increase of transmissivity or, if the slope vanishes, the existence of a recharge boundary. In case the water level recovery after a long pumping $(t_0 \gg t')$ the reduction rate of the water level at the end of the pumping is

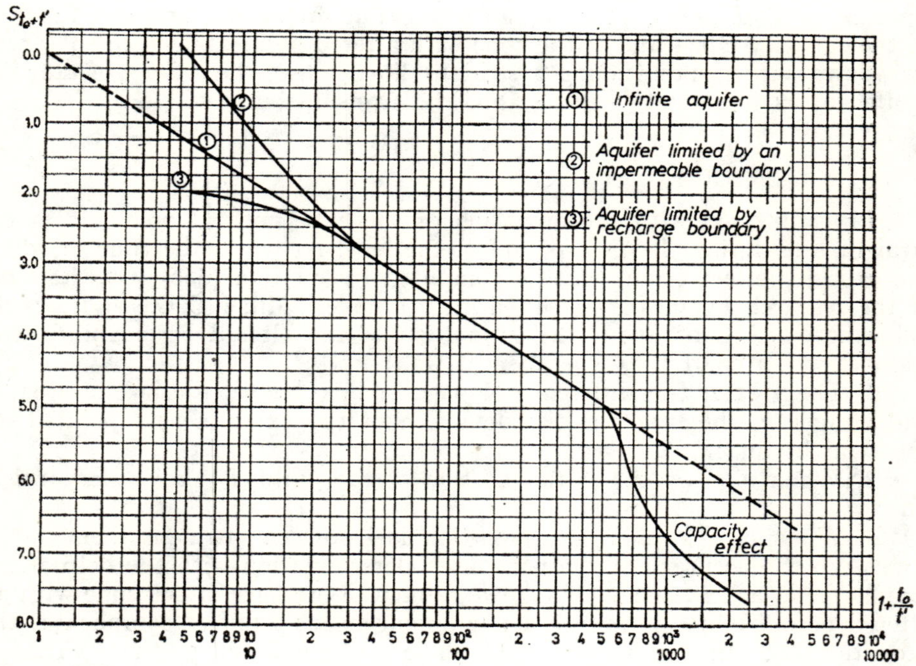

Figure 93. Theoretical curves of the water level recovery in a well (after Forkasiewicz, 1969).

very small in regard to the level recovery rate so that in equation (370) the variation of quantity $-Ei\left(\frac{r^2}{4a(t_0 + t')}\right)$ can be neglected and therefore

$$Q/4\pi T\left[-Ei\left(-\frac{r^2}{4a(t_0 + t')}\right)\right] = S_{t_0+t'}$$

becomes a constant representing the drawdown at the end of the pumping while the relation takes the form

$$\Delta H = S_{t_0} - S_{t_0+t'} = -\frac{Q}{4kM} Ei\left(-\frac{r^2}{4at'}\right) \approx \frac{Q}{4\pi kM} \ln \frac{2.25\,at'}{r^2} \tag{373}$$

From this relation it follows that in the case of long periods of pumping (in steady flow), the water level restoration follows the same law as in the case of lowering during the pumping; therefore, for the processing of the experimental data the method presented in the previous paragraphs can be used by simply substituting of drawdown S, by the level increase (ΔH) with regard to the water level at the end of the pumping, time t' being only considered from the pumping stopage. Logarithmic relation (373) represents the equation of straight line, the co-ordinates of the graph being ΔH and $\log t'$:

$$\Delta H = 0.183\,\frac{Q}{kM} \log \frac{2.25\,a}{r^2} + 0.183\,\frac{Q}{kM} \log t' \tag{374}$$

$$\Delta H = A + m \log t' \tag{375}$$

Since we know the value of the angular coefficient

$$m = \frac{(\Delta H)_2 - (\Delta H)_1}{\log t_2' - \log t_1'} \tag{376}$$

we can compute the transmissivity

$$kM = \frac{0.183\,Q}{m} \tag{377}$$

In this case, the expression of the complex parameter is

$$\log \frac{a}{r^2} = \frac{A}{m} - 0.35 \tag{378}$$

and the expression of the piezoconductivity is

$$\log a = \frac{A}{m} + 2 \log r - 0.35 \tag{379}$$

On this graph also the linear portion only begins at a certain value of the time owing to the same causes, namely the capacity effect and the structure variations of the aquifer in the adjacent zone of the experimental well.

Particularly interesting are the conclusions drawn by Bocever and Orfanidi who, by applying expression (373), computed the hydrogeological parameters k_M, α and ζ_0 for a geological structure with confined aquifers by using the ırement data supplied by both the experimental and the observation The experimental wells were as a matter of fact capture wells having worked uninterruptedly 1.5 — 2 years. On the other hand, let us note the long period of observations about the water level recovery — over 30 days —, which permitted us to draw a longer linear portion of the graph. The practical conclusions are:

— The transmissivity computed with the recovery data in the experimental well and in the observation wells has practically the same value which shows that the non-homogeneousness supplied by the adjacent zone with altered permeability k_m has no influence upon the slope of the linear portion of the graph implicitly upon the transmissivity.

— For the computation of the hydraulic conductivity (a) the observation wells data should be exclusively used with relation (379).

— When the transmissivity and the hydraulic diffusivity are known, relation (373) for the experimental well with the altered adjacent area can be written in the form

$$\Delta H = \frac{Q}{4\pi k M}\left(\ln \frac{2.25\, at'}{r_0^2} + 2\zeta_0\right) \tag{380}$$

thence

$$\zeta_0 = \frac{2\pi k M \Delta H}{Q} - \ln \frac{1.5\sqrt{at'}}{r_0} \tag{381}$$

The global parameter of the seepage resistance (ξ_0) represents the algebraic sum of the two imperfections: the opening degree of the aquifer (ξ^1) (which is always positive) and the kind of opening (ξ_2) (which can be either positive or negative).

The above-mentioned results warrant the periodic checking (by the method of the water level recovery in the well) of the productivity of the capture or drainage wells working during long periods. Even in the case that the investigated well is working in a interfering system, the results are not influenced provided that the stopping period of the well is short enough.

In order to solve quantitatively the problem, two wells should be stopped simultaneously, one of them playing the part of a piezometer.

In the case of a long pumping time, the hydraulic diffusivity can be determined by the point of maximum hydraulic head within the observation well. Owing to the elastic properties of the aquifer, the head variation (by stopping the pumping) is not instantaneously transmitted to the observation well, but only after a certain time interval (t_1), when the water level begins to grow (Figure 94). If this moment of a maximum drawdown (t_m) in the observation well is known and the pumped yield is constant, the hydraulic diffusivity will be expressed by [13]:

$$a = \frac{r^2 t_0}{4(t_{max} - t_0)\, t_{max} \ln \dfrac{t_{max}}{t_{max} - t_0}} \tag{382}$$

Figure 94. Variation of the drawdowns in an observation well.

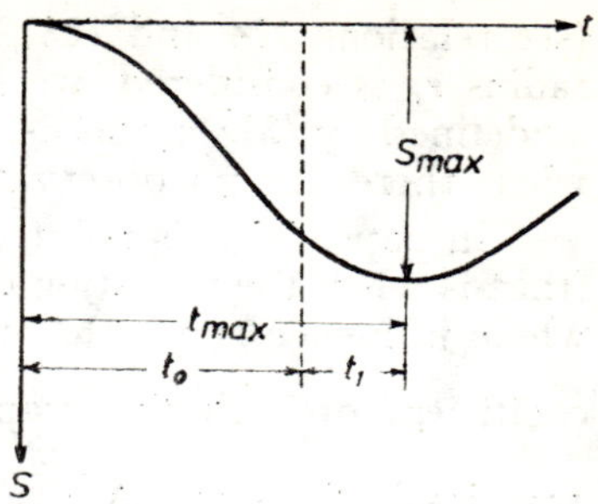

This simple formula offers several advantages:

— the influence of the capacity effect in the experimental well is quite small:

— parameter a characterizes the aquifer between the experimental and the observation well:

— the maximum drawdown point is readily determined by means of the graph.

From the foregoing, there results that, in the case of long periods of pumping, the data interpretation is somewhat easier since at the end of pumping period (t_0), the variation process of the hydraulic head within the whole influence area is practically finished. This type of testing includes first of all the exploitation wells that can be periodically stopped during short intervals for establishing the recovery curve of the piezometric level. The same method of interpretation may however be also used in the case of exploration wells. As a rule, when the pumping period is a good deal longer than the level recovery period ($t_0 \gg t'$) it is considered as a long period pumping. In paper [4] this condition is expressed more clearly by $t' < 0.1\ t_0$; this very stringent condition can sometimes be met with equally in the case of the experimental pumpings with $t_0 = 2$—4 days (when the steady-state flow is reached in the experimental group), the linear portion of the recovery graph begins very shortly after the pumping is stopped. In the case of aquifers with a high permeability and very small ζ_0 parameters, this interval can reach the order of minutes. It is to be emphasized that condition $t_0 \gg t'$ should be appreciated as a function of the considered case characteristics: the hydraulic properties of the aquifer, the seepage resistance of the well, the beginning of the linear portion of the recovery graph.

In case of the experimental well, it is frequently difficult to separate parameters a/r^2; that is why it is recommended to be taken as such. So as to separate hydraulic diffusivity of the aquifer instead of the actual well radius r_0, the reduced well radius r'_0 should be used (see relation 146) which is a function of parameter ζ_0; but, for computing this parameter the data of an observation well will be necessary. An approximate solution is however available by estimating the hydraulic diffusivity supplied by laboratory data lations 124 and 128), from which quantity r_0' is determined. If actual r_0 is considered, we shall obtain the reduced hydraulic diffusivity a_0 as defined by (313); this is, as a matter of fact, the most suitable method when there are no observation wells.

In paper [9] it is pointed out that in the case of long periods of pumping (that is when the variation of the hydraulic head is practically finished in the whole influence zone) the transmissivity computed by means of the angular coefficient of both the graphs $\Delta H - \log t'$ and $S_{t_0+t'} - \log\left(1+\frac{t_0}{t'}\right)$ must be practically the same. A great difference between the two values of the angular coefficient indicates that condition $t_0 \gg t'$ is not fulfilled and therefore that at the end of the pumping no steady or quasi-steady régime is achieved within the influence zone.

If at the end of pumping the régime is unsteady, the hydraulic diffusivity will be supplied by relation (370) by making use of two measurements on the linear portion of the graph

$$\frac{S_{t_0+t_2'}}{S_{t_0+t_1'}} = \frac{-Ei\left[-\frac{-r^2}{4a(t_0+t_2')}\right] + Ei\left[-\frac{-r^2}{4at_2'}\right]}{-Ei\left[-\frac{r^2}{4a(t_0+t_1)}\right] + Ei\left[-\frac{r^2}{4at_1'}\right]} \tag{383}$$

For the graphical solution of this implicit relation, parameter a should be determined first. The relation has no mathematical solution in case of the experimental well ($r = r_0$) since the very small values of the argument (10^{-7}—10^{-8}) practically exclude the dependence on the hydraulic diffusivity.

All these methods assume an instantaneous closing of the well when the pumping is stopped. Actually however, for a certain time afterwards, the water continues to flow in, which influences the recovery curve of the water level along its initial portion. The importance of this capacity effect in the case of deep wells have determined the elaboration of interpretation methods that take into account the water inflow into the well hole after the pumping is stopped. These methods involve a large volume of computation, and are therefore only used in oil and gas exploitation.

There are several possible ways of making a qualitative and quantitative estimation of the state of an experimental well adjacent zone by means of the level recovery curve. For instance, the position of the experimental points in regard to the extrapolated linear portion supplies a qualitative indication about the ratio between the actual permeabilities of the aquifer (k) and of its adjacent zone (k_m). It should be pointed out that the hydraulic conductivity of the aquifer can be obtained on the basis of the linear portion of the graph, independently of the structure alterations occurring in the adjacent zone of the well. Another qualitative indication is supplied by the ratio between hydraulic conductivity k_0 computed by Darcy's relation and hydraulic conductivity k resulting from the level recovery graph by assuming that parameter k_D represents the mean value both of the unaltered aquifer and its adjacent zone:

$$k_D = \frac{0.366\ Q \log R/r}{MS_0} \tag{384}$$

$$k = \frac{0.183Q}{M.m} \tag{385}$$

where m is the angular coefficient of the straight line on the graph and S_0 is the drawdown at the end of the pumping.

Another parameter for estimating the adjacent zone is the productivity rate:

$$P = \frac{Q_d}{Q_0} = \frac{k_D}{k} = \frac{2m \log R/r_0}{S_0} \tag{386}$$

This is a very useful parameter because it also expresses the ratio between the yield of the well (Q_d) which includes degraded zone and the yield (Q_0) of the same well without this zone. The productivity rate of a well can be:

$P = 1$ when there is not an altered permeability zone;
$P < 1$ when a contaminated zone (skin effect) exists;
$P > 1$ in the case of natural development operations.

Another quantitative criterion for the estimation of the structure modifications in the adjacent zone is offered by the reduced radius of the well (r_0') which results from the expression of the water inflow to the well

$$Q = \frac{kMS_0}{0.366 \log R/r_0'} \tag{387}$$

$$\log r_0' = \log R - \frac{kMS_0}{0.366Q} = \log R - \frac{S_0}{2m} \tag{388}$$

From relation (146) the computation formula is deduced for the seepage resistance due to the character of the aquifer opening

$$\zeta_2 = 2.3(\log r_0 - \log r_0') \tag{389}$$

As an example, we worked out the curve of the water level recovery in an exploration well. The tested aquifer is a confined one within a Dacian coal deposit and its initial head H is 65.0 m H_2O. The well was fully penetrated and the aquifer thickness was $M = 9.5$ m. To determine the parameters, we drew the evolution graph of the hydrodynamic elements in the periods of the pumping and of the water level recovery (Figures 95 and 96). From this graph it follows that up to $t' = 25$ hours, the experimental points are not collinear since they are situated above the straight line. According to relation (355), the time in which the capacity effect occurs is

$$t_0' = \frac{\omega S_0}{Q} = \frac{3.14 \times 0.1^2 \times 7.0}{6.4 \times 0.05} = 0.7 \text{ hours}$$

'e can therefore consider that the deviation of the experimental points 'nitial portion of the graph (Figure 96) is chiefly due to the adjacent .vith altered permeability; the position of these points above the straight line means that $k_m > k$ owing to the natural development. The effective permeability of the aquifer is

$$k = \frac{0.183Q}{m.M} = \frac{0.183 \times 154}{2.5 \times 9.5} = 1.2 \text{ m/day}$$

The angular coefficient ($m = 2.5$) was computed for the interval of 25 — 60 hours during which the points are practically collinear.

In this example, the pumping time was $t_0 = 95$ hours, but in the computations we take into account the reduced time (t_{0r}) because of the yield variations during this time. In such cases, Horner [9] recommends for t_{0r} the following calculation

$$t_{0r} = \frac{V}{Q_0} \tag{390}$$

where V is the total volume of the pumped water and Q_0 is the discharge at the end of the pumping.

Since according to the interpretation method the yield should be constant during the pumping, no interruptions can be admitted during this period (t_0) lest the value of the hydraulic conductivity should be affected by errors.

The extension of the straight line in Figure 96 leads to the determination of the initial head which coincides with the measured one, i.e. $H = 65$ m H_2O. It is also to be remarked that the initial ordinate does not coincide with the water level measured at the pumping end. The hypothetical drawdown on the graph (5.9 m) smaller than the measured one in the well ($S_0 = 7$ m) proves the existence around the well of an area with an increased permeability.

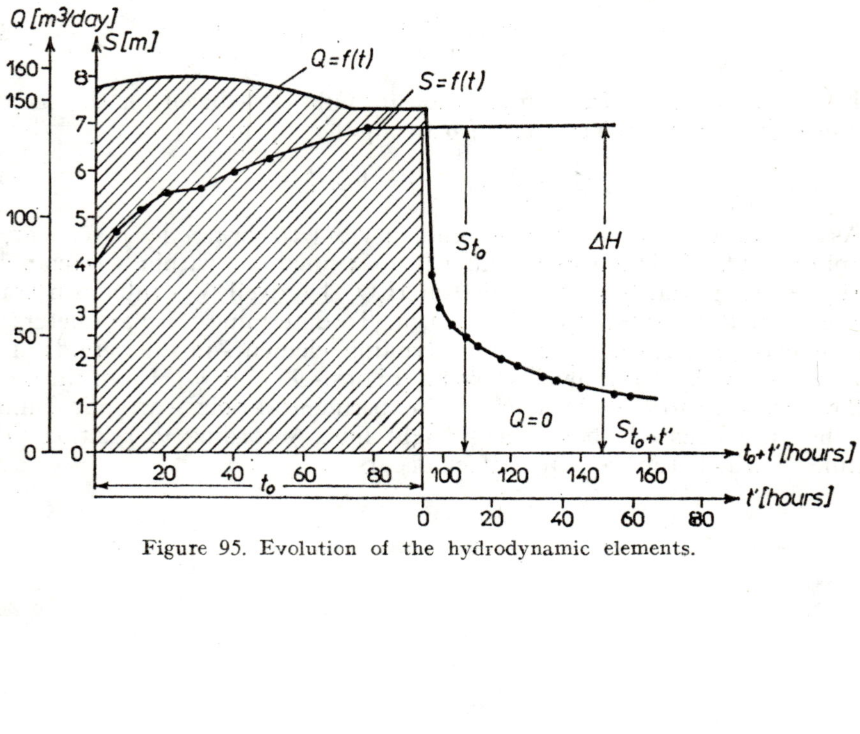

Figure 95. Evolution of the hydrodynamic elements.

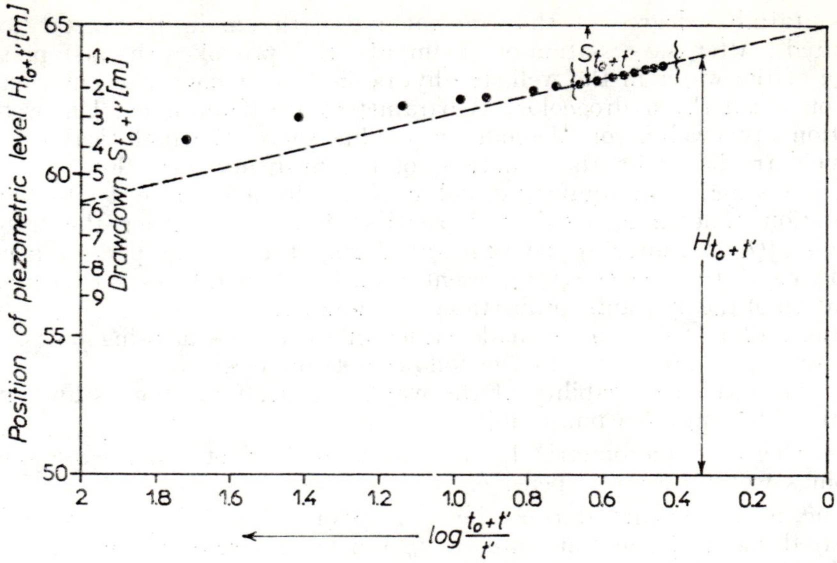

Figure 96. $S - \log \frac{t_0 + t'}{t'}$ graph

So as to ascertain condition $t \gg t'$, the angular coefficient of the straight ine was determined on a graph ΔH — log t' that supplied the value $m = 1.83$. Since this value differs from that determined on the graph (Figure 96) with co-ordinates $S_{t_0+t'}$ and $\log \left(1 + \frac{t_0}{t'}\right)$ it follows that this condition is not fulfilled i.e. at the end of the pumping the flow régime was only practically steady within a limited radius around the experimental well, therefore insufficient to fulfil condition $t_0 \gg t'$. The values of the angular coefficient on the two graphs only coincide when the flow becomes steady in the whole influence area.

2.5.2 The bailing method

The expense reduction of the experimental works and the necessity of individual testing of all aquifers even during the drilling led to the elaboration of quick estimation methods of the hydrogeological parameters, based on the measurements of the water level recovery in the well after repeated bailings. Owing to the short discharging period the extension of the depression area around the well is very reduced. This also appears in the representativity degree of the transmissivity (the main parameter to be computed). After the cessation of the impulse that provoked the fast pressure change of the water in the well, the hydraulic head recovery occurs, on the basis of which the hydrogeological parameters are determined. Under these conditions, this well is considered a vertical line source the analytical solutions of which are based on the equations of the unsteady-state flow.

The testing of an aquifer by bailing is usually made in order to acquire information concerning yielding capacities; however, neither the rate of discharge (Q) (assumed to be constant during the bailing period) nor the specific capacity ($q = Q/S_0$) represents conclusive quantities as concerns the estimation of the hydraulic properties of the aquifer.

The bailing test can be made either in casing or uncasing holes. This procedure is recommended in the following two cases:

— when the permeability of the water bearing formation is low and it is justified to employ a pump unit;

— when the piezometric level is at a great depth and experimental pumpings would be too expensive.

The measurements during the test include both the volume of the evacuated water (V) in time interval t_0 and the residual drawdowns during period t' of the level recovery. In the case of an instantaneous discharge of the well by a single bailing, the residual drawdown is supplied by the following relation proposed by E. Skibitzke [77]

$$S_r = \frac{V}{4\pi T \cdot t'} e^{-\frac{r_0^2}{4at'}} \tag{391}$$

The following transformations are required by the graphical representation of this equation in the form of a straight-line:

$$t'S_r = \frac{V}{4\pi T} e^{-\frac{r_0^2}{4at'}} \tag{392}$$

$$\ln (t'S_r) = \ln \frac{V}{4\pi T} - \frac{r_0^2}{4a} \cdot \frac{1}{t'} \tag{393}$$

$$\log (t'S_r) = \log \frac{V}{4\pi T} - \frac{r_0^2}{1.73\, a} \cdot \frac{1}{t'} \tag{394}$$

On a graph with co-ordinates $\log(t'S_r)$ and $\frac{1}{t'}$, relation (394) is represented by a straight line; the expressions of angular coefficient m and of initial ordinate A permit to determine the hydrogeological parameters

$$m = \frac{r_0^2}{1.73\, a} \tag{395}$$

$$A = \log \frac{V}{4\pi T} \tag{396}$$

In relation (392), since r_0 is very small in regard to time t, exponent $r_0^2/4\,at'$ is nearly equal to zero and consequently term $e^{r_0^2/4at}$ is nearly equal to 1; therefore in the case of a bailed well, this relation is simplified as follows

$$S_r = \frac{V}{4\pi T} \cdot \frac{1}{t'} \tag{397}$$

which represents a straight line (in co-ordinates S and $1/t'$) passing through the origin so that with the help of the angular coefficient we can determine the transmissivity

$$m = \frac{V}{4\pi T} \tag{398}$$

The main error affecting the above method is due to the fact that the usual bailing is not an instantaneous discharge since it lasts several hours. That is why the following changes have to be made in general relation (370):

$$S_r = \frac{Q}{4\pi T} \ln \frac{t}{t'} \tag{399}$$

If the bailing period is t_0 and if t_n represents the time measured at the middle of interval t_0, the following changes must be applied

$$Q = \frac{V}{t_0}; \quad t = t_n + \frac{t_0}{2}; \quad t' = t_n - \frac{t_0}{2} \tag{400}$$

The substitution of these quantities in relation (399) results in

$$S_r = \frac{V/t_0}{4\pi T} \ln \left[\frac{t_n + t_0/2}{t_n - t_0/2}\right] \tag{401}$$

$$S_r = \frac{V/t_0}{4\pi T} \ln \left[\frac{(2t_n/t_0) + 1}{(2t_n/t_0) - 1}\right] \tag{402}$$

With notation $n = 2t_n/t_0$, the logarithmic term can be expressed as a series

$$\ln \left(\frac{n+1}{n-1}\right) = 2\left[\frac{1}{n} + \frac{1}{3n^3} + \frac{1}{5n^5} + \ldots\right] \tag{403}$$

$$\ln \left[\frac{(2t_n/t_0) + 1}{(2t_n/t_0) - 1}\right] = 2\left[\frac{t_0'}{2t_n} + \frac{t_0^3}{3(2t_n)^3} + \frac{t_0^5}{5(2t_n)^5} + \ldots\right] \tag{404}$$

In practice t_0 can be considered smaller than t_n so that only the first term of the series has an appreciable value, while the other terms can be neglected. The substitution of value t_0/t_n in relation (399) will result in

$$S_r = \frac{V/t_0}{4\pi T} \cdot \frac{t_0}{t_n} = \frac{V}{4\pi T} \cdot \frac{1}{t_n} \tag{405}$$

The difference with regard to relation (397) is due to the use of time t_n which is greater than t' by $t_0/2$.

A higher accuracy of the method can be achieved by taking into account each bailing tube within the cycle (repeated bailings with discontinuous discharge):

$$S_r = \frac{V_1}{4\pi T t_1} + \frac{V_2}{4\pi T t_2} + \dots + \frac{V_n}{4\pi T t_n} \tag{406}$$

$$S_r = \frac{1}{4\pi T}\left[\frac{V_1}{t_1} + \frac{V_2}{t_2} + \dots + \frac{V_n}{t_n}\right] \tag{407}$$

where V_n is the water volume evacuated during each bailing cycle and t_i is the time interval between the bailing time and the time of the observed residual drawdown (S_r).

For a uniform evacuated volume at each bailing we have

$$S_r = \frac{V}{3\pi T}\left[\frac{1}{t_1} + \frac{1}{t_2} + \dots + \frac{1}{t_n}\right] \tag{408}$$

$$S_r = \frac{V}{4\pi T}\sum_{i=1}^{n}\frac{1}{t_i} \tag{409}$$

On a graph bearing S_r on the ordinate and $\Sigma \frac{1}{t_i}$ on the abscissa the straight line should pass through the origin; transmissivity T will be computed with the help of angular coefficient $m = V/4\pi T$.

In the case of confined aquifers of infinite or semi-infinite thickness (§ 2.6.2) the hydraulic conductivity can be estimated by relation [70]:

$$k = \frac{0.5\, r_0\, 2 \ln \dfrac{1.47\, l}{\zeta r_0}}{l \,.\, t'} \ln \frac{S_0}{S_r} \tag{410}$$

where S_r is the residual drawdown at time t', S_0 is the drawdown at the end of the pumping ($t' = 0$), l is the screen length, and ζ a coefficient depending on the screen position; $\zeta = 1$ for the infinite aquifer and $\zeta = 2$ for the semi-infinite aquifer (for intermediate positions the value of ζ can be interpolated).

Relation (410) was established on the assumption that, in case of the water level recovery after a short time pumping, the flow to the well is expressed by the equation of the steady régime. So as to ensure a suitable accuracy, the relation can be graphically solved with log S_0/S_r on the ordinate and t' on the abscissa

$$\log S_0/S_r = \frac{lk}{2.65\, r_0^2 \log \dfrac{1.47\, l}{r_0}}\, t' = mt' \tag{411}$$

Of course, in both the initial and the final portions of the graph, certain experimental points will deviate from the straight line owing to the alterations in the well adjacent area and to certain variations within the aquifer itself.

If the angular coefficient m is known, we can determine the hydraulic conductivity

$$k = \frac{2.65\ m\ r_0^2 \log \frac{1.47\ l}{r_0}}{l} \qquad (412)$$

In the case of an unconfined aquifer, the same relations can be used without any change for a well with a flooded screen (completely under the pumping water level). If the screen is not flooded (partially below the pumping level), its active length is

$$l = l_0 - 0.5(S_0 + S_r)$$

This computation method can also be applied during the drilling process when the drill-hole can remain without casing along a small thickness ($l \leqslant 0.1\,M$).

In a fully casing borehole where the water admission takes place only through the bottom, the hydraulic conductivity is supplied by relation

$$k = \frac{1}{4} \frac{r_0}{t'} \ln \frac{S_0}{S_r} \qquad (413)$$

$$\log \frac{S_0}{S_r} + \frac{1.73\ k}{r_0} t' = mt' \qquad (414)$$

The graphical method supplies a suitable value of the hydraulic conductivity.

As an example, the well 2H-Şotînga was considered from which a bailing was made for testing the coal I—II aquifer (Figure 97). Angular coefficient m of the straight line in the graph (Figure 98) being 16.36, it follows that the transmissivity (conformably to relation 372) is

$$T = kM = \frac{0.183\,Q}{m} = \frac{0.183 \times 1152}{16.36} = 12.88\ \text{m}^2/\text{day}$$

According to the procedure established by E. Skibitzke the graph with co-ordinates $\log(S_r \times t')$ and $1/t'$ will supply the value of the initial ordinate $A = \log V/4\pi T = 1.61$, then

$$\frac{V}{4\pi T} = 40.7\,; \quad T = \frac{240}{12.56 \times 40.7} = 0.47\ \text{m}^2/\text{hour} = 11.3\ \text{m}^2/\text{day}$$

The quite close values so obtained show how very reliable the applied methods were.

Since now the angular coefficient of the straight line is known, we can determine the complex parameter $r_0'^2/a$, according to relation (395):

$$\frac{r_0'^2}{1.73\ a} = 1.57\,; \quad \eta = \frac{r_0'^2}{a} = 2.72\ \text{day}$$

Since well 2H-Şotînga presents both the types of imperfections (referring to the degree and kind of the aquifer opening) it is impossible to separate the two components of complex parameter r, i.e. the reduced radius of the well (r_0') and the hydraulic diffusivity.

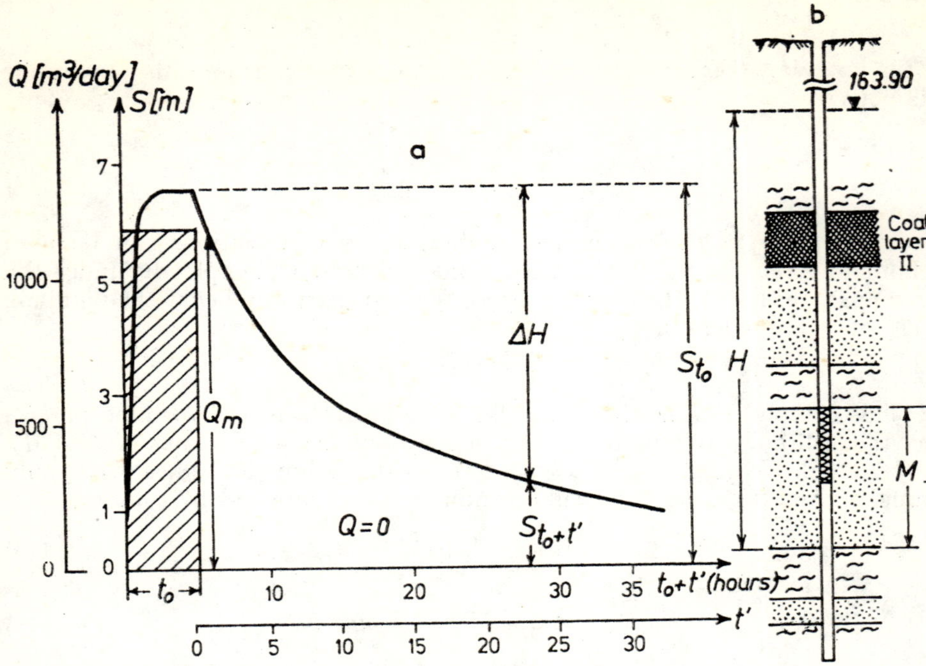

Figure 97. a) Evolution of the hydrodynamic elements during the bailing of well 2H — Șotînga; b) The hydrogeological log — on 1: 500 scale.

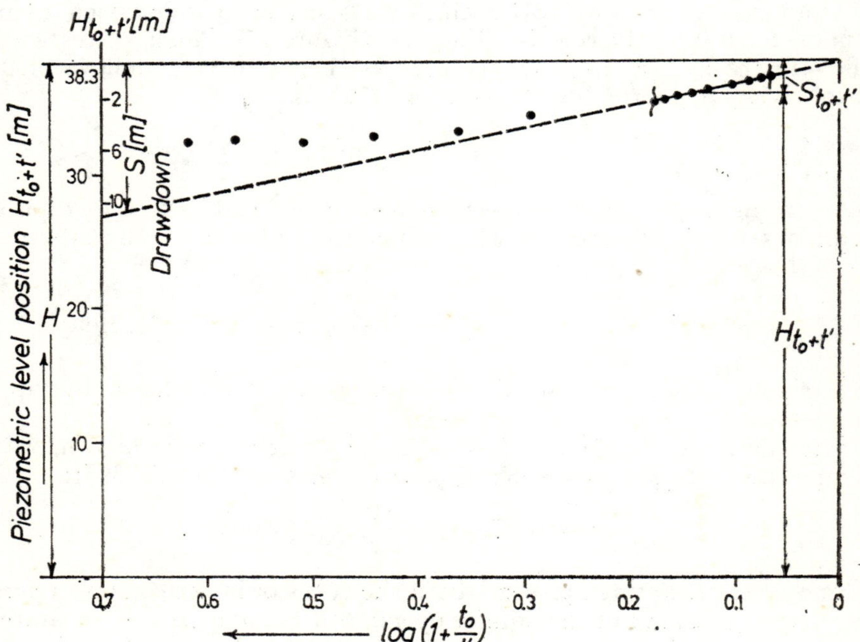

Figure 98. Bailing graph of well 2 H — Șotînga.

The same experimental data served to draw the graphs of relations (397 and 405) (Figure 100). According to the structure of these relations they should be straight lines. Actually however the graph of relation (397) is a very pronounced curve which shows that the simplification introduces gross errors. On the other hand the curve representing relation (405), the radius of curvature of which is very great, can be considered as a straight line in its first portion; transmissivity T computed by this relation at various times of the bailing cycle varies however, from 12.2 to 18.2 m^2/day, which means a noticeable error. This error is due to the great length of the bailing period ($t_0 = 5$ hours), without taking into account the sequence of the evacuated bailers which is expressed by relations (406—409).

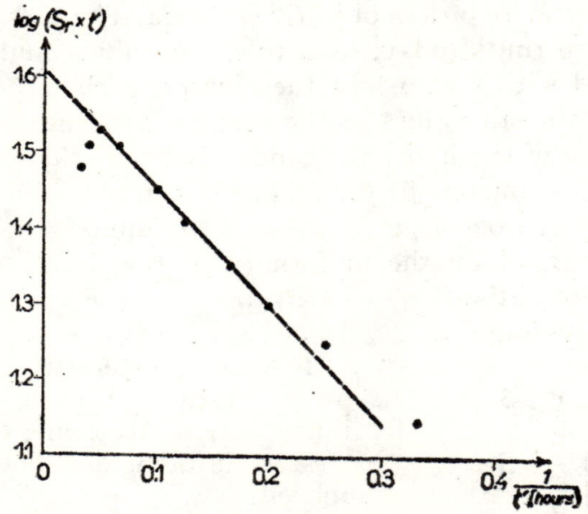

Figure 99. Skibitzke method applied to well 2 H — Șotînga.

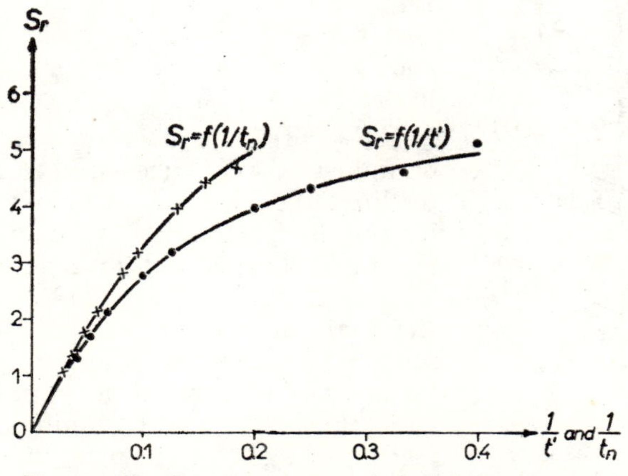

Figure 100. Bailing graphs of well 2 H — Șotînga.

2.5.3 Use of testers

Modern hydrogeology is bound to answer several problems necessitating the study of certain depth structures and of certain sequences of many aquifers. This category of problems includes: the exploration of solid mineral deposits, the prospection of oil and gas deposits and thermal or thermomineral waters, the detection of deep aquifers etc. When the hydrogeological sequence includes many aquifers, the classical method of hydrodynamic investigation by pumping cannot be applied to each aquifer. That is why in every day's practice we proceed to the hydrodynamic test and the collecting of water samples by geological complexes, the selection of which is made by more or less objective criteria. The technology of the hydrogeological wells permits today, as a rule, the tubing and cementing of three casings provided with screens. In these cases, each working hydrogeological complex supplies mean values of the original piezometric surface and of the transmissivity (or of the hydraulic conductivity) as that the estimation of the hydrogeological parameters of each aquifer is impossible. There are of course cases when the hydrogeological research of aquifers belonging to a single complex is warranted by the uniformity of the lithological, structural and hydrogeological conditions, as well as by the problem raised by designing and executing this objective.

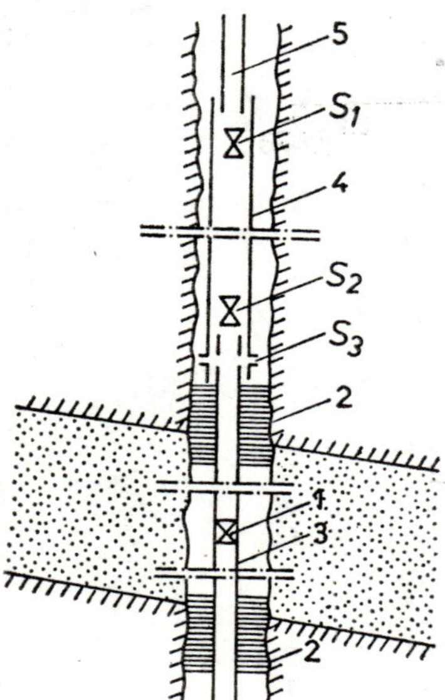

Figure 101. Scheme of a tester: 1 manometer; 2, packer; 3, screen; 4, the tester proper; 5, gear; S_1, relief valve; S_2, check valve; S_3, pressure equalizing jar.

The use of testers with packers allows the separate testing of each aquifer during the drilling process. According to running technique of testers into the drill-hole, the testers can be launched:

— by the drill pipes;

— by a geophysical cable;

— through the annular space of the drill-pipes.

There are various types of testers (with conical or cylindrical packers, tester with foot, hydraulic testers).

Recently the mass production testers of oil industry, were successfull in hydrogeological drilling [3]. The testers are frequently used in the hydrotechnical works where the experimental pouring out and injection tests by sections in the wells are made with the help of the hydraulic tester. Figure 101 represents a scheme of a tester with a double packer. Such testers are used when an uncasing drill-hole goes through several aquifers so that the testing of each of them requires the interval isolation both in the top and in the bottom.

Among the advantages offered by the tester use for testing the aquifers, let us mention:

— The testing is accomplished in short time intervals.

— The experimenting can be made on an uncasing drill-hole without removing the drilling mud.

— The equipment is not complicated.

— The resulting hydrogeological parameters offer adequate accuracy for the preliminary study.

The drawbacks are:

— The drill-hole must be geophysically investigated so that the tester screen could be placed opposite the aquifer and the packer in the impermeable mud.

— The testing should be made shortly after passing through the aquifer, so as to avoid its noticeable contamination by the drilling fluid.

The main objectives of the testing with a tester are: the determination of the initial aquifer head, the collecting of the water sample for determining its physico-chemical properties and the estimating of the transmissivity. To this end, the total volume of the water coming from the aquifer should be known as well as the diagram of the pressure variation by means of a depth manometer fastened below the packer.

On this diagram (Figure 102), each stage should appear of the testing operation made by the tester launched by the drill-pipe. Section *1—2* corresponds to the tester lowering into the drill-hole with empty or partially empty drill-pipes, with both S_1 and S_2 valves closed and the jar of pressure balance (S_3) open. Under these conditions, the pressure increases linearly with time. By means of the drill-pipes, the packer (point *2*), is fixed on the drill-hole wall and jar S_3 is closed thus achieving the isolation of the tested interval. The communication between the aquifer and the tester room is accomplished by opening the back-pressure valve (S_2); the depth manometer shows after a few minutes a decrease of the pressure (segment *2—3*) and then increase of it up to its initial pressure of the aquifer (p_0 corresponds to point *4* on the diagram). By the mechanical opening of the control valve (S_1) the connection is made with the inside of the drill-pipes shown by a sudden pressure decrease (*4—5*). If the aquifer is a freeflowing one the pressure and the rate of discharge with time can be measured at the drilling head. Section *5—6* corresponds to the pressure variation in period t_0 of the water inflow into the drill-pipes, the size of which depends on the technical state of the drill-hole and on the properties of the aquifer; usually, t_0 varies from 20 minutes up to a few hours. The stopping of the water inflow into the drill-pipes (point *6*) is obtained by closing valves S_2 and S_1. Then t'

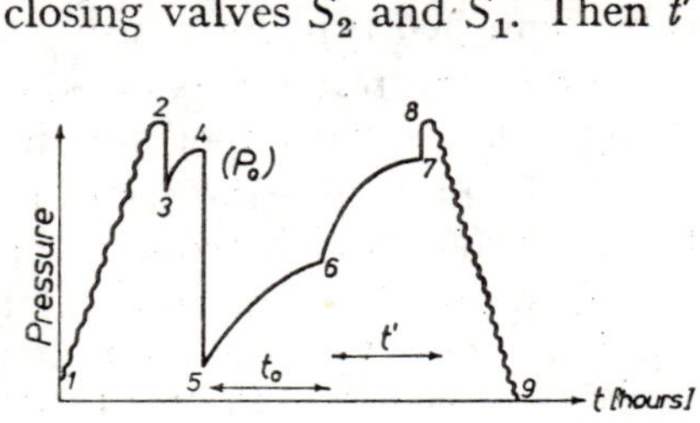

Figure 102. Pressure variation diagram during testing.

is recorded on the recovery curve of the initial pressure (section **6**—*7*). However, a certain increase of pressure (7—*8*) appears as a consequence of handling of the jar S_3. By lifting the tester up to surface, the hydrostatic pressure under, goes a continuous diminution (8—*9*). At the same time as the drill-pipes are lifted, the total volume of collected water is established and the samples are sent for analysis.

The deep hydrogeological wells involve the introduction of testers by means of a geophysical cable, which ensure a good execution and economy of the testing operation. The working principles of these testers are quite different from those launched by drill-pipes: with the help of a skid, a limited sector (and therefore of the tested aquifer) of the drill-hole is sealed up. In Romania, these testers belonging to the oil contractors are rigged up as mobile equipment and possess the following constructing and working characteristics;

— Their main components are: a 12 l — volume vessel, sealing up and fixing skids, control assemblies, pressure transducers and electrical circuits ensuring the recording of the variation diagram of the tested horizon pressure, including the recovery curve;

— They are only useful for depths over 1000 m because the controls are made by means of hydrostatic pressure of the drilling fluid;

— The water inflow from the aquifer into the tester enters by means of a tube that goes through the well within its sealed up section of the skid:

— In the case of a casing hole, the communication orifice is made by means of a jet;

— The location of the testing point is obtained with the help of the spontaneous potential curve recorded concomitantly with the tester launching;

— The testing time proper is 20—30 minutes, followed by the period of the water level recovery;

— In the case of certain testers, the possibility is provided for cementing the communication orifice after the test.

The introduction of tester with the help of geophysical cable presents the following advantages:

— Both their running and their pulling out are made quickly which represents a high economical efficiency of the well testing process;

— There is no eruption danger in the case of overflowing aquifers since the test is made at a single point of the well while the rest of its area is under the back-pressure of the drilling fluid;

— The sample waterproofing (the holding out of the field pressure in the reservoir) allows for the accurate study of the gaseous factor.

Of late, in USSR, hydrogeological testers were made of the geophysical cable launched type for a usual depth range of 50—1000 m.

Both the tester types — those launched by drill-pipes or geophysical cables — have two important drawbacks:

1. To introduce the tester it is necessary to bring the drilling string out to the surface which reduces the drilling rate (chiefly at the great depths).

2. When the time interval between the drilling and the testing of the aquifer is long, the contamination of the adjacent zone by the drilling fluid is likely to appear within an important radius and so the water sample will not

display the genuine quality of the water. This drawback can however be avoided by using a suitable drilling fluid or the water instead of the usual drilling mud.

In order to avoid these deficiencies, in some countries (USSR, Canada) testers were made that, though launched within drill-pipes, do not require the interruption of the drilling process so that the testing follows close on the introduction and consequently a good determination of the aquifer's natural permeability can be obtained.

Whatever the tester type, its transmissivity is established by processing the curve of the water level recovery. Since, under the working conditions of the tester, an instantaneous vertical source appears, the transmissivity can be computed by means of the previously described grapho-analytical methods. All the previously mentioned reasons concerning these processes are also valid for these tests. In formula (377), the rate of discharge is not constant during the testing, but decreasing. When this is the case, a mean value of the discharge rate is recommended which results from the division of the water volume (V) by the testing period (t_0).

To conclude, the application of the testers to hydrogeological research by deep drillings is a way of increasing the technical and economical efficiency at least in the preliminary research stage.

2.5.4 Determination of the geometrical elements of a confined aquifer

When designing capture and drainage systems, the great values of the radius of influence can determine a complicated (in the plane) geometrical configuration of the depression area where it includes one or more recharge or impermeable boundaries.

It is appreciated for exemple that if the density of the impermeable boundaries is higher than 0.4 km/km^2, the drainage will be very difficult even if the density of the works is high owing to the lack of a direct hydrodynamic connection. For the unconfined aquifers, these boundaries can be established easily enough by geological methods. On the other hand, for the deep and confined aquifers, the estimation of the position and of the permeability of the boundaries can only be made by hydrodynamic methods the investigation radius of which is great around the experimental well.

The influence of external impermeable boundaries on the extreme portion of graph $s - \ln t$, emphasized in § 2.3.1., where graph is drawn on the basis of the pumping datta in unsteady flow régime.

The degree of permeability or of hydrodynamic communication of a fault can also be determined by pumpings in interference or in experimental groups with many observation wells on the alignment crossing the fault plane.

An efficient hydrodynamic method, based only on the experimental well data, is the *interception method* elaborated by Houpeurt [61]. This method makes use of the final portion of the graph of aquifer pressure recovery

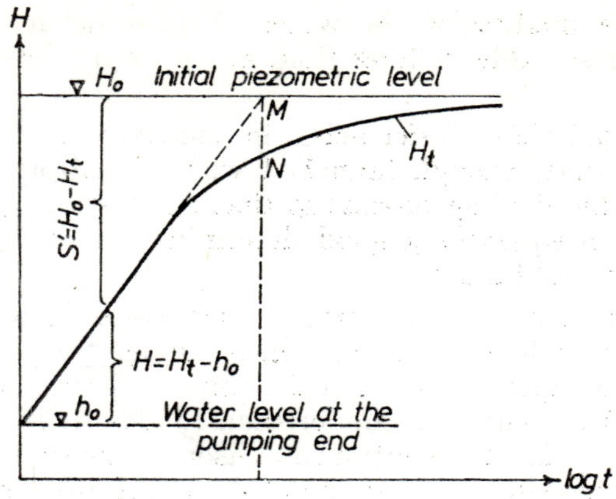

Figure 103. Theoretical diagram of the interception method.

after a previous pumping in permanent régime ($t_0 \gg t'$) at a constant rate of discharge. Houpeurt studied three cases of the confined aquifers: 1) with an infinite extension, 2) with an open circular boundary (of water supply) and 3) with a closed circular boundary (impermeable). Figure 103 represents the theoretical diagram of the method that essentially consists in the determination of segment MN. Point M is the intersection of the prolonged linear portion of the graph with the initial piezometric level, while N is the intersection of the vertical passing through M with the recovery curve.

For an infinitely extended aquifer, the value of segment M_1N_1 is supplied by relation

$$M_1N_1 = \frac{Q}{4\pi kM} \ln 2 = \frac{0.183\,Q}{kM} \log 2 \tag{415}$$

Relation (372) showed that quantity $0.183\,Q/kM$ represents the angular coefficient of the linear portion of the graph and therefore $M_1N_1 = m \log 2 = 0.301$ m.

Thence, the value of segment M_1N_1 being known (by measurement) as well as that of angular coefficient m (by computation), the numerical coefficient is determined; this value being equal to 0.301, there results that the extension of the aquifer is infinite because of the depression area such as they are created by pumping.

The problem is similarly solved for the other two cases but then there appear also the relations for the estimation of the distance from the experimental well to the boundary (r_c). Thus for an aquifer with an open circular boundary:

$$M_2N_2 = 0.084 \text{ m} \tag{416}$$

$$r_c = 1.5\sqrt{at} \tag{417}$$

For an aquifer with a closed circular boundary, the computation relations are

$$M_3N_3 = 0.173 \text{ m} \tag{418}$$

$$r_e = 3.26\sqrt{at} \tag{419}$$

In these relations, r_e is the very radius of influence; to determine it, we must also know the hydraulic diffusivity (a), time t resulting from the graph by protracting segment MN up to its intersection with the abscissa.

In practice, there are however known aquifer types with other than circular boundaries, for which no elaborate solutions are known, but the interception method offers qualitative data as is shown by the solution supplied for well 5H-Șotînga. Figure 104 represents graph H—$\log t$ after the pumping in steady flow régime in this well. The angular coefficient is $m = 16.22$ and segment $MN = 2$ m, hence $NM = \eta m$; $\eta = \frac{NM}{m} = \frac{2}{16.22} = 0.123$.

This excludes the case of the "infinite" aquifer but, according to this numerical value (η), this case can be assimilated to one of the two diagrams studied in the case of those with circular boundaries. From hydrogeological exploration documentation it follows that the aquifer tested by well 5H is

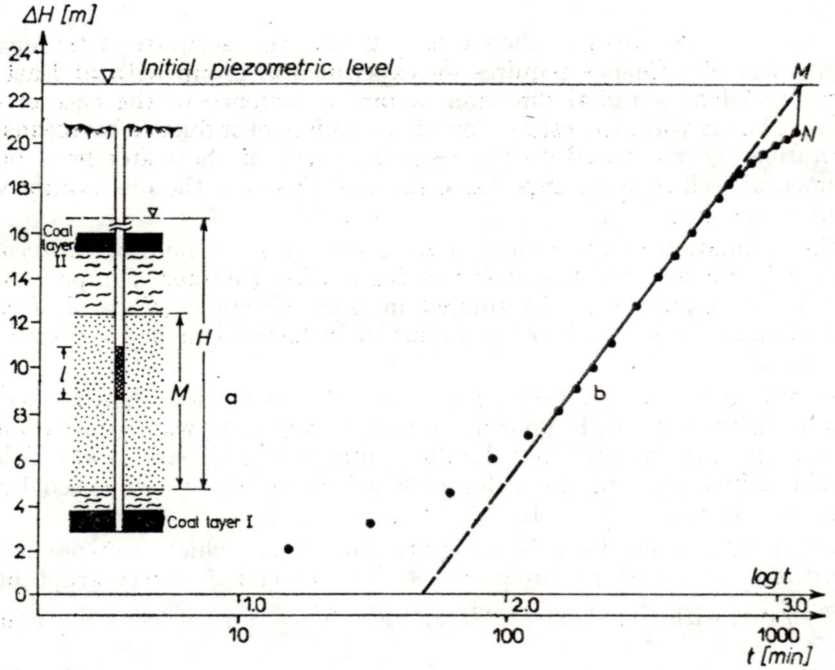

Figure 104.
a) $\Delta H - \log t$ graph in the level recovery period; b) Hydrogeological log of well 5 H — Șotînga(1: 500 scale).

a typical case of limited aquifer; its width occupies approx. 600 m on the well 5H area and at about 600 m towards the north the aquifer is also partitioned by fault F_1. Consequently, the interpretation of the data obtained by the method of interception becomes complicated when the value of numerical coefficient η differs from its three analytical determined values; it is however possible to reach qualitative conclusions by assimilating the aquifer to one of the typical geometric schemes.

The method of interception is advantageous when there is a good homogeneity of the aquifer (uniform values of the thickness and of the transmissivity within the cone of depression area). Nevertheless, on the assumption that η is the proportionality factor between segment MN and angular coefficient m, the method of interception can also be applied to cases of the aquifer non-homogeneity, provided the results are only considered as qualitative. In this case, η can be called a parameter of the drainage potential. A map with the isolines of this parameter (the values of which vary from 0.084—0.301) can be very useful for locating the capture wells (in the zones of minimum values) and for establishing the density of the drainage wells.

2.5.5 Estimation of the reduced radius of influence

As already shown in § 2.1.8, the accurate determination of the radius of influence requires an experimental group with at least two piezometers along a radial direction, a rare occurrence in the case of deep aquifers. That is why, the estimation of the radius of influence by means of a computation scheme based on the recovery curve of the water level in the experimental well is quite important, since in this case the observation wells are not necessary.

The estimation of the radius of influence with the help of the recovery graph can be made in the following two cases: after the short pumpings when the radius of influence is determined in unsteady flow (R_t) and after the long pumpings ($t_0 \gg t'$) when the radius of influence can be estimated in a steady flow (R).

According to relation (323), parameter R_t can be determined when the hydraulic diffusivity (a) is known; in fact, however, as was pointed out in § 2.3.1.2 the experimental well data are sufficient to determine the reduced hydraulic diffusivity (a_0) the value of which is moreover influenced by the seepage resistance of the well.

Under these conditions the computation scheme which does not include the hydrological diffusivity are preferred. To this end, recovery graph of the level is drawn with co-ordinates ΔH and $\log \frac{t_0 + t'}{t'}$ for the confined aquifers and with co-ordinates $H^2 - h_t^2$, and $\log \frac{t_0 + t'}{t'}$ for the unconfined aquifers. The linear portion on these graphs is taken into account in the computations.

The radius of influence (R_t) is computed by the relations established by Kerkis [80]:

— for the confined aquifers:

$$\log R_t - \log r_0 = \frac{0.5\,S_0}{\Delta H_2 - \Delta H_1} \log \frac{t_2(t_0 + t_1)}{t_1(t_0 + t_2)} \tag{420}$$

— for unconfined aquifers:

$$\log R_t - \log r_0 = \frac{0.5(H^2 - h_0^2)}{h_2^2 - h_1^2} \log \frac{t_2(t_0 + t_1)}{t_1(t_0 + t_2)} \tag{421}$$

where S_0 and h_0 are the drawdown and the water level at the end of the pumping; ΔH is the level increase with regard to the water level at the end of the pumping at time t' (for confined aquifer); H and h_t are the hydrostatic level and the water level in the well at time t' with regard to the lower impermeable layer (for unconfined aquifers); ΔH, h and t (with indices 1 and 2) correspond to the beginning and end of the linear portion of the graph.

There are several methods for estimating the radius of influence in steady régime (R). One of them is based on the graphs the co-ordinates of which are ΔH and $\log t'$, respectively $H^2 - h_t^2$, and $\log t'$ and on which the linear portion is delimitated; formulae 420 and 421 are used where the logarithm sign is applied to ratio t_2/t_1.

Another method, elaborated by I. A. Tcharnyi, makes use of the final stage of the level recovery period. The radius of influence of a confined aquifer is then determined by relation

$$R = 1.59 \sqrt{\frac{a(t_2' - t_1')}{\log S_1' - \log S_2'}} \tag{422}$$

where S_1' and S_2' are the residual drawdowns at times t_1' and t_2' from the end of the pumping. The following relation, easier to apply since it does not include the hydraulic diffusivity (a), can equally be used:

$$\log(\log R - \log r_0) = \log \frac{S_0}{S_1'} - \frac{t_1}{t_2 - t_1}(\log S_1' - \log S_2') + 0.243 \tag{423}$$

In the case of an unconfined aquifer, relation (422) can be used; it includes however the hydraulic diffusivity as defined by formulae (129), where H_m is the mean thickness of the aquifer at the end of the recovery period: $H_m = H - 0.25\,(S_1' + S_2')$.

All the formulae comprising the well radius (r_0) are valid provided the imperfection of the experimental well is not significant ($\zeta_0 = 0$); otherwise, the reduced radius (r_0') must be used. Of course, it is not always possible to compute r_0', but it should be remembered that in most of the formulae for the determination of the transmissivity the radius of influence is included in expression $\log R - \log r_0$ so that the separation of the two parameters is unnecessary.

2.6 Aquifer testing under intricate hydrogeological conditions

The possibility of a minimum alteration of the natural situation when schematizing the hydrogeological conditions depends on the hydrogeological complexity of the area of influence as well as on the boundary conditions. The ideal situation requires a hydrodynamic research made under relatively simple hydrogeological conditions so that the errors introduced in the processing and interpretation of the observation data should be reduced to a minimum. But this cannot always be achieved, since the hydrogeological conditions are complicated by natural factors; on the other hand since the subjective factor can also have an influence it is necessary that in designing a hydrodynamic investigation the hydrogeological conditions should be simplified within the limits allowed by the research objectives.

The simple hydrogeological conditions correspond to the following cases;

— confined aquifer with infinite extension;

— upper and lower confined beds consisting in practically impermeable deposits;

— aquifer with uniform thickness and homogeneous permeability;

— perfect experimental well as concerns the degree and the kind of aquifer opening;

— existence of observation wells.

When these conditions are simultaneously fulfiled the choice of the interpreting models is facilitated. But it is not always possible to achieve them either because of the natural factors or the experimental conditions required.

In the previous paragraphs, certain schemes with complex hydrogeological conditions (leakage, variable thickness of the aquifer, etc.) were studied. Hereafter, some other factors will be presented liable to complicate still more the designing of the investigation and the choice of the interpreting model.

2.6.1 Unconfined aquifer and fully penetrating experimental well

In the case of pumping from an unconfined aquifer, the radial-plane motion is distorted owing to the reduction of the saturated thickness from H (at the external limit of the depression area) to h_0 which represents the pumping water level position in the experimental well with regard to the impermeable bed. The deformation of the radial-plane motion results in the appearance of a vertical component of the seepage velocity, the value of which increases as the experimental well is nearer and as the drawdown increases. To obtain the hydrodynamic solution of the problem, we start from the differential equation of the water inflow to the well by a radial-plane motion $\left(Q = 2\pi\, kxy \frac{dy}{dx}\right)$ which introduces therefore an error due to the omission of the above-mentioned vertical component. The formulae supplied by the solution of this differential equation can as a matter of fact be deduced from the corresponding formulae of a confined aquifer where the saturated thickness is introduced as an arithmetic mean of quantities H and h_0 (183—186). For the steady flow pumpings, this problem was given as an example in § 2.2.2.3.

To the unsteady flow pumpings, all computation schemes in § 2.3.1 can be applied provided product $M.S$ is modified according to formulae (183—186). In this case, when processing the data from an observation well (n), formula (293) becomes

$$S_n(2H - S_n) = \frac{0.366\,Q}{k}\log\frac{2.25\,a^*}{r_n^2} + \frac{0.366\,Q}{k}\log t \qquad (424)$$

If we substitute C to $\frac{0.366}{k}$ and A to $C\log\frac{2.25\,a^*}{r_n^2}$, the relation becomes

$$S_n(2H - S_n) = A + C\log\log t. \qquad (425)$$

Since this is the equation of a straight line, it can be solved with the help of a graph where quantity $S_n(2H - S)$ is on the ordinate while $\log t$ is on the abscissa. On this graph, quantity A is the ordinate and C is the angular coefficient of the linear portion

$$C = \frac{S_2(2H - S_2) - S_1(2H - S_1)}{\log t_2 - \log t_1} \qquad (426)$$

$$k = \frac{0.366\,Q}{G} \qquad (427)$$

$$\log a^* = 2\log r_n - 0.35 + \frac{A}{C} \qquad (428)$$

Now parameters k and a^* being known, the superficial specific yield (n_s) can be estimated (129).

The mean thickness (H_m) of the aquifer current at the end of pumping period, for the aquifers with a large expansion area is estimated by the following approximate relations:

— when there are no observation wells:

$$H_m = \frac{2H + h_0}{3} \qquad (429)$$

— when there is an observation well:

$$H_m = \frac{h_1 + h_0}{2} = H - 0.5(S_0 + S_1) \qquad (430)$$

— when there are two observation wells:

$$H_m = \frac{h_1 + h_2}{2} = H - 0.5(S_1 + S_2) \qquad (431)$$

Relation (424) can also be used for the experimental well (r_0 and S_0) but according to the indications in § 2.3.1.2 the reduced hydraulic diffusivity

(a_0*) which includes moreover the seepage resistance of the well (ζ_0) will result.

In connection with the use of the relations established for a radial-plane motion in the case of the pumpings from the unconfined aquifers, work [20] makes some actual recommendations. For instance, if the drawdowns are very small in regard to the initial thickness of the aquifer ($S < 0.1\ H$) these relations can be directly used. If the drawdowns are important ($0.1\ H \leqslant S \leqslant 0.3\ H$) the relations valid for the confined aquifers can be used provided that the computations make use of corrected drawdowns (S'):

$$S' = S - S^2/2H \tag{432}$$

where S is the measured drawdown.

This correction permits us to compute the hydrogeological parameters in the case of both steady and unsteady flow. So as to estimate the hydraulic conductivity, transmissivity T so obtained will be related to mean values H_m, computed by one of the previous relations. Coefficient n_s is

$$n_s = [(H - S)/H]\ n'_s \tag{433}$$

where n'_s is the value supplied by the computation.

When $S > 0.3\ H$, the motion is considered as three-dimensional and therefore the use of the relations established for confined aquifers is not recommended any more. In practice, however, this recommendation is not applied because the literature does not offer suitable solutions.

Therefore, in the case of pumpings from unconfined aquifers, the drawdown is recommended to be $S_0 < 0.3\ H$; if this condition cannot be fulfilled, at least two observation wells are necessary with $S < 0.3\ H$, that could be used for the computation of the hydrogeological parameters.

In § 2.2.2.3, the total well loss (Δh) was computed for experimental station Plopu. This value includes however component Δh_1 due to the curvilinear character of the depression surface and to the noticeable deformation of the flow lines in the neighbourhood of the fully penetrating experimental well. The parameters determining this well loss are: the reduced rate of discharge $\left(\bar{Q} = \frac{Q}{k}\right)$, the pumping water level in the well (h_0) and the well radius (r_0). By synthesizing the experimental data supplied by several researchers [63], V. M. Shestakov obtained the following relation:

$$\Delta h_1 = \sqrt{\left(0.73 \log \frac{\sqrt{\bar{Q}}}{r_0} - 0.51\right) \bar{Q} + h_0^2} - h_0 \tag{434}$$

For experimental well Plopu, there results:

$$\Delta h_1 = \sqrt{\left(0.73 \log \frac{\sqrt{4.91}}{0.71} - 0.51\right) 4.91 + 5.86^2} - 5.86 = 0.20 \text{ m}$$

where

$$\bar{Q} = \frac{993.6}{202.2} = 4.91 \text{ m}^2$$

This reduced value is due to the very high permeability of the aquifer.

The difference $\Delta h_2 = 2.68 - 0.20 = 2.48$ m represents the well loss due to the imperfection according to the kind of opening (the presence of the clogged up area) and to the constructive characteristics of the screen.

2.6.2 Imperfect wells according to the opening degree

The imperfection according to the opening degree (partially penetrating wells) of the aquifer can be conditioned, i.e. the position and the length of the screen both in the experimental well and in the observation wells can be established as a function of its purpose as well as of the necessity of using a simple computation scheme for interpreting the data.

The method of the hydrodynamic study of the partially penetrating well is theoretically based on its replacing by a system of punctiform sources with a constant linear intensity. The difference between the various computation methods consists in the kind of appreciation of the effective potential according to the system of sources. A method, proved to supply good results, was proposed by H. L. Hain and N. N. Verighin for the vertical wells in the confined and homogeneous aquifers; in this method, the effective potential is considered to be equal to the mean potential along the well outline.

A frequently used method is that of the hydraulic resistance. The principle of this method consists in dividing the whole aquifer current in the zone of influence into fragments and to estimate in each of them the local head loss. The total head loss is equal to the sum of the local losses. The hydraulic head can be replaced by the hydraulic resistance the value of which is directly proportional to the head loss. The hydraulic resistance (Φ) is by definition the ratio of the drawdown (S) to the discharge (Q) and its dimensions are therefore day/m^2. To make it easier, a dimensionless parameter of the resistance is used, the expression of which depends on the structure of the aquifer current.

As a rule, the addition of the hydraulic resistances is the same as that of the electric resistances. It is characteristic of the imperfect well that the deformations of the flow lines (with regard to the radial-plane motion) occur around the well within a radius equal to the aquifer thickness. The important local hydraulic resistance of this area depends moreover on the well characteristics. At great distances, the hydraulic resistance is only determined by the water inflow towards the well whether it is perfect or imperfect; in this portion, the motion is radial-plane.

For wells that are imperfect according to their opening degree, many interpreting models are elaborated: some of them for particular conditions, other ones for a higher degree of generality. Hereafter, we present a few models of the last category.

Confined aquifers. When the thickness of aquifer (M) is known for any position of the screen, the seepage resistance of the imperfect well can be

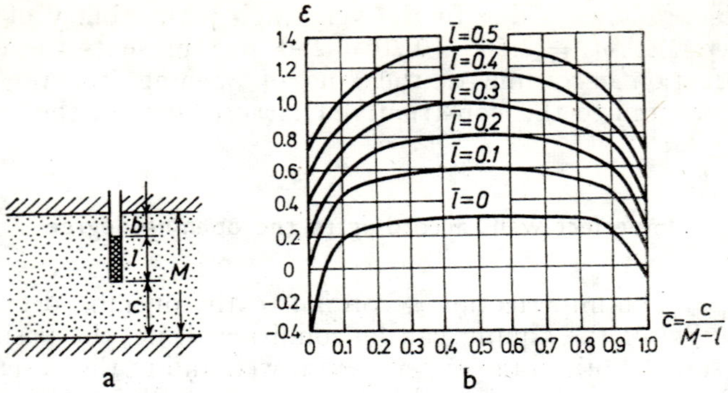

Figure 105. Partially penetrating well in a confined aquifer (after Shestakow, 1965): *a*, Diagram; *b*, Graph for determining quantity ε.

determined by V. M. Shestakov's method synthesized in the following simplified formula [63]:

$$\zeta_1 = \frac{1-\bar{l}}{\bar{l}}\left(\ln\frac{l}{r_0} - \varepsilon\right); \quad \bar{l} = \frac{l}{M} \tag{435}$$

where ε is a quantity depending on parameters c/l and l/m. i.e. on the position of the screen in the aquifer. The values are determined from the graph up to $\bar{l} = 0.5$ (Figure 105). For some particular cases, analytical formulae are supplied; thus, for $\bar{l} > 0.3$

$$\varepsilon = 1.8\,\bar{l} - 0.27 + 1.38\sqrt{\bar{c}(1-\bar{c})} \tag{436}$$

and, when the screen is in contact with the lower impermeable bed ($c = 0$) whatever the value of l,

$$\varepsilon = 1.9\,\bar{l} - \frac{0.4}{1+\sqrt[4]{\bar{l}}} \tag{437}$$

The simplified formula 435 is affected by errors of 3—4% in regard to the exact analytical solution when $l/r_0 > 10$ which is the case of the experimental well. If seepage resistance ζ_1 is known, instead of the imperfect well, a perfect one (fully penetrating) can be used which differs from the actual one by its smaller diameter (the dynamic level and the rate of discharge are the same in both these wells).

The expression of the reduced radius of the equivalent well (r'_0) results from the formula of a well discharge as established by the method of the hydraulic resistances:

$$Q = \frac{2.\pi.k.M.S_0}{\ln\frac{R}{r_0} + \zeta_1} = \frac{2\pi k.M.S_0}{\ln\frac{R}{r_0 e^{-\zeta_1}}} \tag{438}$$

$$r'_0 = r_0{}^{-\zeta_1} \tag{439}$$

Once this quantity is known, the hydrogeological parameters will be determined by the formulae for the fully penetrating wells, unconditionally in the case of pumpings in steady flow and conditionally in the case of pumpings in the unsteady flow. This condition was established by A. L. Hain as follows:

$$\zeta_1 = \zeta_{10} - \zeta_{1t} \tag{440}$$

where ζ_{10} is the additional resistance independent of time and ζ_{1t} the dependent on time one.

Quantity ζ_{1t} diminishes rapidly with time and so it can be neglected for $\alpha = \frac{r^2}{4at} < 5.10^{-5}$, i.e. almost always when the computation point corresponds to the experimental well or is near enough to it.

When the observation wells are partially penetrating, their seepage resistances ζ' should be taken into account as pointed out by relations (141, 142). The value of this parameter can be determined with the help of Table 66 if quantities l/M and M/r are known. From this table it follows that when the distances between the observation wells and the experimental well are greater than the aquifer thickness, the seepage resistance due to imperfection is negligible.

Table 66

$\frac{l}{M}$	M/r									
	0.5	1	3	10	30	100	200	500	1000	2000
0.1	0.00391	0.122	2.04	10.4	24.3	42.8	53.8	69.5	79.6	90.9
0.3	0.00297	0.0908	1.29	4.79	9.2	14.5	17.7	21.8	24.9	28.2
0.5	0.00165	0.0494	0.656	2.26	4.21	6.5	7.86	9.64	11.0	12.4
0.7	0.00055	0.0167	0.237	0.879	1.69	2.67	3.24	4.01	4.58	5.19
0.9	0.00005	0.0015	0.0251	0.128	0.3	0.528	0.664	0.846	0.983	1.12

The hydrogeological literature gives many hydrodynamic schemes of pumpings in steady flow régime from imperfect wells with or without observation wells. They should only be used very discriminately, because they are elaborated for particular cases. A basic criterion of the elaboration and classification of these schemes in the screen position in both the experimental well and in the piezometers, with regard to the upper and lower impermeable beds. If, at both the extremities the condition $b > 3$ to $4\,l$ and $c > 3$ to $4\,l$, is fulfilled, the relief of the pressures will be symmetric around the experimental well, the equipotential surfaces will have the form of spheres and the aquifer will be considered as infinite (as to the thickness); from this, the other condition follows, namely $l < 0.3\ M$. The aquifer is considered as semi-infinite if $b < 3$ to $4l$ or $c < 3—4\ l$; in this case the pressure relief will be influenced by the upper and lower impermeable limits. The aquifer has a finite thickness when $l > 0.3\ M$ and the distance from the screen ends to the upper and lower impermeable beds is smaller than 3—4 l; in this case the form of the equipotential lines is influenced by the two impermeable limits. In the case of the semi-infinite and infinite aquifers, the computation relations do not include thickness M. The scheme corresponding to an infinite aquifer is indispensable if the aquifer consists of several layers; such a scheme permits the hydrodynamic study of each layer.

In the case of pumpings in unsteady flow from imperfect wells and from aquifers with an infinite thickness, the determination of the hydrogeological parameters can be made by the relations valid for the finite aquifers where the seepage resistances (ζ) are considered equal to zero and the aquifer thickness (M) is replaced by l (the length of the experimental well screen).

Unconfined aquifers. The kinematics of the flow in the neighbourhood of the experimental well depends on the screen position. In this respect, two cases must be considered: the well with a non-flooded screen, when the depression curve crosses the well within the screen limits, and the well with a flooded screen when the latter is placed below the pumping level in the well (Figure 106)

In the case of the *non-flooded screen* (Figure 106 *b*) the deformations of the flow due to its imperfection occur around the experimental well along a distance approximately equal to h_0 (the position of the pumping level with regard to the impermeable bed). The problem of the seepage resistance due to this imperfection is not theoretically solved; that is why the method established for the confined aquifers (435—437) is recommended. To this end, the computation values of the current thickness (h_M) and of the screen length (l) should be estimated. The thickness of the aquifer current at a distance equal to h_0 can be determined by the simplified formula [4]:

$$h_M = \sqrt{h_0^2 - 2\bar{Q}_0 f}$$

$$f = 0.366\ \log \frac{h_0}{r_0}; \quad \bar{Q}_0 = \frac{Q_0}{k} \tag{441}$$

where Q_0 is the rate of discharge computed for the considered well on the assumption that it is fully penetrating

$$Q_0 = \pi k \frac{(h_M - c)^2 - (h_0 - c)^2}{\ln \frac{h_M}{r_0}} \tag{442}$$

$$l = l_0 + h_M - h_0 \tag{443}$$

$M = h_M$, l and $\bar{c} = 0$ being known, parameters ζ_1' and r_0' can be determined by the formula elaborated for the confined aquifer. The well loss (Δh_n) can be estimated by relation (434), which may also be applied with an admissible approximation to the partially penetrating wells provided l_0 is substituted to h_0.

This computation scheme was verified in practice and can be applied both to the experimental wells and to the designing of working wells.

In the case of an experimental well, the reduced radius (r_0') is first determined without making use of the observation well data. With this quantity the hydraulic conductivity is then computed with the help of one of the perfect well schemes; the existence of an observation well at a distance $r_1 > h_0$ is recommended. In this case, the formula is:

$$k_{0-1} = 0.73 \frac{Q \ln r_1/r_0'}{h_1^2 - h_0^2} \tag{444}$$

where Q is the experimental pumping rate of discharge in steady flow.

Quantity r_0' may also be used in the formulae valid for pumpings in unsteady flow. When designing exploitation wells, if k is known, rate of discharge Q and well loss Δh_0 will be computed for various known values of parameters r_0, h_0 and l_0 and the best version will be adopted.

In case of the *flooded screen*, the kinetics of the flow in the neighbourhood of the experimental or observation well is more intricate than in the case of the non-flooded screen, which explains why the theoretical solution is only approximate. In such wells, a "suspension" of the depression curve in contact with the plain casing above its pumping level appears (Figure 106 *b*). Another characteristic is the horizontal shape of the depression surface in contact zone with the plain casing of the well.

So as to determine the reduced radius of the well (r_0') the computation thickness of the flow is determined just as in the case of the non-flooded screen by formula (441). Hereafter, to compute the seepage resistance ζ_1(435), the following quantities are used:

$$\bar{l} = \frac{l_0}{h_M}; \quad \bar{c} = \frac{c}{h_M - l_0}; \quad l = l_0 \tag{445}$$

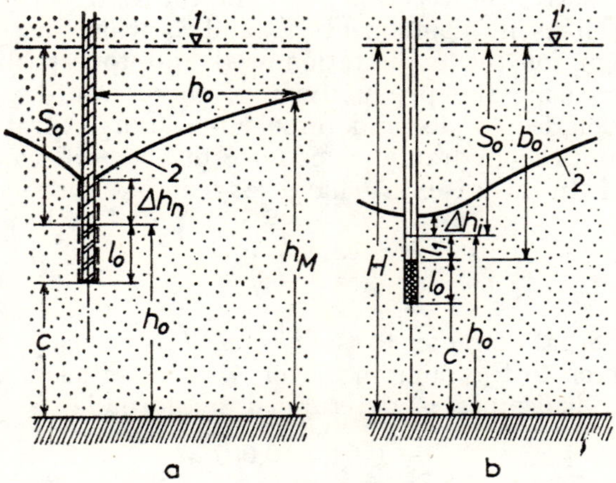

Figure 106. Partially penetrating well diagrams in unconfined aquifers;

a, with unflooded screen; *b*, with flooded screen; *1*, hydrostatic level; *2*, depression curve.

Lastly, for computing the rate of discharge or the hydraulic conductivity the perfect well is considered with r_0' and pumping level h_0. The estimation of the "suspension" height (Δh_1) still constitutes a difficult problem from the analytical standpoint.

When drawdowns S_0 are relatively small, the free level aquifer can be treated as a confined one by considering a conventionally impermeable top, the elevation of which is the average between the hydrostatic and the pumping levels of the experimental well. In this case, the aquifer thickness will be:

$$M = H - 0.5\, S_0, \tag{446}$$

and the computation screen length will be:

— for the non-flooded screen

$$l = l_0 - 0.5\; S_0 \tag{447}$$

— for the flooded screen

$$l = l_0; \; b = b_0 - 0.5 \; S_0 \tag{448}$$

These parameters allow to use Shestakov's method for computing the seepage resistance (ζ_1) and then the rate of discharge or the hydraulic conductivity.

2.6.3 Imperfect wells according to the opening kind

In the previous paragraphs, we presented, supported by examples, the computation method for seepage resistance ζ_2, the value of which depends both on the type and construction of the screen and on the variations of the aquifer structure and permeability in the neighbourhood of the well. On the basis of the experimental data, the total seepage resistance ζ_0 is determined as well as its components ζ_1 and ζ_2 which permits us to make a resonable design for the exploitation well and to establish the required measures for increasing the productivity of these aquifers.

In the case of deep wells and important discharge rates, we have also to take into account the hydraulic resistance of the well casing to the water flowing through it. In this case, the charge loss can be expressed [64] by relation

$$\Delta h_c = \lambda \frac{L}{d} \cdot \frac{Q^2}{2g} \tag{449}$$

where d is the internal diameter of the casing; L is the casing length from the middle of the screen to the pumping level; λ the coefficient of friction which for the metallic drill-pipes is determined by formula

$$\lambda = 0.04 - 0.002 \;\; d'' \tag{450}$$

d'' being expressed in inches.

This charge loss should be subtracted from the measured drawdown in the well.

2.7 Estimation of the optimum volume of experimental data as a function of the permeability heterogeneity

So that the hydrodynamic parameters could become significant and fit for use, their hydrogeological synthesis must be achieved. This synthesis is indispensable in the case of heterogeneous aquifers the hydrodynamic behaviour of which in disturbed régime is hard to appreciate. On the other hand, the hydrogeological synthesis is also imposed by the fact that in a heterogeneous system many measurements — chiefly pumpings — are necessary so as to give a good knowledge of the aquifer characteristics. Lastly, the hydrogeological synthesis helps to define a model which takes into account the irregular distribution of the desired parameter and permits us to obtain additional information about the development and management of the aquifer. Therefore the veracity of a model depends on the coherence of the experimental data as well as on their volume.

The accuracy of the mean hydraulic conductivity is influenced by the method of its determination and of the character of its distribution. Three groups of such methods are known: in laboratory, geophysical and hydrodynamic. In most cases, the values supplied by these three groups are not coherent and the selections cannot be merged so as to obtain a single mean value.

Laboratory methods give a high enough accuracy but the smallness of the sample entails a punctual character of the result. The supplied values are therefore characterized by a considerable variation which means that only the aquifer micro-heterogeneity could thus be estimated after the permeability.

The interpretation of the geophysical logs supplies mean values depending on the investigation radius of the applied method. Obviously, under these conditions the coefficient of variation is much smaller than the laboratory one.

In the case of experimental pumpings the average character of the hydraulic conductivity is even stronger since the values are influenced by the aquifer heterogeneity in the whole influence area.

Even on the assumption that the errors of all three methods are reduced to a minimum, the obtained values will only be close and comparable to one another in the case of a homogeneous and isotropic aquifer, which is rarely exhibited by nature.

The direct comparison between the results supplied by the three groups of methods may therefore lead to erroneous conclusions. Naturally as the volume of data increases, so the mean values tend to grow nearer, but the differences are still appreciable; the frequency curves are also different.

Assuming that the behaviour of an aquifer as a whole is hardly influenced by the local heterogeneities the permeability data obtained on the basis of the experimental pumpings are important.

As a rule, according to their permeability, the formations are divided into *homogeneous*, *heterogeneous* and *very heterogeneous*, in the cross-section or in the plane [24].

A homogeneous aquifer is characterized by relatively constant values of the hydraulic conductivity. The heterogeneity on vertical cross-section of a lithological complex should only be quantitatively estimated when the same capture or drainage is employed for several aquifers. In this case, if criterion $\frac{k_1}{k_2}, \frac{k_2}{k_3}, \ldots < 20$ (where k_1, k_2, k_3 ... are the hydraulic conductivity of neighbouring aquifers) is fulfiled, then the complex is considered as heterogeneous in the cross-section and the computation value of the hydraulic conductivity is the weighted mean as a function of the elementary thicknesses (M). If the criterion is greater than 20, the complex is considered very heterogeneous in cross-section and the computation scheme must take into account each aquifer.

The heterogeneity in the plane of an aquifer is a usual problem of hydrogeological exploration. The character of the heterogeneity is estimated by the ratio between the hydraulic conductivity of two wells next to each other. If the difference is small $\left(\frac{k_1}{k_2}, \frac{k_2}{k_3}, \ldots < 20\right)$ the aquifer may be considered as heterogeneous in computations by taking as k value its weighted mean with regard to the single areas of the influence zones (F):

$$k_m = \frac{k_1F_1 + k_2F_2 + \ldots + k_nF_n}{F_1 + F_2 + \ldots + F_n} \tag{451}$$

When the values of the hydraulic conductivity present great variations, i.e. where $\frac{k_1}{k_2}, \frac{k_2}{k_3}, \ldots > 20$ the aquifer is considered as very heterogeneous and the considered zones are taken singly into account in the hydrogeological computations.

E. E. Kerkis [34] points out that the use of the arithmetic or weighted mean leads to underestimated values of the hydraulic conductivity which is unsatisfactory chiefly in the case of drainage problems. In these cases, the statistical methods appear quite suitable, as they can supply more truthful results. In most cases, the small number of experimental hydrogeological wells implies a small volume selection of the hydraulic conductivity ($n < 30$), i.e. a particular method of processing the collectivity.

As an example, we shall give hereafter the hydrogeological synthesis of the filtration coefficient of coal II—III aquifer within the area of lignite fields Cuzap-Voivozi south. There are 19 experimental wells and groups separated from one another by distances of 40—1700 m (Figure 107). The highest ratio between the hydraulic conductivities is on alignment 27H—24H $\left(\frac{1}{0.06} = 17\right)$, the permeability variation in the plane is not too high and consequently it is possible to use an average hydraulic conductivity.

The arithmetic mean of k is:

$$\bar{k} = \frac{\Sigma k_i}{n} = \frac{12.26}{19} = 0.65 \text{ m/day} \tag{452}$$

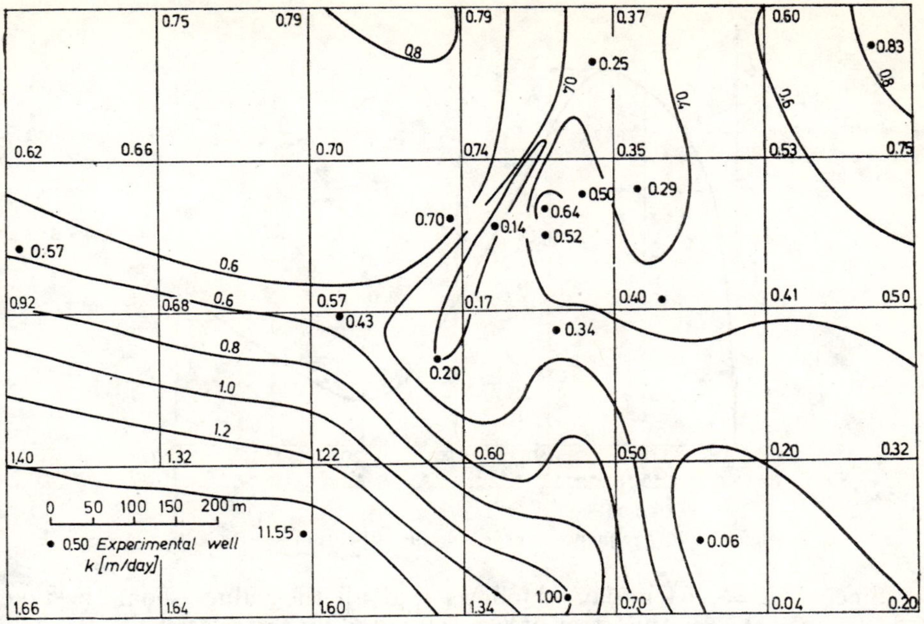

Figure 107. Hydraulic conductivity isolines in perimeter Cuzap-Voivozi.

The standard deviation (σ) can be computed with the help of the scattering amplitude (R) (for small volume populations):

$$R = k_{max} - k_{min} = 1.64 - 0.06 = 1.58 \text{ m/day}$$

$$\sigma = \frac{R}{d_n} = \frac{1.58}{3.69} = 0.43 \text{ m/day}$$

where d_n is a quantity depending on the population volume [54]

The coefficient of variation is:

$$V = \frac{0.43}{0.65} 100 = 66.2\%$$

Once these parameters are computed, we can ascertain the existence of the gross errors to be excluded from the population. For small populations F. Grubs' criterion can be used:

$$x_{(n)P} = \bar{k} + \sigma V_p \tag{453}$$

where parameter V_p is a function of n and P [54].

For $P = 0.99$ we have $x_{(n)P} = 0.65 + 0.43 \times 2.93 = 1.91$.

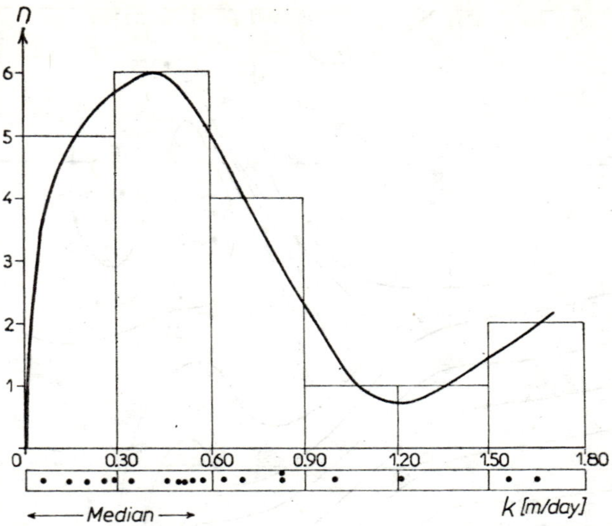

Figure 108. Frequency curve of the filtration coefficient.

Since $k_{max} = 1.64$ m/day it follows that all the values should be taken into account. The relative error of k is estimated by the relation:

$$\varepsilon_k' = \pm \frac{t_q V}{\sqrt{n-1}} = \frac{1.73 \times 66.2}{\sqrt{18}} = 27\,\% \qquad (454)$$

where Student's t_q value corresponds to $P = 90\,\%$. Taking into account that the relative error should be smaller, it follows that for a detailed exploration the volume population should be larger ($n > 30$). Before drawing any conclusions the two principal deficiencies of this statistic model must be pointed out:

— the arithmetic mean is not sufficiently substantiated to be taken as a computation parameter:

— relation (454) can only be used in the case of normal distributions because of the connection between probability P and value t_q.

In Figure 108, the distribution of the hydraulic conductivity appears to have an asymmetric character; if we add the fact that the population is a small volume one, it is obvious that the result $\varepsilon_k' = 27\,\%$ is erroneous.

A solution to this problem was given by E. E. Kerkis who proposed, for the fissured-karstic aquifers, to draw the dispersion graph and the frequency curve with quantity $\sqrt{k}$. This manner of solving the problem is supported by the fact that the hydraulic conductivity of these aquifers is proportional to the third power of the fissure opening. It is thus possible to find the symmetric frequency curve. In our case — of the laminar flow — the hydraulic conductivity is proportional to the square of the grains or pores diameter; the frequency curve can therefore be drawn with the help of quantity $\sqrt{k}$ (Table 67 and 68). This led to a normal distribution and to results devoid of the errors of the previous computations (Figures 108 and 109).

Table 67

Values of hydraulic conductivity (k) and of $\sqrt{k}$ according to pumping test data

Well No	k [m/day]	$\sqrt{k}$	Well No	k [m/day]	$\sqrt{k}$
24 H	0.06	0.25	32 H	0.53	0.73
35 H	0.14	0.37	66 H	0.57	0.75
33 Ha	0.20	0.45	36 He	0.64	0.80
44 H	0.25	0.50	34 He	0.70	0.84
38 He	0.29	0.54	29 H	0.83	0.91
31 H	0.34	0.58	39 H	0.83	0.91
30 H	0.46	0.68	27 H	1.00	1.00
37 H	0.50	0.71	7 H	1.21	1.10
36 Ha	0.52	0.72	1 H	1.55	1.25
			10 H	1.64	1.28

Table 68

Distribution of k and $\sqrt{k}$ quantities

k [m/day]	n	n%	$\sqrt{k}$	n	n%
0 —0.30	5	26.3	0.25—0.43	2	10.5
0.30—0.60	6	31.6	0.43—0.61	4	21.1
0.60—0.90	4	21.0	0.61—0.79	5	26.3
0.90—1.20	1	5.3	0.79—0.97	4	21.1
1.20—1.50	1	5.3	0.97—1.15	2	10.5
1.50—1.80	2	10.5	1.15—1.33	2	10.5
	19	100.0		19	100.0

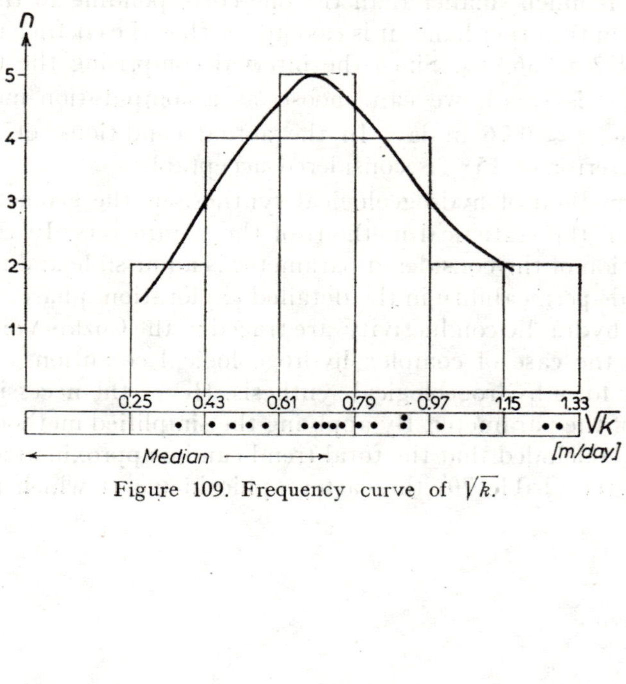

Figure 109. Frequency curve of $\sqrt{k}$.

For comparison, the typical values are given — the mean, the mode and median — of the two distributions (Table 69) from which it follows that in the case of a normal distribution the difference between them diminishes.

Table 69

Distribution		Mean	Median	Mode
Asymmetric	k	0.65	0.53	0.44
Normal	$\sqrt{k}$	0.76	0.73	0.69
	k	0.58	0.53	0.48

If quantity $\sqrt{k}$ is taken into account, all the typical values shall be calculated again

$$R = \sqrt{k_{max}} - \sqrt{k_{min}} = 1.28 - 0.25 = 1.03 \tag{455}$$

$$\sigma = \frac{R}{d_n} = \frac{1.03}{3.69} = 0.28$$

$$V = \frac{0.28}{0.76} 100 = 36.8\,\%$$

$$\varepsilon' = \pm \frac{1.73 \times 36.8}{18} = 15.0\,\%$$

This result proves that the adopted model is close to reality, since the relative error is much smaller than the one corresponding to the asymmetric distribution. On the other hand, it is easy to see that the coefficient of variation drops from 68.2 to 36.8 %. Since the interval comprising the typical values of quantity $\sqrt{k}$ is small, we can choose as a computation mean the value $\sqrt{k} = 0.76$, i.e. $k = 0.76$ m/day. In the actual conditions of the pumping techniques an error of 15 % is considered acceptable.

Another method of hydrogeological synthesis is the geometrical analysis (combined with the statistical method) of the parameters. In this procedure, a linear variation of the considered parameter is admissible and this is the case of Dacian sands permeability in the detailed exploration phase. In Figure 107 the isolines of hydraulic conductivity are traced in the Cuzap-Voivozi southern perimeter. In the case of complex hydrogeological conditions, such a map is unsatisfactory for a hydrogeological synthesis. Hence the necessity of drawing a trend map of the parameter. By applying the simplified method of regression (§ 1.6) it was concluded that the total trend can be approximated by a surface of the IV degree (Table 70) the cartographic image of which is represented

Table 70

Computation of total trend of the hydrualic conductivity

Polynomials contribution / On columns / On rows					Degree I	+0.357	+0.238	+0.119	0	−0.119	−0.238	−0.357
					Degree II	+0.035	0	−0.021	−0.028	−0.021	0	+0.035
					Degree III	−0.110	+0.110	+0.110	0	−0.110	−0,110	+0.110
					Degree IV	+0.018	−0.042	+0.006	+0.036	+0.006	−0.042	+0.018
Mean	Degree I	Degree II	Degree III	Degree IV	SUM	+0.300	+0.306	+0.214	+0.008	−0.244	−0.390	−0.194
0.727	−0.170	+0.144	+0.002	−0.013	0.690	0.99	1.00	0.90	0.70	0.45	0.30	0.50
0.727	−0.085	−0.072	−0.004	+0.052	0.618	0.92	0.92	0.83	0.63	0.37	0.23	0.42
0.727	0	−0.144	0	−0.078	0.649	0.95	0.95	0.86	0.66	0.41	0.26	0.46
0.727	+0.085	−0.072	+0.004	+0.052	0.796	1.10	1.10	1.01	0.80	0.55	0.41	0.60
0.727	+0.170	+0.144	−0.002	−0.013	1.026	1.33	1.33	1.24	1.03	0.78	0.64	0.83

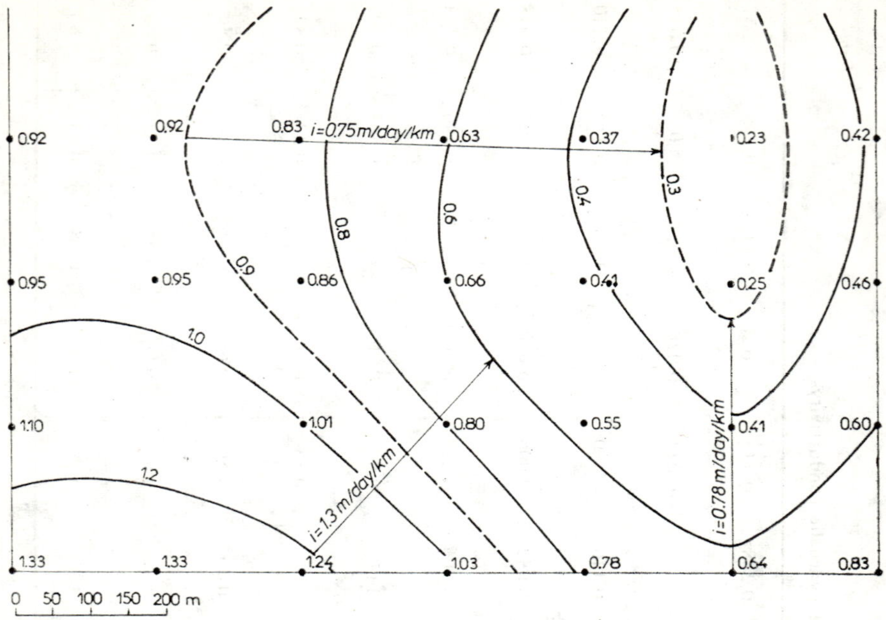

Figure 110. Total trend of the hydraulic conductivity [m/day] in perimeter Voivozi-Cuzap.

by Figure 110, where the gradients of the hydraulic conductivity (i) for values going from 0.74 — 1.3 m/day/km also appear. The residual mean is small: 0.03.

The use of the map with the isolines of the hydraulic conductivity permits us to determine the weighted mean value of k by formula

$$k_m = \frac{\Sigma k_i F_i}{\Sigma F_i} \tag{456}$$

where

F_i is the elementary surface comprised between two neighbouring isolines

k_i is the mean hydraulic conductivity of the elementary surface.

The computation result for the considered perimeter was $k_m = 0.65$ m/day which is close to the value supplied by the statistical method.

We can therefore state that the estimation of the heterogeneity as a function of an aquifer permeability is a quite important problem of applied hydrogeology, on the solution of which depends the efficiency of the capture or drainage works designed. The complex and very complex hydrogeological conditions involve great difficulties in establishing a rule of estimation for the optimum volume of experimental works. The application of the statistic model to solving this problem should lead to the discarding of the subjective criteria still used. As to the necessity of the experimental groups, the experience showed that the processing of the data from the sole experimental well cannot supply true values of the hydrogeological parameters, especially in the case of a reduced permeability ($h < 10$ m/day). Whence, the necessity of using experimental groups with at least two observation wells along a radius. For economy's sake, this requirement cannot be generalized to the deep aquifers. In this case, the optimum solution is that to some of the experimental wells observation wells should be added, so as to permit us to determine the correction factor $\eta = k_0/k_c$ (where k_0 is the actual hydraulic conductivity as deduced from the observation wells data and k_c is deduced from the experimental well data).

This correction factor (estimated by means of several experimental groups) will improve the erroneous hydraulic conductivity determined on the basis of the data supplied by isolated experimental wells. So that this method can be more efficient, the imperfection due to the kind of opening must be quite small in order that the permeability of the adjacent area of the experimental well would be less altered. Therefore natural development will be prohibited while the physico-chemical properties of the drilling fluid should be quite uniform.

2.8 Hydrodynamic study of jointed and fissured aquifers

The practical problems connected with the drainage of underground waters located in fissured and karstic rocks are particularly important for establishing dam sites as well as for the draining of mineral deposits. The vast groundwater resources should be added existing chiefly in the calcareous formations, which constitute an important source for water supply with large yields and of good quality. In order to simplify the terminology, the notion of *fissured rocks* is used; here the term *fissure* includes all the openings in the massive whatever its geological origin (schistosity plane, stratification plane, faults, karstic openings etc.).

In the hydrodynamic research of these aquifers it should be taken into account that the geometry of the fissure network results in two types of aquifers with neatly different régimes. There is ground with a regular and homogeneous fissuration consisting in independent fissures or openings. Such geological formations do not generate proper aquifers since there is no practical hydraulic connections between openings. This type of aquifer could not have a single piezometric level; at a certain level between two aquiferous fissures (each with its piezometric level), there is an impermeable area devoid of water. Such cases are characteristic of certain limestone formations including systems of joints developed at great distances on the horizontal. In the case of formations with irregular and heterogeneous fissuration and with very many orientations there is a possibility of hydraulic continuity and thence of a proper aquifer.

The notion of continuity (or of discontinuity) of a fissured terrain depends on the relative scale of the phenomenon as well as on the fissuring modulus and therefore on the mean distance between two elementary fissures. No general rules are available on this question.

Besides these two principal cases, an intermediate one could exist: a network of parallel aquiferous fissures with the same piezometric level due to the hydraulic connection between them by the intermediate of an open fault or of permeable formation in contact with it.

The investigation "in situ" of these aquifers characterized by a great hydraulic anisotropy would be likely to lead to the determination of some parameters such as: the opening and its relative variation on an elementary fissure, the clogging, the degree of separation or discontinuity of the fissures etc. Since it is practically impossible to determine separately each of these parameters, attempts were made to determine the cummulated effect of them, which would be a representative parameter easy to determine in the field. Its definition starts from the law of the flow through an elementary fissure:

$$v = kI^{1/n} \tag{457}$$

$$\Delta H = S = \frac{L}{k^n} \cdot v^n \tag{458}$$

$$S = \frac{L}{k^n \omega^n} Q^n \tag{459}$$

$$K_g = \frac{k^n \omega^n}{L} \tag{460}$$

$$Q^n = K_g S \tag{461}$$

where

v is — the flow velocity of ground water;
k — the hydraulic conductivity of the elementary fissure which depends on its opening and rugosity, on the physical properties of the water, etc.
I — the hydraulic gradient
n — the exponent of the flow régime (1 ... 2)
ΔH — the head loss equal to the drawdown S
L — the length of the stream-line
Q — the flow rate

In this way, a synthesis parameter K_g is found, called *global hydraulic conductivity*; it can be determined by a system of hydraulically bound fissures [42].

So as to determine the two unknowns K_g and n of the relation (461), at least two tests are required for which Q and S are measured in steady flow. The dimensions of parameter K_g result both from the defining equation (460) and from the equation of yield-drawdown diagram (461); for turbulent flow (when $n = 2$), the parameter K_g can be expressed by m^5/h^2.

The global hydraulic conductivity can also be deduced from Bernoulli's equation written for a flow-line between the surface of the aquiferous current in the fissure and the emergency level (spring or experimental well), corresponding to the reference plane:

$$h_1 = \frac{p_2}{\gamma} + \frac{\alpha v^2}{2g} + \tau_{1-2} \tag{462}$$

where

h_1 is the position energy;

$\frac{p_2}{\gamma}$ — the pressure energy;

$\frac{\alpha v^2}{2g}$ — the kinetic energy;

τ_{1-2} — the head loss.

The head loss is due both to the longitudinal resistances along the rectilinear paths with a uniform cross-section and to local resistances resulting from the cross-section variation. In case the flow occurs through fissured rocks, the separation of these two head losses is impossible. Undoubtedly, the local head losses are very great when the water flows through fissures. Bearing in mind that both the head losses are directly proportional to the kinetic term $\left(\frac{\alpha v^2}{2g}\right)$ multiplied by several resistance factors. Bernoulli's equation is simplified:

$$h_1 = \frac{p_2}{\gamma} + \frac{\alpha v^2}{2g} + \eta \frac{v^2}{2g} + \zeta \frac{v^2}{2g} \tag{463}$$

where η and ζ are the resistance coefficients corresponding to the longitudinal and local head losses

$$h_1 = \frac{p_2}{\lambda} + (\alpha + \eta + \zeta) \frac{v^2}{2g} \tag{464}$$

$$h_1 = \frac{p_2}{\gamma} + \lambda Q^2 \tag{465}$$

$$\lambda = \frac{\alpha + \eta + \zeta}{2g \cdot \omega^2} \tag{466}$$

where λ is the global resistance parameter, i.e. the inverse of the global hydraulic conductivity $\left(\lambda = \frac{1}{K_g}\right)$

$$\lambda Q^2 = \frac{Q^2}{K_g} = S \tag{467}$$

$$h_1 = \frac{p_2}{\gamma} + S \tag{468}$$

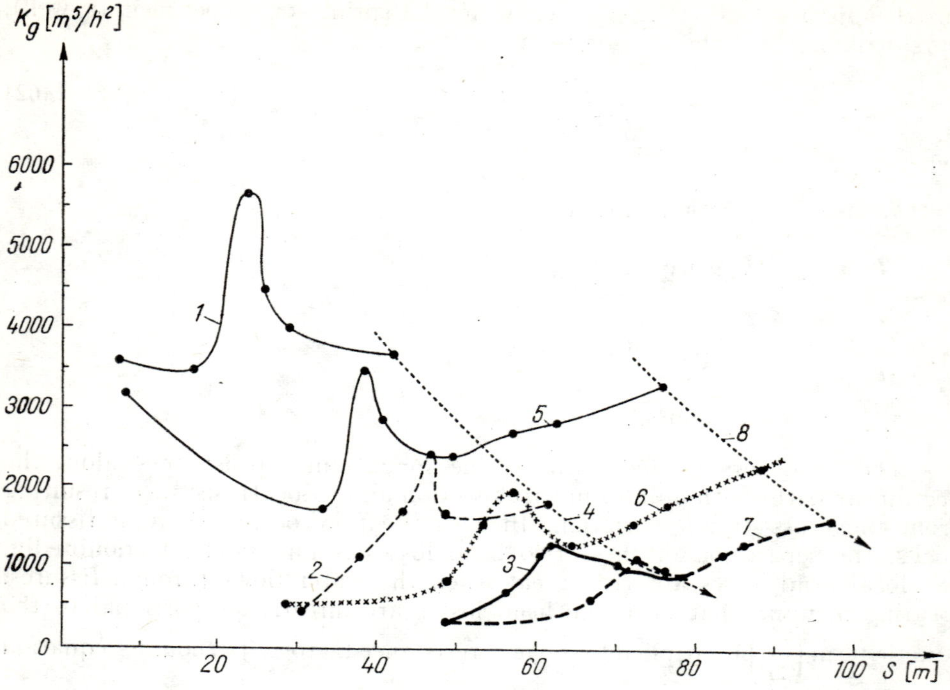

Figure 111. Variation of the global hydraulic conductivity (K_g as a function of drawdown S) at Kipushi ore-deposite (Zair Republic):
System I: *1*—Test A; *2*—Test B; *3*—Test C; *4*—Variation with time of K_g for the system I.
System II: *5*—Test A; *6*—Test B; *7*—Test C; *8*—Variation with time of K_g for the system II.

Relation (467) is practically the same as (451) and is valid for the turbulent flow (K_g). Figure 111 shows the variation of global hydraulic conductivity within the frame of each test of two systems of aquiferous fissures (I and II) drained by two groups of underground wells (1150 m-deep) the ore deposit of complex sulphides Kipushi in the Zair Republic. The different hydraulic behaviour of the two tectonic fissure systems that affect the limestones in footwall of ore body are conspicuous by the variable reduction of parameter K_g in the 14-day interval between tests *A* and *C*:

— for system I, parameter K_g is 4 times smaller;

— for system II, parameter K_g is only 2 times smaller.

After a time, a similitude between K_g values of the two fissure systems were noticed. For example, in the case of test C, when $S = 80$ m, K_g values are practically the same. This is due to an equalization of hydraulic régime in Kakontwe calcareous complex (where the two fissure systems are met) when the drawdown is high.

The reduction with time of parameter K_g is favourable to the efficiency of the mining drainage.

2.9 The network of piezometers

The control of the evolution of the ground water resources in a working aquifer or of the drainage efficiency can be achieved by a continuous supervising of the piezometric surface by means of a network of observation wells (piezometers); this method is also called *piezometry*. It is based upon the fact that the piezometric surface represents the final effect of all the natural or engineering factors likely to affect the aquifer.

With the help of the systematic measurements of the piezometric level, periodic piezometric maps can be drawn on which the hydraulic gradients can be defined, flow lines can be traced and areas can be delimited where the working conditions of the aquifer are not conformable to the forecast.

The piezometry is the most adequate method for the aquifers characterized by a great heterogeneity (e.g. those located in calcareous formations or affected by tectonic dislocations) because the classical hydrogeological parameters are very difficult to determine by the experimental pumping method. In such aquifers, the network of piezometers should not follow a regular geometry since the locations are determined by the general hydrogeological conditions and by the technical conditions of the capture or of the drainage. The processing of the piezometric data with a view to determining the hydrogeological parameters can be made by means of the mathematical models; to this end in order to have a higher efficiency, the method of finite differences or the method of the finite element can be used.

During the drilling of each well-piezometer, a programme of hydrogeological research must be accomplished (pumpings, piezometric level recovering, chemical sampling, etc.) and the resulting data must be directly processed and interpreted. By so doing, these data can point out immediately the location and the drilling depths for the subsequent piezometers. The final diameter of the screen can be reduced down to 2 inches, but for the hydrogeological study of the well the minimum diameter should be 6 inches. The borehole can be provided with a gravel pack that ensures a long working period for the piezometer.

When the piezometric control must be extended to a group of 2—3 superposed aquifers the same piezometric hole can be fitted with 2—3 distinct screens providing a good hydraulic isolation between them.

The piezometry is a very effective method for the control of a rational exploitation and of the aquifer drainages because it permits to apply immediate measures with a view to comply with the technical design.

3

THE INFLUENCE OF THE PETROPHYSICAL FACTORS ON THE DRAINAGE CAPACITY OF SANDS

The term "drainage" describes the specific drainage works which accomplish the evacuation of the ground waters from the saturated formations. The groundwater drainage works permit the reduction of the level of the ground waters with a certain purpose (mining exploitation, civil engineer works etc.).

The Romanian technical vocabulary also uses the term "dewatering" which in fact expresses the same drainage processes applied to the drainage of agricultural areas and surface building construction areas; therefore this term is only restricted to phreatic horizons (unconfined aquifers). In the case of agricultural areas, the dewatering process could also include the zone of aeration area when the humidity reduction is required as well.

Dewatering is also the drainage of the useful mineral fields both in open pits and underground working. In this case the useful domain is obviously much larger since it includes both the unconfined (phreatic) and confined aquifers at great dephts. The dewatering degree can widely vary up to the total exhausting of an aquifer.

In field hydrodynamic research the mean characteristics of the rocks must first be studied. While in the capture of the ground waters the mean hydrodynamic characteristics of the aquifer are sufficient to solve the drainage problems some quantitative aspects of the influence of the petrophysical factors on the capacity of sand drainage should be tackled.

The importance of the petrophysical factors grows as the sand grain size diminishes, which means that to estimate the drainage capacity of the fine to silty and medium sands (very frequently met with) these factors have to be taken into account. The considerable influence of the petrophysical factors upon the drainage capacity of these types of sand grain-size often determines the non-concordance between the hydrodynamic forecasts and the actual behaviour of the formation. For this reason, the choice and the sizeing of certain drainage works (wells for underground dewatering, well points system etc.) is often made on empirical bases only by using the acquired experience. But a suitable treatment of these cases, i.e. the achievement of a realistic design of the drainage works can only be made by a harmonious combination of the petrophysical factors with the acquired experience about similar objectives from a hydrogeological standpoint.

The knowledge of the petrophysical factors influence is particularly important on choosing the size of the dewatering system of the coal or other mineral deposit, especially of those worked in underground conditions. Sometimes the exploitation of useful minerals requires the elimination of a large part or entire quantity of the water hold up in the pores (this even could ultimately lead up to the exhausting of the aquifer), which would thus result in a sandy massive with improved physico-mechanical properties. The discussion of the petrophysical factors is made on the basis of the experimental data supplied by sand samples in the Şotînga-Doiceşti lignite field.

The hydrodynamic research might emphasize a certain statistical homogeneity. Actually, however, the grain-size and the parameters present a greater variation that should be taken into account chiefly in the case of the underground dewatering systems. An important contribution to hydrogeological research can be made by the geological exploration drillings from which the sand samples required for the petrophysical study can be gathered. Similar data can be furnished also by the sand samples gathered during the mining work drifting; such a research, at a very minute scale, constitutes very valuable documentary material. By the hydrogeological interpretation of the petrophysical parameters (the values of which are in fact punctiform) the degree of uniformity of the aquifer drainage capacity will be found. When the number of samples is high, the processing and interpretation of the petrophysical data may make use of the mathematical statistic methods.

3.1 Grain-size composition

The drainage capacity of a sandy formation is chiefly determined by the parameters of the porous space: the cross-sections of the pores and of the interconnection channels between them. The determination of these parameters implies however an intricate working technique not yet to be found in the usual programme of the laboratories. But on the other hand these parameters are determined by the characteristics of the grain-size distributions: the composition by fractions, fraction sizes, non-uniformity of the composition. The laboratory technique for determining the grain-size distributions is very simple so that they can constitute an important indicator of the drainage capacity of the sands.

Of late, many endeavours were made to establish the various types of drainage on the basis of the grain-size composition. Thus, on the diagram of Figure 112 an area of the slow gravitational drainage is separated, corresponding to a portion of the floury sand with $d = 0.05 — 0.1$ mm. Theoretically, this fraction can also be gravitationally drained but over a long time interval. The lower limit of the vacuum drainage goes down even in the clays domain

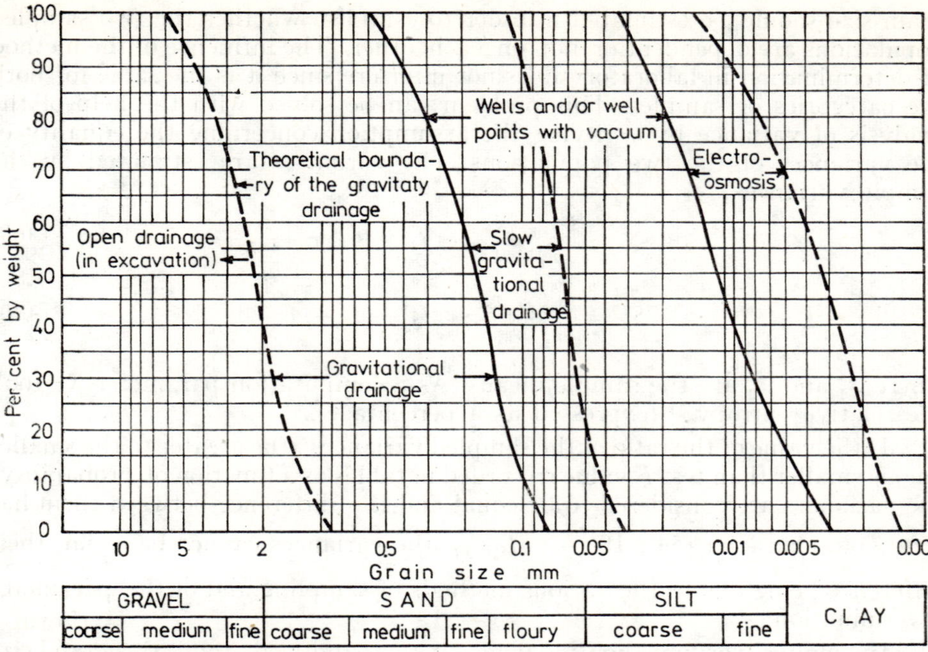

Figure 112. Classification of the dewatering systems according to the grain-size (after Moretrench Corp.).

to $d = 0.003$ mm provided that the proportion of this fraction is $< 15\ \%$. The interpretation of the grain-size data solely by means of a simple ternary diagram constitutes a simplistic method and cannot form a basis of hydrogeological interpretation. In the same way the following indicators may be tackled: the grain-size domain, the mean grain-size distribution, the grain-size coefficients (mean diameter and sorting degree, asymmetry coefficient, sorting degree and sorting coefficient), and the entropy.

The interpretation of all these indicators can be made as a function of the minuteness degree and of the data volume for the whole aquifer or separate by structures or fields.

3.1.1 Statistical estimation of the influence of the sample collecting conditions

Besides the grain-size analyses on core-samples of the exploration drillings the grain size determinations could equally be made on the samples collected from drilling performed in underground works.
In such a case, it is obvious therefore that both the populations should be within the same statistical complex, which would lead to a higher degree of certainty of this interpretation. However, before using these data, the

grain size should be estimated in order to establish whether the two sampled populations are dependent or not on each other. The influence of the method of determination in laboratory does not interfere since it is the same for both the categories of samples. This problem can be solved with the help of the analysis of variance by verifying the assumption concerning the equality of the variances of the two populations. The variances are estimated by the sample variances:

$$S_1^2 = \frac{(x_{i1} - \bar{x}_1)}{n_1 - 1}; \quad f_1 = n_1 - 1 \tag{469}$$

$$S_2^2 = \frac{(x_{i2} - \bar{x}_2)}{n_2 - 1}; \quad f_2 = n_2 - 1 \tag{470}$$

where $\bar{x}_1$ and $\bar{x}_2$ are the sample means. As a computation parameter we took the relative entropy [1] (expressed as a percentage).

If the value of the ratio of the sampled variances (the greater to the smaller one) is smaller than test F value (recorded in tables as a function of probability) the variances are considered to be equal and the difference between them has a random character [54]. If $\frac{S_1^2}{S_2^2} > F_{(1-P)}$ the variances cannot be equal, their difference being due to the various methods of sampling and of the parameter determination.

In many working lignite fields it was noticed that the grain size composition determined by core samples taken from exploration wells is somewhat different from that determined by core samples from the underground drillings. This is due to the fact that the exploration drillings (both geological and hydrogeological) are executed by the rotary-hydraulic system while the underground ones are usually made by hand and without a drilling fluid. In order to apply the statistical principle, the first sand bank was considered in the top of coal layer II. Each of these populations consists of $n_1 = 37$ samples for the exploration wells and $n_2 = 30$ samples for the underground wells. The results were:

$$S_1^2 = \frac{\Sigma(x_{i1} - 71.9)^2}{36} = 198.7$$

$$S_2^2 = \frac{\Sigma(x_{i2} - 51.6)^2}{29} = 52.8$$

$$\frac{S_1^2}{S_2^2} = \frac{198.7}{52.8} = 3.76$$

From the tables it follows that value $F_{0.001} = 3.10$. The ratio of the variances being greater than that, it is considered that the difference is no more random but due to the sampling factor. The probability of this hypothesis is higher than 0.999.

---

[1] The entropy will be computed in § 3.1.4.

In order to verify this the mean grain-size distributions of both the populations were marked on the probabilistic diagram (Figure 113). The mean grain size distribution in the underground drillings is well represented by a straight line, which proves the lognormal character and implicitly the accuracy of these distributions. On the same diagram the mean grain-size distribution corresponding to the exploration drillings is represented by a broken line at the point with $d = 0.1$ mm, which shows a different character from the lognormal one. Up to $d = 0.1$ mm, both the mean grain-size distributions coincide. This deviation can be due to:

— the core-sample contamination by the drilling fluid;
— the non-accuracy of the core-sampling

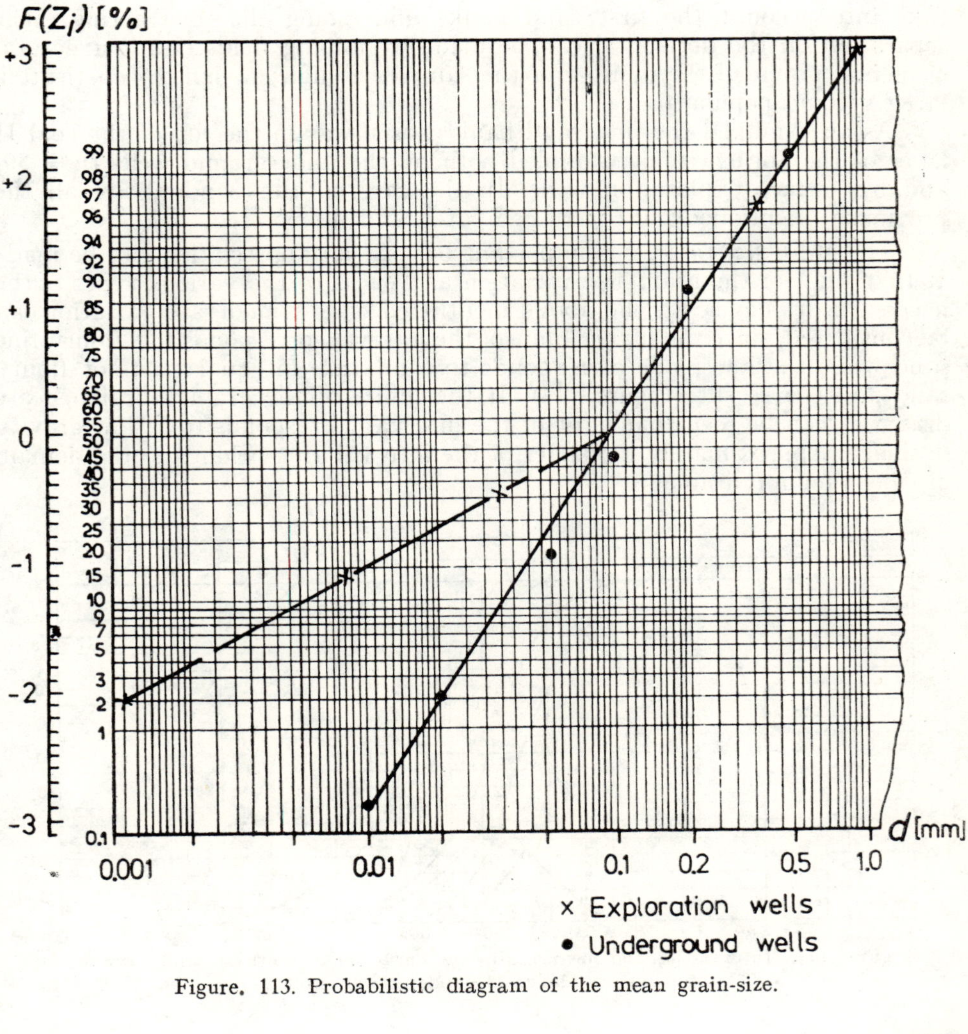

Figure. 113. Probabilistic diagram of the mean grain-size.

While sample collecting appears as a random factor, the sample contamination is a permanent and therefore a principal factor of the deviation of the grain-size distribution from the lognormal character. This factor should be considered, because the hydrogeological interpretation is only made on the basis of core-samples from the exploration well and when it is neglected erroneous conclusions can arise.

Grain size of the sand bank in the top of coal layer II (Şotînga). The systematization of the grain size data and their hydrogeological interpretation should be made by horizontally and vertically delimited unities. The delimitation on the vertical is made as a function of the interval to be dewatered, by the same dewatering works. For instance, in the case of an underground dewatering by means of penetrating screens, it is sufficient to take into account the first sand banks undergoing the dewatering. The separation on the horizontal can be made by mining fields or by structural elements, provided the number of the samples is high enough to constitute a large volume population.

As an example, the first sand bank chosen was in the top of the coal II layer of the Şotînga mine, sampled both by the underground wells ($n = 30$) and by surface exploration wells. In the case of this sandy body all the previously mentioned data were computed and interpreted.

The *grain size interval* is important because an appreciation can be made at first sight of the grain size variation as well as of the various types of the necessary drainages. In the above mentioned case a moderate variation can be considered, since the higher limit of the interval corresponds to a floury-fine sand while the lower one corresponds to a mixture in equal parts of floury sand and coarse silt (Figure 114). If the limits of the gravity drainage are drawn, it can be seen that most of the interval corresponds to a slow gravity drainage. Approximately a quarter of the interval area belongs to the domain of the vacuum drainage.

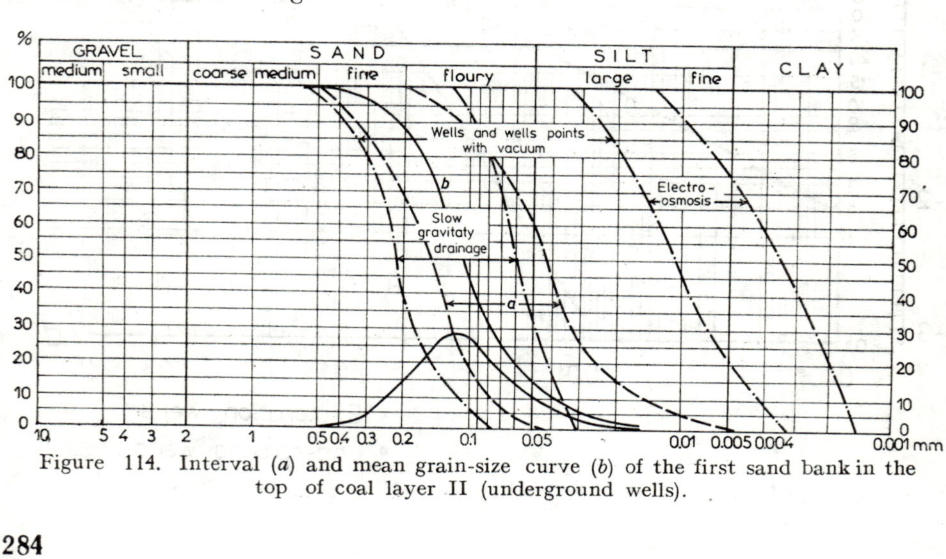

Figure 114. Interval (*a*) and mean grain-size curve (*b*) of the first sand bank in the top of coal layer II (underground wells).

The *mean grain size curve* is obtained by the mean arithmetic computation of all the fractions. It is necessary because it estimates the granulometric trend. In our case, it slightly deviates from the median line of the interval towards the higher limit of the latter. Practically, owing to this position it is completely framed within the domain of the slow gravity drainage.

The granulometric domain corresponding to the drainage by vacuum shall be scrutinized so as to be able to specify whether it is due to small residues of the pelitic fraction or to the latter having an important weight.

Sample	8	12	14	34	36	37	54	55
Content (%)	7	3	15	25	12	8	85	30

Except sample 54 and to a certain extent samples 34 and 55, it turns out that the small contents of pelitic material do not determine the drainage type. Sample 54 represents a special case; it indicates in fact the lower limit of the grain size interval. If this sample is excluded, the grain size domain corresponding to the vacuum is much reduced (about an eighth of the whole interval area), the effect of which could be the generalization of the slow gravity drainage.

The extent of the domain of drainage by vacuum in the grain size interval established on the basis of the exploration wells (Fig. 115) — approximately half of the interval area — emphasizes once more the deviation of these granulometries from the actual ones since the experiment proved the efficiency of the gravitational dewatering.

The *character and the type of the grain-size distributions* are determined by the sorting degree of the water bearing formation. The sorting degree determines the geometry of the porous space and therefore the water yielding properties. All the frequency curves drawn at a logarithmic scale emphasized unimodal and lognormal distributions which shows the relatively constant character of the sedimentation régime in the whole area of the bank under study. W. Engelhardt [18] studied the geometric models of certain distributions

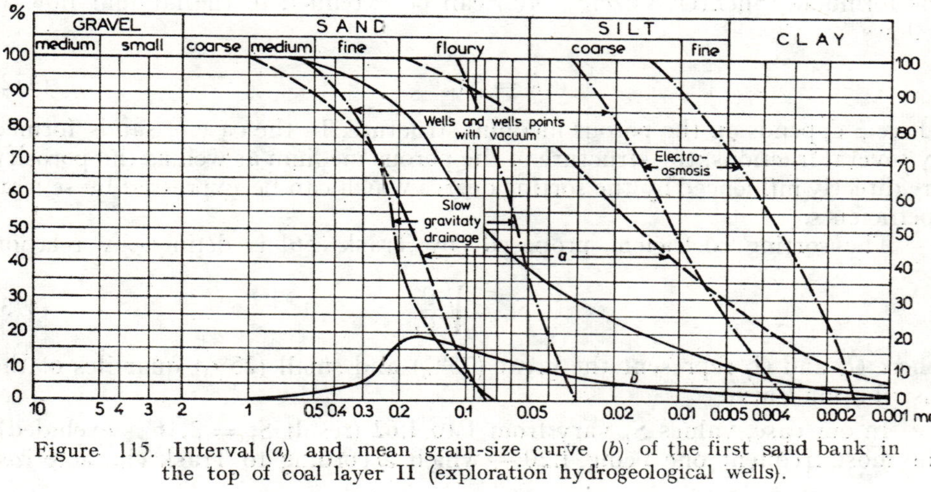

Figure 115. Interval (*a*) and mean grain-size curve (*b*) of the first sand bank in the top of coal layer II (exploration hydrogeological wells).

of natural sands and drew the conclusion that the symmetrical distribution with a well defined maximum corresponds to the highest porosity. A clear definition of the maximum appears in the case of a small number of fractions. Evidently, a lognormal distribution alters this conclusion as a function of the asymmetry sign. In the above-mentioned case, all the grain-size distributions present a positive asymmetry which contribute to increa se the porosity. It is thus possible to appreciate that the packed sands present porosities of 40—50% which leads to the increase of the static water resource. The porosity of most of the samples collected from underground of the Șotînga mining are within these limits, though a few samples have porosities of $n < 40\%$ and $n > 50\%$. According to Engelhardt, the sands with a high porosity are also well sorted. It is to be noted that all the documentation of hydrogeological exploration about the lignite fields emphasized the lognormal character of the grain size distributions; the rather high variation of the porosity (usually $n = 35$—55%) are due to: the existence of a prevailing fraction, the great number of fractions and the distribution asymmetry.

The *grain size coefficients* are particularly useful when comparing the results. The various handbooks propose many coefficients determined by graphical and analytical methods, the latter making use chiefly of methods of mathematical statistics. The choice of these coefficients as well as their interpretation are made as a function of their purpose.

The mean grain size, expressed by the mean diameter (d_m), is an important indicator of the unimodal distributions. This indicator is computed as a weighted mean of the granulometric fractions the total of which is $\sum g_i = 100$

$$d_m = \sum (g_i\, d_i)/\sum g_i \tag{471}$$

The concentration of two-thirds of the results within a small interval ($d_m = 0.10 - 0.12$) brings into relief the reduced character of the fluctuations of the available energy on the erosion areas and in the transport systems. The mean diameter is used for estimating the hydraulic conductivity since all the formulae (Slichter, Cozeny, etc.) can be extended to the laminar flow as follows

$$k = \beta d_m^2 \tag{472}$$

where β depends on the porous medium structure. In the case of sands formed by several fractions, the structure of the porous medium as well as the porosity are directly influenced by the sorting degree which can be expressed by several coefficients.

The sorting coefficient proposed by Trask [56] is defined by relation

$$S_0 = \sqrt{\frac{Q_3}{Q_1}} \tag{473}$$

where Q_3 and Q_1 represent the great (75%) and small (25%) quartiles of the distribution curve.

In our case, values S_0 vary from 1 to 1.62 (result $S_0 = 2.16$ is excluded), the most frequent one being 1.20 — which according to Trask characterizes the sands as well sorted ($1 < S_0 < 2.5$). This coefficient is not however significant for dewaterings since it disregards the very finest fractions with a cumulative content of 25% and the presence of which are likely to alter considerably the structure of the porous space.

The asymmetry of the grain size distributions is estimated with the help of the asymmetry coefficient [56] as defined by formula

$$S_a = \frac{Q_1 Q_3}{M_d^2} \tag{474}$$

where M_d is the median diameter (50%).

Since the computed values are 0.68—1.25 (Figure 116), they characterize the sands as well sorted (according to Trask this quality goes up to $S_a = 2.5$). The drawback of this coefficient is theoretically the same as that of the sorting coefficient since both of them are computed on the basis of the quartiles. In some cases, the quartiles are replaced by deciles 10 and 90% on the very purpose of considering the extremities of the distribution. This new coefficient of asymmetry (S'_a) is in some samples smaller while in some other ones greater than S_a: its higher limit increases however considerably and reaches such values as 7—8. This wide variation of the asymmetry coefficient is due to an important diminution of the slope of the distribution curve in its extreme portions: the lower portion (< 25%) reveals a decrease and the higher part (> 75%) an increase of S'_a in regard to S_a. Therefore the degree of objectivity of the asymmetry coefficient (as well as that of S_0) is small because the punctiform values taken into account for their computation are arbitrarily chosen on the distribution curve.

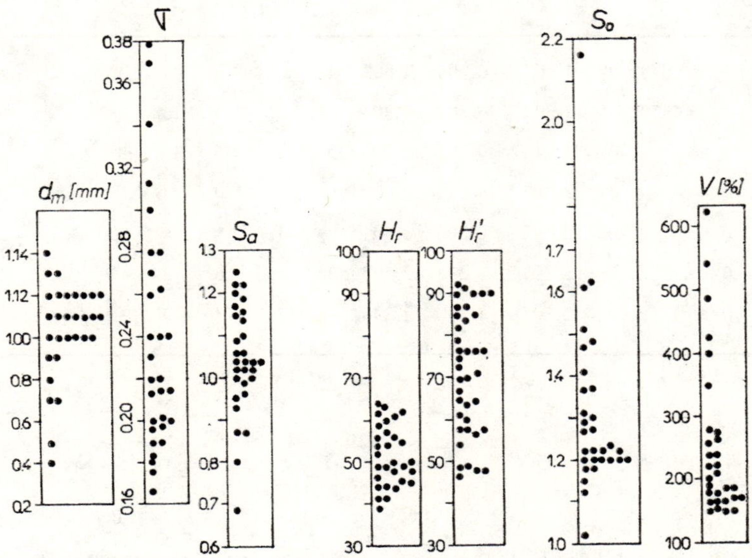

Figure 116. Dispersion graphs of the grain-size coefficients (first bank in the top of Șotînga coal layer II).

The estimation of the sorting degree can be made by the statistical computation of the distribution deviation from the mean diameter. The greater the deviation, the smaller the sorting is. Consequently, the standard deviation (σ) is also given the name of *sorting coefficient* and is expressed as a logarithmic value when the distributions have a lognormal character. The sorting degree can equally be expressed as a function of the coefficient of *variation* defined as the ratio of σ to d_m (in %).

For a quick calculation of parameters d_m and σ the method of the conventional moments (Table 71) was applied by using the logarithms of the finite dimensions of the grain size classes. Here are the computation relations:

$$\log d_m = x_0 + b_m V_1 \tag{475}$$

$$\sigma = b_m \sqrt{V_2 - V_1^2} \tag{476}$$

where x_0 is the mean diameter of the grain size class with the conventional number $\alpha = 0$, b_m is the mean value of the grain size classes, and V_1 and V_2 are the first and second conventional moments.

Let us note that the σ values are not concentrated within a very small interval (as the case is with the mean diameter) but uniformly enough spread

Table 71

Examples of grain size computation (sandy bank, above coal layer II Șotînga)

Crt. No	Grain size [mm]		Log interval		P4				P5			
	mm	log	b	x	$\Delta g(\%)$	α	$(\Delta g)\,\alpha$	$(\Delta g)\,\alpha^2$	$\Delta g(\%)$	α	$(\Delta g)\alpha$	$(\Delta g)\alpha^2$
1	0.5 ÷0.32	$\bar{1}.699 \div \bar{1}.505$	0.194	$\bar{1}.602$	1	+3	3	9	1	2	+ 2	+ 4
2	0.32 ÷0.20	$\bar{1}.505 \div \bar{1}.301$	0.204	$\bar{1}.403$	4	+2	8	16	4	1	+ 4	+ 4
3	0.20 ÷0.13	$\bar{1}.301 \div \bar{1}.114$	0.187	$\bar{1}.208$	35	+1	35	35	40	0	0	0
4	0.13 ÷0.08	$\bar{1}.114 \div \bar{2}.903$	0.211	$\bar{1}.008$	48	0	0	0	37	−1	−37	+37
5	0.08 ÷0.05	$\bar{1}.903 \div \bar{2}.699$	0.204	$\bar{2}.801$	8	−1	−8	8	12	−2	−24	+48
6	0.05 ÷0.031	$\bar{2}.699 \div \bar{2}.491$	0.208	$\bar{2}.595$	3	−2	−6	12	5	−3	−15	+45
7	0.031÷0.02	$\bar{2}.491 \div \bar{2}.301$	0.190	$\bar{2}.396$	1	−3	−3	9	1	−4	− 4	+16
					$\Sigma + = 46$ $\Sigma = 89$				$\Sigma + = 6$ $\Sigma = 154$			
					$\Sigma - = 17$				$\Sigma - = 80$			
					$\Sigma = +29$				$\Sigma = -74$			
	V_1				0.29				−0.74			
	V_2				0.89				1.54			
	$\log d_m$				$\bar{1}.066$				$\bar{1}.06$			
	d_m mm				0.12				0.11			
	σ				0.18				0.200			
	$V(\%)$				150%				182%			

and within a larger interval ($\sigma = 0.16 — 0.38$). This behaviour should be attributed to the moderate variations of the hydrodynamic and morphological conditions on the sedimentation area.

A diminution of the sorting degree while the mean diameter remains unchanged will obviously lead to a diminution of the permeability. In this connection, W. C. Krumbein pointed out that a well sorted fine sand can display the same permeability as a coarse sand poorly sorted. In the case under study, since the variation interval is rather small ($\Delta\sigma = 0.22$), the permeability variations due to the sorting degree will present the same character.

Most values of the coefficients of variation are from 150 to 250%, which means a small to average variability for grain size of the sand bank.

3.1.2 Use of the relative entropy as a measure of the sorting degree

An important application of the relative entropy is found in the study of the grain-size distributions where it is considered as the true measure of the sorting degree. Value $H_r = 0$ corresponds to a very good sorting (all the grains are of a single class), while the maximum value ($H_r = 100\%$) corresponds to a very poor sorting (the contents of all the grain-size classes are equal).

By the relations in § 1.4 the relative entropy was computed of the grain-size samples from the underground wells (Table 72). From the dispersion graph (Figure 116), it follows that the values are compactly grouped ($H_r = 39—64\%$). In the literature, the sorting degrees are not differentiated

Table 72

Examples of relative entropy computation H_r(sandy bank above II coal layer Şotinga); sample collected from underground drillings

Crt. No	Grain size [mm]	P4			P5			P7			P8		
		$\Delta g(\%)$	$-p \ln p$	H_r	$\Delta g(\%)$	$-p \ln p$	H_r	$\Delta g(\%)$	$-p \ln p$	H_r	$\Delta g(\%)$	$-p \ln p$	H_r
1	2.0 —1.0	—	—	—	—	—	—	—	—	—	—	—	—
2	1.0 —0.5	—	—	—	—	—	—	—	—	—	—	—	—
3	0.5 —0.25	1	0.046	—	2	0.084	—	4	0.129	—	1	0.046	—
4	0.25 —0.13	34	0.366	—	42	0.364	—	31	0.364	—	29	0.359	—
5	0.13 —0.06	60	0.307	—	45	0.360	—	52	0.340	—	52	0.340	—
6	0.06 —0.032	4	0.129	—	9	0.216	—	8	0.202	—	11	0.243	—
7	0.032—0.016	1	0.046	—	1	0.046	—	5	0.150	—	4	0.129	—
8	0.016—0.006	—	—	—	—	—	—	—	—	—	3	0.106	—
9	0.006—0.002	—	—	—	—	—	—	—	—	—	—	—	—
10	TOTAL	100%	0.894	0.62	100%	1.070	0.49	100%	1.185	0.54	100%	1.223	0.62

as a function of the H_r value. It is appreciated that for small H_r (25—55%), the sorting is considered as a good one and for a 50—75% H_r, as a moderate one.

So as to estimate the influence of the sampling conditions, the relative entropy was moreover computed for the grain size of the samples from the exploration wells. The dispersion graph shows a great variation of the sorting degree ($H'_r = 47 - 92\%$) on the considered area going down to a very poor sorting; this emphasizes the granulometric curve errors due to samples from the hydrogeological or geological exploration drillings.

3.2 Capillary potential

The classical hydrodynamic analysis of the drainage process is based on the assumption that the soil and the rock are saturated and the interaction between the water and the solid tectonic framework is only due to the friction between water and particles. The underground measurements in the case of a free dewatering (without pumping) shows however that the drainage dynamics is more intricate and could not be simply reduced only to a continuous advance of the drainage front that would divide permanently the two hydraulically distinct areas — the drained zone and the saturated zone. The prevalence of the fine and floury sand fractions and the small contribution of the dynamic ressources to the making up of the flow rate — in most of the Romanian lignite fields — are the main factors which complicate the dynamics of the drainage process.

The moisture distribution curves, experimentally established in laboratory [1] on similar terrains, demonstrated that the continuous advance phase of the drainage front is rather short-lived and during it the water free or very weakly retained is drained. In the second phase of the drainage the terrain is unsaturated and the drainage rate diminishes continuously owing to the increase of the interaction between the water and the solid frame. So as to understand this interaction, the forces and potentials at stake have to be considered. Potential Φ of the water in the terrain is supplied by expression

$$\Phi = -\Phi_1 - \Phi_2 + \Phi_3 + \Phi_4 + \Phi_5 \qquad (477)$$

where

- $-\Phi_1$ is the adsorption potential;
- $-\Phi_2$ is the surface tension potential;
- $+\Phi_3$ is the osmotic potential (the value of which can be neglected)
- $+\Phi_4$ is the Newtonian potential ($= +gh$)
- $+\Phi_5$ is the potential due to pressure p.

The retention or capillary potential is supplied by expression

$$\Phi_r = \Phi_1 + \Phi_2 \qquad (478)$$

If the soil is saturated, $\Phi_r = 0$,thence

$$\Phi = \Phi_4 + \Phi_5 \quad \text{or} \quad \Phi = gh + \frac{P}{\gamma} \tag{479}$$

Physically, the capillary potential is the moisture potential resulting from the moisture difference between two points of the terrain. Therefore, flow rate q by unit cross-section is equal to the capillary potential difference between two points at distance x from each other:

$$q = -k_w \frac{d\psi}{dx} \tag{480}$$

where k_w is the coefficient of the capillary hydroconductivity.

The capillary potential can be measured by means of the suction ($\Phi_r = H$) which represents the pressure decrease of the water in the pores with regard to the atmospheric pressure due to the interaction between the water and the mineral frame. Practically this decrease is determined by the height to which a gram of water goes up in the terrain within the static gravitational field. Since the variation of the capillary potential is very large (H is expressed in cm), the logarithm of Φ_r denoted by pF is usually used, i.e. $pF = \log H$.

The methods of determining the suction depend on pF value. Since the sands in the lignite deposits have a $pF < 3$, the method of devices with a suction plate may be used. Using this method, the geotechnical laboratory INCERC-Bucharest constructed the suction-moisture (moisture retention) curves of three sand samples taken from the Șotînga area, namely samples N and Z from the outcrops and sample C from the underground. They were chosen so that they could include an interval of large grain size range permitting to appreciate the dependence of the capillary potential on the grain-size composition. The principal geotechnical parameters of the three samples are recorded in Table 73 and the grain-size curves are represented by Figure 117.

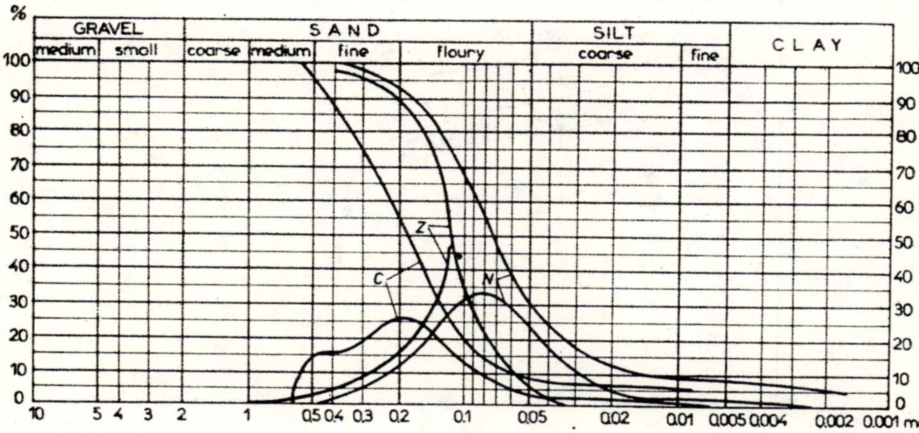

Figure 117. Grain-size distribution and frequency curves of sand samples C, N and Z Șotînga.

Grain size (%)	Sample *C*	Sample *Z*	Sample *N*
Coarse sand	5	3	—
Medium sand	25	3	—
Fine sand	50	59	27
Floury sand	11	29	38
Silt	5	6	23
Clay	4	—	7
Porosity (n%)	42.3 ... 45.3 / 45	47.5 ... 50.7 / 49.4	49.2 ... 52.5 / 50.7
Specific gravity γ_s[to/m³];	2.65	2.65	2.65

Figure 118 represents the suction-moisture curves of the three samples. Since $pF < 3$ (1 at), the suction will also be given in the diagrams in cm H_2O. These curves permit us to estimate the moisture decreases by applying various dewatering techniques.

The differences between the capillary potentials of the three samples agree with the grain-size composition. Thus, sample *N* (fine-floury sand) has the highest capillary potential, owing to its high silt fraction; samples *C* and *Z* offer equally a palpable difference between their capillary potentials, owing to the various grain size of the medium sand fractions (25% for sample *C*) against that of the floury sand (29% for sample *Z*). There results from the analysis of the suction-moisture curves that the work spent by the water

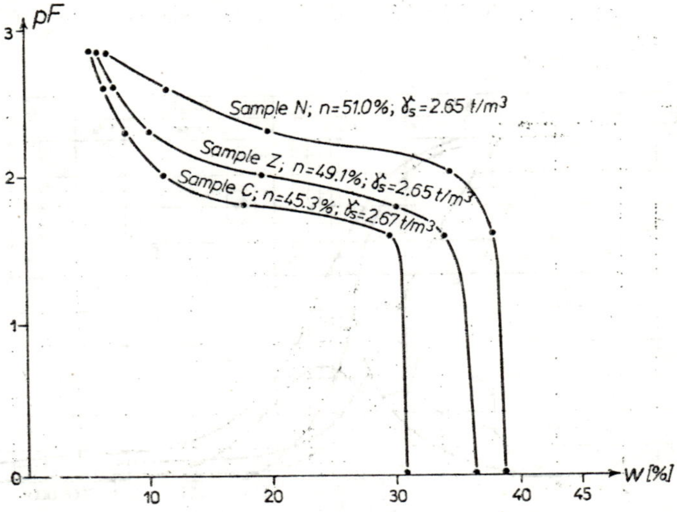

Figure. 118. Suction-moisture curves of the sand samples from the Șotînga Dacian.

displacement within the rock is a continuous function of the moisture content. So that the capillary water could be drained, the draining force should reach a certain value of the capillary potential; only in this case an active drainage takes place, accompanied by the corresponding moisture. To exemplify this, the degree of saturation and the coefficient of surface storage were computed for several steps of the capillary potential (Table 74).

Table 74

pF(H)	Humidity w%			Saturation degree S			Surface storage coefficient n_s[%]		
	C	*Z*	*N*	*C*	*Z*	*N*	*C*	*Z*	*N*
0	31.1	37.0	39.0	1.00	1.00	1.00	0	0	0
1.7 (= 50 cm)	27.0	33.5	37.6	0.87	0.90	0.97	5.85	4.94	1.52
2.0 (= 100 cm)	11.5	20.0	35.0	0.37	0.54	0.90	28.35	22.72	5.07
2.30(= 200 cm)	8.2	9.8	20.0	0.26	0.26	0.51	33.30	36.56	25.86
2.48(= 300 cm)	7.0	8.2	14.0	0.23	0.22	0.36	34.65	38.53	32.45
2.60(= 400 cm)	6.5	7.4	11.5	0.21	0.20	0.30	35.55	39.52	35.49
2.70(= 500 cm)	6.0	6.9	9.3	0.19	0.19	0.24	36.45	40.01	38.53
2.78(= 600 cm)	5.7	6.3	7.4	0.18	0.17	0.19	36.90	41.00	41.07
2.83(= 675 cm)	5.5	5.8	6.3	0.18	0.16	0.16	36.90	41.50	42.59

Beside these parameters, we can also take into consideration the dewatering degree (A), defined as the ratio of the yielded water volume to the pores volume. Numerically $A = 1 - S$.

The data recorded in Table 74 allow us to draw the variation diagrams of the degree of saturation and of the coefficient of surface storage as a function of the capillary potential (Figure 119 and 120). Note that the storage coefficients at $H < 400$ cm H_2O are smaller for sample N than that for samples Z and C: the ratios change for $H > 400$ cm H_2O and finally, for $H = 675$ cm H_2O, sample N has the highest coefficient of storage.

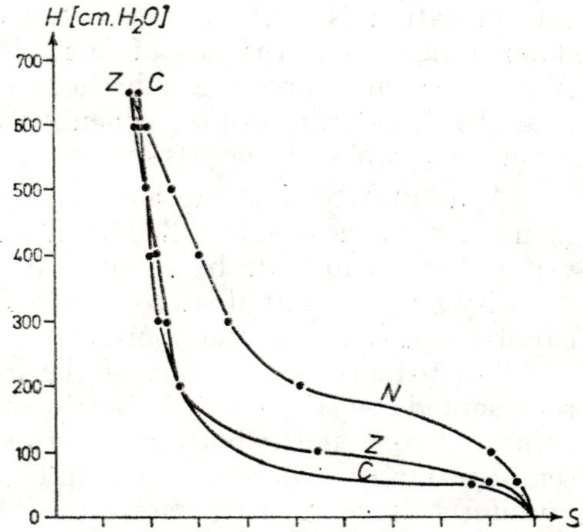

Figure 119. Suction-moisture degree curves of the sand samples from the Șotînga Dacian.

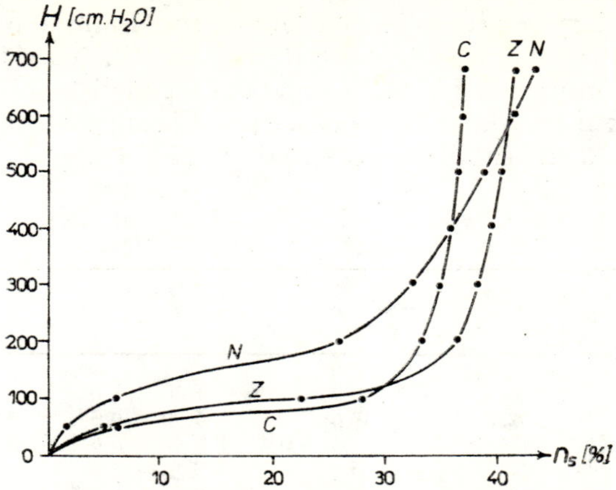

Figure 120. Suction — coefficient of surface storage of the sand samples from the Șotînga Dacian.

S. Andrei [1] pointed out that for pedology's sake the gravitational drainage corresponds to a suction $H = 300$ cm H_2O ($pF = 2.5$). To this suction, the following reductions correspond of moisture w and of degree of saturation S:

	w[%]	S
— sample C	31.1 — 7.0 = 24.1;	1.0 — 0.23 = 0.77
— sample Z	37.0 — 8.2 = 28.8;	1.0 — 0.25 = 0.75
— sample N	39.0 — 14.0 = 25.0;	1.0 — 0.34 = 0.66

From Figure 120, it follows that samples C and Z should not necessarily reach such values since $n_s = 30\%$ (which corresponds to 60—70% of the water elimination) can be obtained for $H = 125$ cm H_2O. This is difficult in the case of sample N (floury-fine sand) because $n_s = 30\%$ corresponds to a high enough capillary potential: $H = 250$ cm H_2O. A practical conclusion of this interpretation is that rock formation of types Z and C can be gravitationally dewatering even in the case of small drawdowns. The case is different for type N sands because pressure reductions above 250 cm H_2O can only be achieved near the dewatering works; when this is the case, a higher density of the dewatering works is necessary or else a forced dewatering.

A frequently used method of studying the capillarity of sands consists in using a vertical tube filled with sand and immersed into a vessel full of water. If the tube can be divided into several sections, the moisture and porosity can be individually determined, and finally the dependence of the capillary pressure on the moisture content is obtained.

The balance distribution of the moisture contents in the type C sand is represented by Figure 121 which shows that the practically completely saturated zone at the bottom is very small (5—10 cm). The zone of capillary semi-continuity (where only the fine pores are saturated) is characterized by a uniform decrease of the moisture. It was impossible to also bring into relief the upper zone where the uniform moisture corresponds to the pellicular

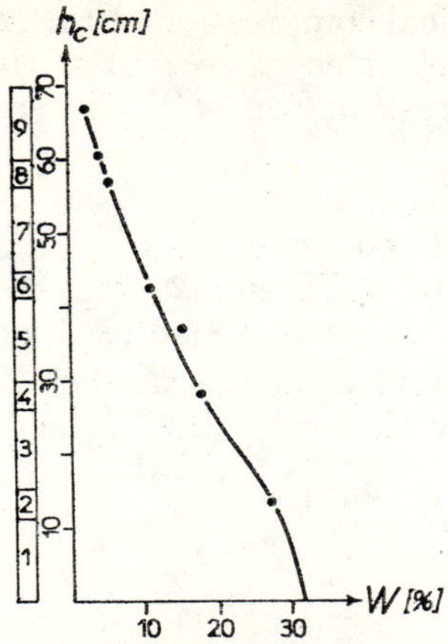

Figure 121. Moisture content distribution in a capillary tube (type *C* sand).

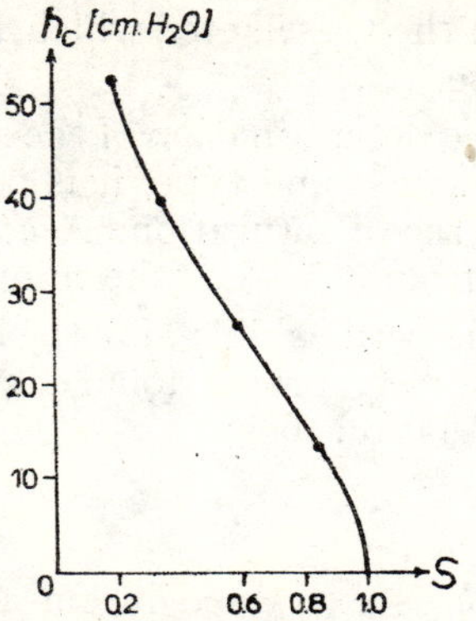

Figure 122. Degree of saturation distribution in a capillary tube (type *C* sand).

water. The variation diagram of the degree of saturation in the capillary tube (Figure 122) does not show anything significant, in contrast with the suction which presented an obvious apron for $S = 0.35$. As to the accuracy of the determination in the laboratory of the capillary height, in paper [65] is pointed out that owing to the various forms and sizes of the capillaries, the capillary pressure in the field (h_r) and in a tube (h_c) cannot be the same. By experimental research, it was possible to establish the approximate relation between these pressures

$$h_r = Bh_c$$

where coefficient B varies for various terrains from 20—300. The various results supplied by the two methods — the suction and the vertical tube — confirm partially these appreciations. As was shown in work [14] the study of the capillarity phenomenon in the vertical tube assumes certain schemes

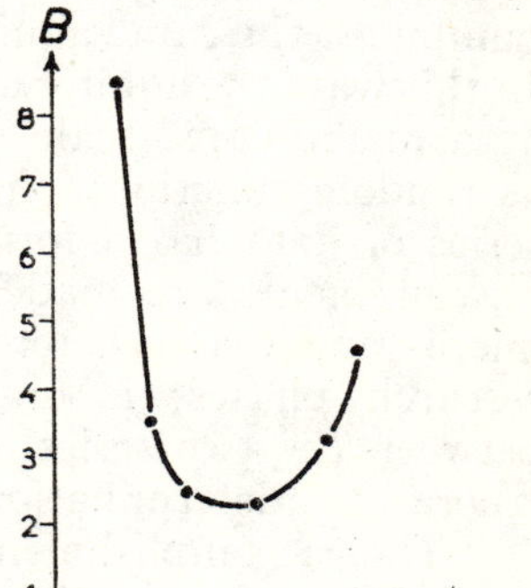

Figure 123. Variation of coefficient B as a function of S.

and approximations (the neglecting of the vertical compression of the materials and the consideration that atmospheric pressure is equal to zero at the water level in the vessel, etc.) which lead to an underestimation of the intergranular pressure.

The determination of the suction and of the capillary height being operated on the same sand type, it is possible to estimate coefficient B as a function of the degree of saturation. According to the graph in Figure 123, this coefficient varies from 4.5—8.5, the minimum recorded value corresponding to $S = 0.5$.

The permeability, i.e. the hydraulic conductivity and the capillary pressure, are inversely proportional, as was analytically shown by D. Taylor [65], the final relation being

$$\frac{0.7}{\sqrt{k}} < h_c < \frac{2.4}{\sqrt{k}} \tag{481}$$

where h_c is expressed in cm and k in cm/s.

By transformation, this inequality becomes

$$\frac{0.5}{h_c^2} < k < \frac{5.8}{h_c^2} \tag{482}$$

There results thence the important influence of the capillary rise upon the hydraulic conductivity.

3.3 Permeability

Usually, an apparent lack of coherence exists between the permeabilities determined by means of the experimental pumpings and those obtained in the laboratory. G. Matheron points out that the estimation of the homogeneity (or heterogeneity) degree of the permeability is determined by the scale of the observations. For instance, an aquifer may be characterized as heterogeneous from the punctiform permeabilities (determined in laboratory), while by changing the observation scale, for instance by the experimental pumpings, the same aquifer can present a homogeneity of a statistical nature. In this way, the aquifer can be considered as a homogeneous statistical medium at a macroscopic scale while the punctiform permeabilities can be treated as random functions. Consequently, the lack of coherence between the two series of data has a formal character determined by the scale at which the measurements are made. In the physics of the oil deposits, the notions of micro- and macro-heterogeneity of the permeability correspond to the two research phases: laboratory and experimental pumpings. The divergence between the two series of data increases in proportion, as the volume of the laboratory determinations increases.

The accomplishment within the framework of the hydrogeological exploration of a certain number of laboratory determinations of the hydraulic conductivity, is first required by the fact that the number of experimental pumpings is relatively small and therefore the work is made with sampled population S based on a small hydraulic conductivity. But even in the case of a great number of experimental pumpings, the laboratory determinations are useful because of the need to know the micro-heterogeneity of the aquifer in order to establish the size of the drainage system. For instance, the achievment of an efficient dewatering (especially in the case of the underground method) is conditioned by the knowledge of the permeability structure and composition in the various areas of the aquifer. These elements cannot always be appreciated on the basis of the hydraulic conductivity supplied by the pumpings, but on the basis of the statistical processing of a great volume of laboratory data. The necessary samples are easily enough obtained from core taken in the geological and hydrogeological drillings.

So as to obtain in the laboratory an effective permeability, the determination should take into account the influence of all the peculiarities likely to appear in the natural conditions. The hydraulic conductivity is not a constant since it depends on the: temperature, time and hydraulic gradient. In the drainage problems certain conditions are simplified because the depths are relatively small, while the groundwater has only a small mineralization; there remains however to estimate the influence of the hydraulic gradient on the hydraulic conductivity in the case of both the vertical and horizontal flow. To this end two permeameters were used, with constant head, namely with vertical and horizontal flow; the determinations were made for the fine-medium sand of sample *C*. Both diagrams $k = f(I)$ offer a maximum point corresponding to $I = 1$ for the horizontal flow and $I = 1.75$ for the vertical one (Figures 124 and 125).

The diminution of the hydraulic conductivity in the case of large hydraulic gradients is explained by the high hydrodynamic pressures that in the process of flow reduce the porosity by: the rearrangement of the particles, the clogging of certain pores, etc. Since in the application of the drainage systems the hydraulic gradients are not great enough, only the mounting branch of the

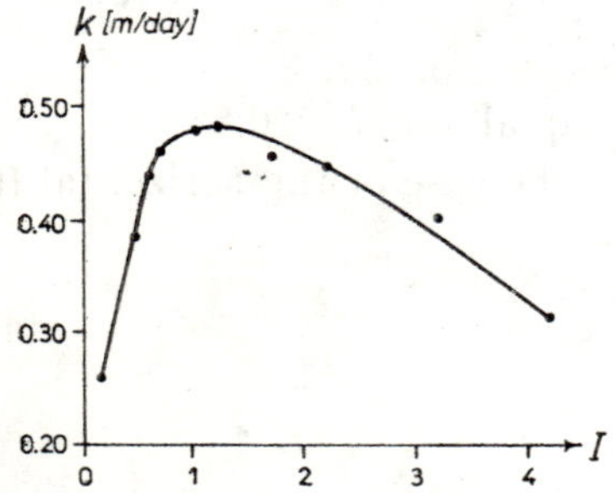

Figure 124. Variation of hydraulic conductivity *k* as a function of the hydraulic gradient in the horizontal flow.

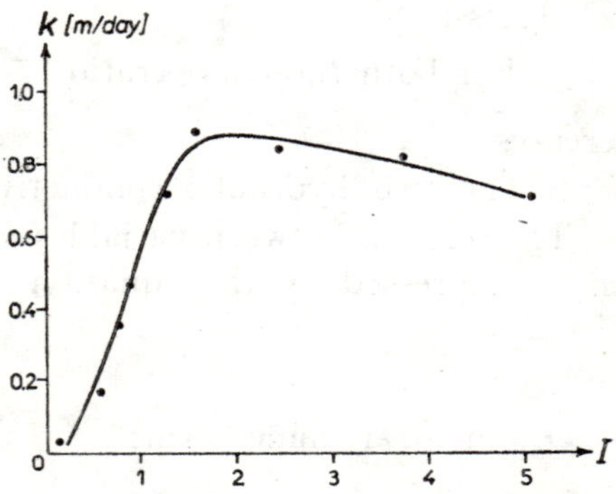

Figure 125. Variation of hydraulic conductivity *k* as a function of the hydraulic gradient in the vertical flow.

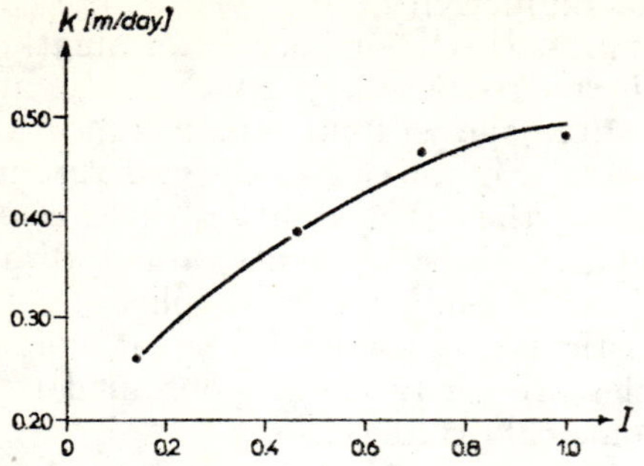

Figure 126. Dependence $k = f(I)$ in case of the horizontal flow.

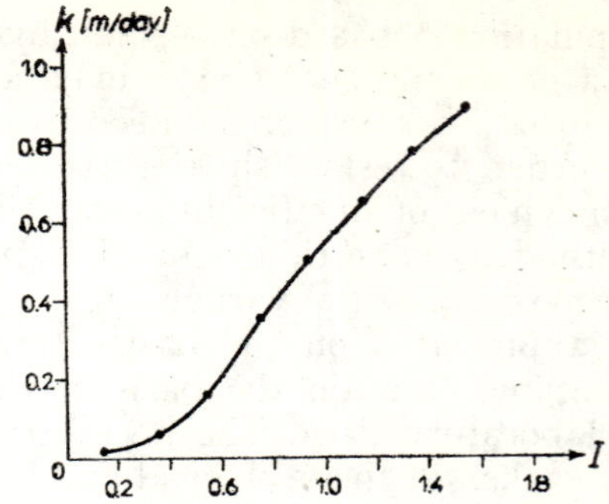

Figure 127. Dependence $k = f(I)$ in case of the vertical flow.

diagrams was thoroughly studied. The comparison of the two curves (Figures 126 and 127) throw into relief the following peculiarities:

a) The initial hydraulic resistance of the sand is higher in the case of the vertical flow than in the case of the horizontal one;

b) The mean angular coefficient is $\alpha = 0.54$ in the case of the vertical flow curve while being only $\alpha = 0.26$ in the case of the horizontal flow curve, which proves a change of the ratio of the hydraulic resistances as compared with the initial state;

c) The decrease of the hydraulic resistance of the vertical flow with regard to the horizontal one is also proved by the maximum value of k for $I = 1.75$;

d) The structure changes of the sample are more important in the case of the vertical flow because the curve presents an inflexion point (at $I = 0.5$) while in the case of the horizontal flow the concavity of the curve faces the abscissa;

e) For both the curves ratio $\frac{\Delta k}{\Delta I}$ decreases when the hydraulic gradient increases;

f) The two hydraulic conductivity are equal for $I = 0.85$.

The relation between variables k and I in the case of the horizontal flow can be expressed by the equation

$$y = ax^b \tag{483}$$

Or, in logarithmic form:

$$\log y = \log a + b \log x \tag{484}$$

By using the graphical criterion (Figure 128) there results that the variation law of parameters k and I is followed by the proposed equation.

Figure 128. Verification of the $k - I$ dependence.

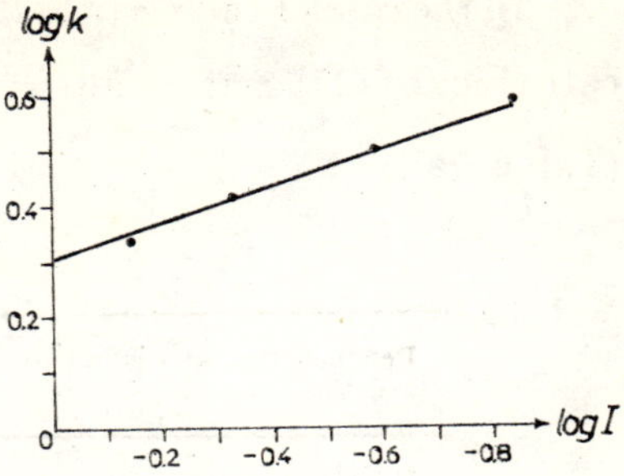

To estimate parameters a and b, the method of least squares was applied (Table 75):

$$\log y = n \log a + b \Sigma \log x \tag{485}$$

$$\log y \log x = \log a \Sigma \log x + b \Sigma(\log x^2) \tag{486}$$

$$b = \frac{n\Sigma \log y \log x - \Sigma \log y \Sigma \log x}{n \Sigma(\log x)^2 - (\Sigma \log x)^2} \tag{487}$$

$$\log a = \frac{\Sigma \log y - b \Sigma \log x}{n} \tag{488}$$

$$b = \frac{(4 \times 0.679) - (1.656 \times 1.320)}{4 \times 0.837 - 1.742} = 0.330$$

$$\log a = \frac{-1.656 - 0.330(-1.320)}{4} = 0.305$$

$$a = 2.02$$

The empirical equation is therefore:

$$k = 2.02\ I^{0.33} \tag{489}$$

where k is expressed in m/day and I in fractions of unit.

Table 75

Crt. No	x	y	$\log x$	$\log y$	$\log x \cdot \log y$	$(\log x)^2$	Computed y values
1	0.145	0.260	−0.839	−0.585	0.491	0.704	0.262
2	0.462	0.385	−0.335	−0.415	0.139	0.112	0.384
3	0.715	0.460	−0.146	−0.337	0.049	0.021	0.444
4	1.000	0.480	0	−0.319	0	0	0.495
Σ			−1.320	−1.656	0.679	0.837	

In the case of the vertical flow, the curve of function $k = f(I)$ is more intricate (Figure 127) as there also results from the variation of angular coefficient $\frac{\Delta k}{\Delta I}$ (Table 76).

Table 76

Determination (No)	1	2	3	4	5	6	7	8
k [m/day]	0.02	0.06	0.17	0.36	0.51	0.66	0.78	0.89
I	0.15	0.35	0.55	0.75	0.95	1.15	1.35	1.55
$\frac{\Delta k}{\Delta I}$	0.28	0.55	0.95	0.75	0.75	0.60	0.55	

The intricate form of the curve did not allow us to establish the empirical formula by means of the levelling method nor by the comparison of the graphs. Consequently, the parabolic interpolation was applied by using Newton's interpolation polynomial. Taking into account the monotone character of the curve in certain of its portions, experimental points 3, 5 and 7 were given up and the result was a polynomial of the fourth degree. For the unequal intervals, Newton's interpolation polynomial takes the form [66]

$$y = Ln(x) = f(x_0) + (x - x_0) f(x_0;\ x_1) + (x - x_0)\ (x - x_1)\ f(x_0;\ x_1;\ x_2) +$$

$$+ \ldots + (x - x_0) \ldots (x - x_{n-1}) f(x_0 \ldots x_n) \tag{490}$$

where $f(x_0 \ldots x_n)$ represents the divided differences which can be computed by Table 77.

Table 77

Point No	x (I)	y (k m/day)	Divided differences of order : first	second	third	fourth
0	0.15	0.02				
			+0.20			
1	0.35	0.06		+0.92		
			+0.75		−0.92	
2	0.75	0.36		0		+0.527
			+0.75		−0.182	
3	1.15	0.66		−0.218		
			+0.575			
4	1.55	0.89				

Finally, the divided differences entering into the general relation according to this table will have the following values:

$$f(x_0;\ x_1) = +\ 0.2;\quad f(x_0;\ x_1;\ x_2) = +0.92;$$

$$f(x_0;\ x_1;\ x_2;\ x_3) = -0.92;\ f(x_0;\ x_1;\ x_2;\ y_3;\ x_4) = +0.527$$

With these values, we can write the polynomial:

$$Ln(x) = 0.02 + (x - 0.15)\ 0.2 + 0.92(x - 15)\ (x - 0.35) - 0.92$$
$$(x - 0.15)\ (x - 0.35)\ (x - 0.75) + 0.527(x - 0.15)$$
$$(x - 0.35)\ (x - 0.75)\ (x - 1.15)$$

The final expression of the polynomial is:

$$y = 0.53x^4 - 2.18x^3 + 3.05x^2 - 0.93x + 0.1$$

$$k = 0.54\ I^4 - 2.18\ I^3 + 3.05\ I^2 - 0.93\ I + 0.1 \qquad (491)$$

All the experimental results were verified by this polynomial.

The quantitative analysis emphasized the important influence of the flow type as well as of the hydraulic gradient upon the hydraulic conductivity which can therefore no longer be considered as a constant for the given rock and fluid. In laboratory practice, these two elements are not taken into account, the determinations being often made for $I = 1$. Under these conditions, the statistical processing and the interpretation of the laboratory data do not give a real image of the effective permeability; the determinations should be made under hydrodynamic conditions (flow type and hydraulic gradient) closest to the actual conditions of the drainage. While the flow is practically horizontal in most of the cases the choice of the most suitable hydraulic gradient is still more difficult since it is determined by the actual conditions of the drainage. Thus in the case of underground dewaterings by means of rows of wells, the slope of the piezometric surface of the top dewatered aquifer reaches its maximum at the beginning of the drainage after which it decreases with time and finishes by being close to the impermeable bed slope.

Empirical relation $k = f(I)$ can be equally used for estimating the hydraulic gradient permitting the highest hydraulic conductivity, hence an increased discharge. This is the case of the drainage wells with a great diameter, which can work at great drawdowns owing to the gravel pack.

The hydrodynamics of the underground drainage processes (by underground wells) is characterized by a continuous decrease of the discharge rate. This is due to the active exhaustion of the static resources (the dynamic resources being unimportant), which leads to the use of Boussinesq's equation for sources during draught periods. These equations were elaborated on the assumption that the higher limit of the aquifer is the depression surface which undergoes a uniform lowering while the hydraulic conductivity is a constant of the rocks. As a matter of fact, however, the rocks situated above the depression surface cannot be considered promptly drained, owing to the interaction between the water and the mineral grains (§ 3.2). Therefore, the contribution of the upper zone to the formation of the water inflow is important, even decisive, towards the end of the dewatering period when on all the adjacent area the drainage is a free-level one. Then, the problem arises of the permeability estimation of the upper zone the moisture of which decreases with time. L. A. Richards proposed also to generalize Darcy's law to the ground water movement within unsaturated rocks:

$$v = k_w \frac{\Delta u}{\Delta l} \tag{492}$$

where v is the water movement velocity, Δu the water pressure in the pores and Δl the path length.

The values of the k_w coefficient are a continuous function of the moisture state; as the moisture decreases, its values decrease much more rapidly. Obviously, the complete solution of the problem would require the establishment of a variation law of k_w as a function of the moisture degree and of the drainage time. Up to now, the problem related to the water motion within the unsaturated grounds do not have a complete theoretical solution; it is only tackled by the pedologists. Therefore, for hydrogeologists the solving of the problem is now limited to the laboratory determination of the interdependence between coefficient k_w on the one hand and the moisture state or suction on the other hand. Determinations made on the fine sand of Văleni, Romania (siliceous sands) by S. Andrei [1] established a logarithmic dependence between k_w and the moisture: at $w = 26\%$ (saturation degree $S = 1$) corresponds $k_w = 6 \times 10^{-3}$ cm/s, while at $w = 3\%$, corresponds $k_w = 1.7 \times \times 10^{-5}$ cm/s. It should, however, be remarked that owing to their prevailing siliceous composition, the hydraulic resistance of these sands is small.

The percolation coefficient can also be estimated by some empirical relations, such as the following one elaborated by S. F. Averianov for the aeration zone:

$$k_w = k\left(\frac{w_v - w_{0v}}{e - w_{0v}}\right)^{3.5} \tag{493}$$

where k is the hydraulic conductivity (when the medium is completely saturated), w_v is the volumetric moisture content, w_{0v} the maximum pellicular water content and e the void ratio (= 0.83).

By this relation the decrease was determined of ratio $\frac{k_w}{k}$ as a function of the degree of saturation, for sample C (fine-medium sand) analysed by suction. Volumetric moistures w_v and w_{0v} were recalculated by relations:

$$w_v = w.\gamma_s \tag{494}$$

$$w_{0v} = w_0\gamma_s = 5.5 \times 2.67 = 15\% \tag{495}$$

where w_0 is the maximum pellicular water content appreciated by the suction curve at 5.5%.

Figure 129. Relation between k_w/k and the degree of saturation.

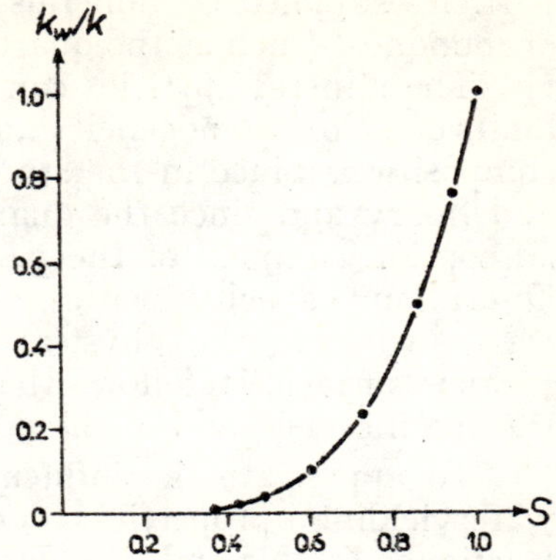

In order to draw diagram $k_w/k = f(S)$ the degree of saturation was calculated by the relation

$$S = \frac{w_v}{e} \tag{496}$$

The resulting variation curve (Figure 129) is very like the one proposed by S. Irmay, though constructed by means of another empirical relation [6]. The only difference is that in our curve there is practically no water motion for S values lower than 0.30, while on the S. Irmay's diagram this corresponds to $S < 0.20$. Of course, this interval of the degree of saturation is determined by the intensity of the reaction between the water and the mineral grains as there resulted from the capillary potential analysis. Ratio k_w/k is also called *coefficient of the relative permeability* and it is a dimensionless parameter.

3.4 The influence of the mineralogical composition upon the wateryielding properties

A decisive factor of the wateryielding properties of the sands is the mineralogical composition. This dependence is as yet hardly studied: much more attention is granted to the relation between the mineralogical composition and the sand grain size. The grain size distribution can however be indirectly used for interpreting the wateryielding properties. The influence of the mineralogical composition upon the filtering properties can also be analysed from the point of view of the mineral properties as well as of the adsorption capacity of the mineral particles.

The mechanical properties of the minerals — cleavage, fracture, hardness, etc. — determine the form of the particles and thence the geometric characteristics of the porous space. The sand particles formed by minerals with a perfect cleavage (such as the muskovite) supply smaller pores than the particles supplied by minerals with an imperfect cleavage and a high degree of roundness (such as the quartz).

The mineral particles can be considered as electrically neuter within the limits of a close homogeneity and when quartz is prevailing in the mineralogical composition. Since in the studied coal complexes most of the sands are fine and floury and since the quartz is not prevailing in their composition, the adsorption complex of the particle should be taken into consideration. The absorption capacity varies, however, even within the same grain-size class, owing to the physical and chemical properties of the monomineral particle surface. It follows that for sands with similar grain-size the values of the specific yield and of the permeability are different.

To appreciate the influence of the mineralogical composition upon the wateryielding properties, determinations should be made by grain-size fractions. To this end, the data supplied by 15 samples taken from the first bank of the top of Șotînga coal II and III layers were used. As a function of the hydrogeological behaviour, the minerals were divided into five classes: rocks and coal fragments, quartz, felspars, micas (muskovite and subordinately chlorite) and heavy minerals (including metallic oxides). The melanocratic minerals and the carbonates were not taken into account owing to their small and insignificant contents as to the purposes of the researches. So as to facilitate the estimation of the trend, on the dispersion graphs (Figure 130) the frequency mean curves were approximated too. All the graphs emphasize a greater or smaller trend of increase or decrease of each mineral content towards the gross or fine fractions. Thus, fragments of rocks or coals begin to appear in the $d = 0.20$ mm fraction (fine sand) after which a sudden increase is marked so that the $d = 0.50$ mm (medium sand) prevails now over the other minerals. The quartz distribution presents two distinct fields: the small contents field (as an average less than 20%) for fractions $d = 0.20$—1.00 mm (fine and medium sands) and the large contents field (a 50% mean) corresponding to the floury sands. The large contents correspond to the $d = 0.12$ mm fraction; they increase up to 75% while their mean (the maximum of the mean curve) is 60%. In the case of felspars too, it is possible to separate two content fields delimitated similarly to those of the quartz. The contents difference is however smaller: a few per-cents for $d = 0.25$—0.75 mm fractions (fine and medium sands) and about 24% for the floury sand. Micas are present up to $d = 0.75$ mm. The maximum content occurs at $d = 0.25$ mm i.e. between the fine and floury sands; the mode of the frequency curve is 38%. Within the floury sand class the mica content diminishes down to the mean of 10%.

The heavy mineral fraction is present in the $d < 0.5$ grain-size class. The largest contents correspond to $d = 0.25$ mm (18% average). Floury sands get up to about 10%.

The hypotheses related to the variation laws of the mineralogical composition by fraction are quite varied, because the mineral distribution by fractions is due to several factors, the separate influence of which is difficult to appreciate (the physical resistance and the chemical stability of the minerals, the physico-geographical conditions in the erosion area, the primary material

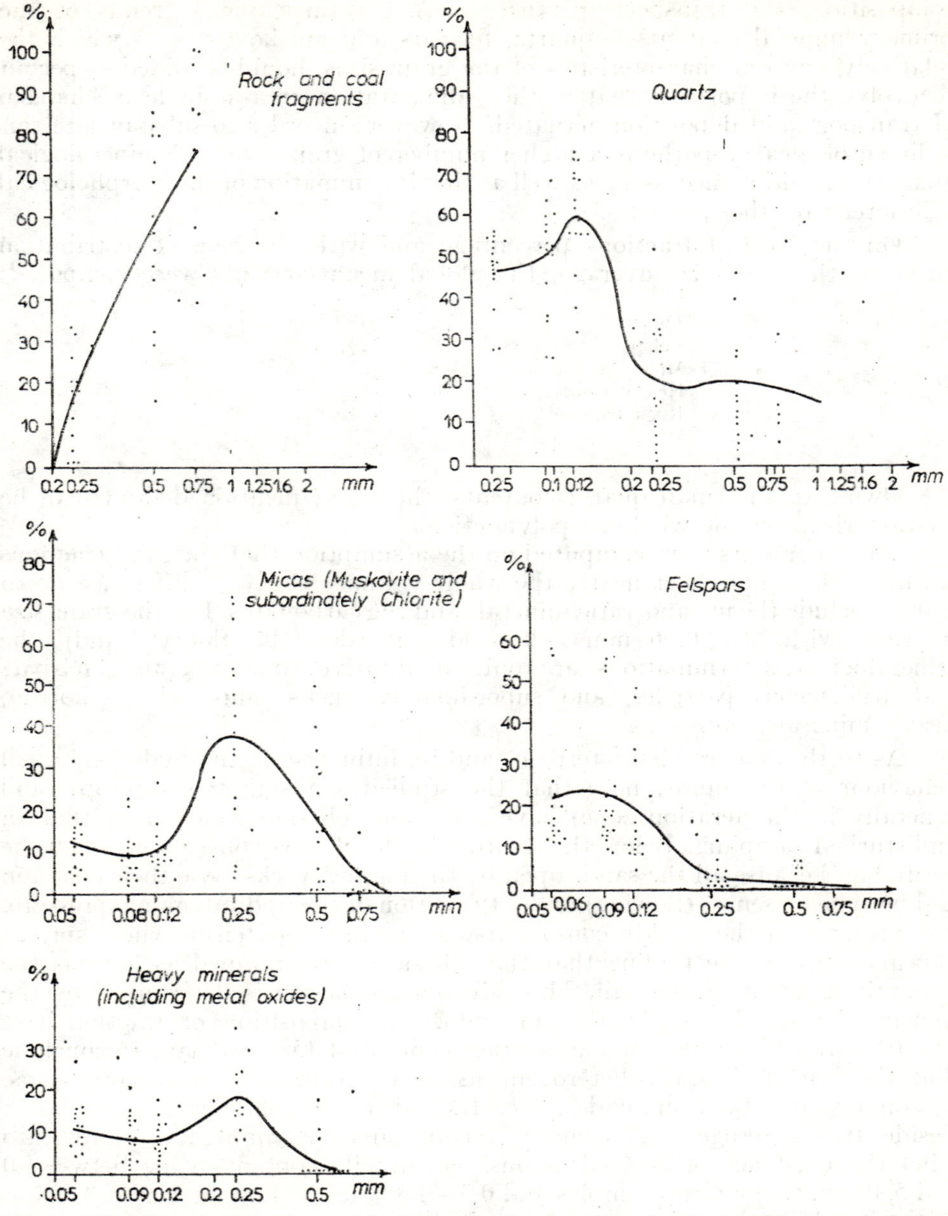

Figure 130. Mineral dispersion graphs by grain-size fractions (sands in the top of Șotînga coal layers II and III).

composition, the transport distance, etc.). Certain obvious trends of the primary mineral contents — quartz, felspars and muskovite — to which the relatively uniform characteristics of the grain-sizes should be added — permit to evolve the hypothesis that in the sedimentation area a single mechanism of transport and deposition actuated. However, in order to substantiate this sedimentological hypothesis a higher number of grain-size and mineralogical analyses should be necessary as well as the determination of the morphological characters of the particles.

On the base of fractions percentage and with the help of distribution curve of the grain-size average, the global mean contents were computed:

Quartz	50%
Felspars	20%
Micas	12%
Heavy fraction	11%
Rock fragments	3%
	96%

Owing to the small quartz contents, the above-mentioned sands can be characterized on the whole as polymictic.

These contents were computed on the assumption that the sand fractions with $d > 0.06$ mm cover nearly the whole volume; the 4% (difference up to 100%) include the melanocratic minerals and the carbonates. For the grain-size fraction with $d < 0.06$ mm (silt and subordinately floury sand) the mineralogical determinations are only qualitative: quartz grains, felspars and argillaceous particles, and subordinately micas (muskovite, chlorite), heavy minerals, etc.

As to the cement characteristics and its influence on the hydrogeological behaviour of the sands, note that the studied sands in the outcrops (and generally in the aeration zone) have a certain cohesion which permitted an undisturbed sampling. Below the hydrostatic level, this cohesion ceases to be verifiable because all the sands open by the mining works loose their cohesion and become looseness (flowing) under the action of the hydrodynamic pressure. The presence of the weakly consolidated sands in the outcrops where springs appear confirm the finding that the cohesion is maintained as long as the hydraulic gradients are small. This adherence cohesion is determined by the cement characteristics. At the mineralogical composition of fraction $d < < 0.05$ mm, which represents an average content of 15%, we have to conclude that the cement is quite heterogeneous, being formed by micaceous clays, carbonates, quartz, iron oxides, etc. Its adherence cohesion is very small. Beside this heterogeneous cement, certain analyses brought equally into relief the existence of the calcareous cement; its contents vary between 0 and 5.4% in the outcrop samples and 0.3—4.8% in the mining works samples. Consequently, the above mentioned heterogeneous cement gives the sand its weakly consolidated character; the adherence cohesion due to this cement is small and in cases of high hydraulic gradients it vanishes.

The results of the experiments made by V. V. Ochotin with the primary minerals — quartz with rounded and angular particles, orthoclase and muskovite — are quite useful for estimating the influence of the mineralogical composition upon the water-yielding properties (Table 78).

In order to facilitate the comparison, the hydraulic conductivity of the muskovite was conventionally considered as equal to 1 for each grain-size fraction. The increasing order of the filtration coefficient is: muskovite < orthoclase < angular quartz < rounded quartz. From Table 78 a reduction of the mineralogical composition influence on the hydraulic conductivity results in proportional to the reduction in size; thus, while for a coarse sand the ratio k_{max}/k_{min} is 20, for the floury sand the same ratio diminishes to 5. As regards variation of the capillary rise the highest recorded values are shown by muskovite.

Table 78

Influence of mineralogical composition on the hydraulic conductivity k [m/day], porosity in the compact state n [%] and capillary rise h_c [cm]

Grain size fraction	Muskovite			Orthoclas			Angular quartz			Rounded quartz		
	k	h_c	n	k	h_c	n	k	h_c	n	k	h_c	n
2–1	1	12.3	80.5	15	6.8	44.5	—	5.5	37.4	20	9.0	33.4
1 –0.5	1	20.8	75.2	17	13.3	46.3	12	13.2	38.8	35	14.6	34.1
0.5 –0.25	1	32.7	72.2	5	23.3	47.2	6	25.7	40.1	12	26.3	33.8
0.25–0.1	1	65.8	66.3	5	49.2	49.2	7	56.0	42.6	11	61.8	33.7
0.1 –0.06	1	122.4	66.0	7	99.3	51.4	5	100.3	43.1	15	82.7	35.9
0.06–0.01	1	—	65.3	—	—	—	3	205.0	39.6	5	127.7	—
0.01	1	—	—	2	—	—	3	—	—	—	—	—

For the same grain-size fraction, the porosity varies as a function of the particle shape. The experiments showed that whatever the grain-size fraction, the muskovite possesses the greatest porosity and the rounded grains of quartz the smallest porosity. On the other hand, the more the grain size diminishes, the more the muskovite porosity decreases while the orthoclase and quartz porosity increases.

On the basis of Ochotin's experimental data an estimation has been made of the sands in the Șotînga area (Romania), taking also into account that in the analysed thin sections the grains with $d > 0.1$ mm are slightly rounded, subangular or even angular.

The determinant part of water-yielding properties are played by the quartz and the felspars grain, the largest percentage of which being of $d = 0.09$ to 0.12 mm; this interval corresponds to the mode of the grain-size curve. The position of the mode interval within the limits of the floury sand fraction determines similar behaviour of the felspars and of the angular quartz. The angular quartz grains influence negatively the water-yielding properties, owing to the decrease of the hydraulic conductivity, thence an increase of the capillarity of the $d < 0.10$ mm fractions in regard to the rounded grains of quartz. In packed form, the angular shaped quartz has greater porosities than the rounded one, but since a good part of the pore channels are capillary or subcapillary the hydraulic conductivity is finally reduced.

F. A. Trebin also emphasizes the aspect of the hydraulic resistance of the particles surface: the quartz resistance is lower than that of the felspar. But, since the influence of this factor is incomparably smaller than that of the other ones, it cannot practically be expressed quantitatively.

In conclusion, the mineralogical composition of the sands is an important element for appreciating the hydraulic conductivity; it permits us to explain the variation limits of the various parameters, such as porosity, capillary rise, hydraulic conductivity, etc. Let us, however, state that to consider only the grain-size composition constitute a one-sided approach of the problem likely to lead to an erroneous interpretation of the water yielding properties.

3.5 The structure of the porous space

When solving the practical hydrogeological problem of the sandy rocks reservoirs, particular importance should be granted to the structural features of the porous space among which the most significant are the size and form of the pores, the quantitative ratios between pores of various sizes and forms, as well as the characteristics of the prevailing pores. But no functional and quantitative dependence exists between the permeability and the total porosity, only a qualitative approximate relation. A functional dependence exists however between the permeability and the effective porosity (specific yield) but only if the structural factors of the porous space are taken into consideration. Therefore, a thorough characterization of the filtering properties can only be made when the porous space structure is known. A. A. Hanin [28] established such a correlation between the permeability and the mean diameter of the pore channels, for the coarse sands reservoirs. On the other hand, D. Taylor [65] pointed out that the seepage velocity is directly proportional to a coefficient of the cross-section form (C_s) of the channel, which analytically can only be determined for simple forms of the cross-sections though not depending on their sizes. In an ideal case (channels with a circular cross-section) $C_s = 1/3$. Actually however, these cross-sections are very varied so that an analytical solution of the problem capable of also taking into account the structural parameters of the porous space is impossible. For this reason, the use of the correlation method is quite frequent.

When classifying the pores by their sizes, we must consider the interaction between the water and the mineral particle; The water mobility and consequently the water yielding properties depend on this interaction. However, the separation of the various pore types by size constitute a difficult problem, dealt with by the researchers in various ways. All the classifications assume the existence of three pore types: supercapillary pores (or macropores), capillary pores (or micropores) and subcapillary pores. But their limits are debatable. Thus, the limit between macropores (not to be confounded with the macropores microchannels in the loess soils) and the capillary pores varies from 0.5 to 0.1 mm. Size $d = 0.5$ mm (or more accurately 0.508 mm) was established by theoretical computations. It is however, estimated that for sizes over 0.1 mm the influence of the surface tension is practically negligible. The lower limit of the capillary pores varies from 0.2 to 1 μ.

Taking into account that the porosity given by the subcapillary pores corresponds to the highest hydroscopicity, and the largest thickness of the hygroscopic water film is (after B. V. Deriaghin) 1 μ, we can assume it as the lower limit of the capillary pores. There results that the capillary porosity corresponds to pores of $d = 1 - 100$ μ. Within this capillary porosity, we can separate the large capillary pores ($d > 20$ μ), the most active ones from the standpoint of the groundwater flow, and the small capillary pores with $d < 20$ μ. The separation of the great capillary ($d > 20$ μ) is justified by the existence of the "suction without surface tension"; since this section is situated in the central capillary thread, i.e. outside the attraction force of the solid phase, it can more easily yield the water, called in this case *mobile capillary water.*

The quantitative study of the pore types and of porosities is called *porometry*. For a clearer presentation of the method of processing and interpreting of the porometry data the problem will be treated on the basis of the experimental results supplied by the three samples of Dacian sands taken from the Șotînga area (Romania) (§ IIII.2): samples N and Z from the outcrops and sample C from the underground.

An application of the suction phenomenon to the porous materials that do not show noticeable volume variations when moisture varies (e.g. the sands) is the computation of the pore distribution by sizes. Within this experimental capillary model, the "non-wetting" phase is the air and a certain value of the suction corresponds to a certain diameter of the pores under the action of the suction phenomenon motion. By so doing, curve $pF - S$ (capillary potential — degree of saturation) represents the cumulative curve of the pore distribution by sizes while the degree of saturation corresponds to the porosity fraction occupied by the water at a certain suction H. By equating the moisture suction to the height of capillary rise (Laplace's formula) we can determine the diameters of the capillaries:

$$H = \frac{2\sigma \cos \alpha}{r \gamma_a} \tag{497}$$

where H is the suction (cm H_2O); α the wetting angle (for full wetting $\alpha = 0$, and therefore $\cos \alpha = 1$); σ is the surface tension of the water (dyn/cm); r the capillary radius (cm); γ_a the water specific gravity (dyn/cm²).

At $t = 20$°C, the tables give $\gamma_a = 72.53$ dyn/cm and $\sigma = 979$ dyn/cm so that relation (460) becomes

$$H = \frac{0.148}{r} \tag{498}$$

or:

$$d = \frac{0,30}{H} \tag{499}$$

where d is the capillary diameter.

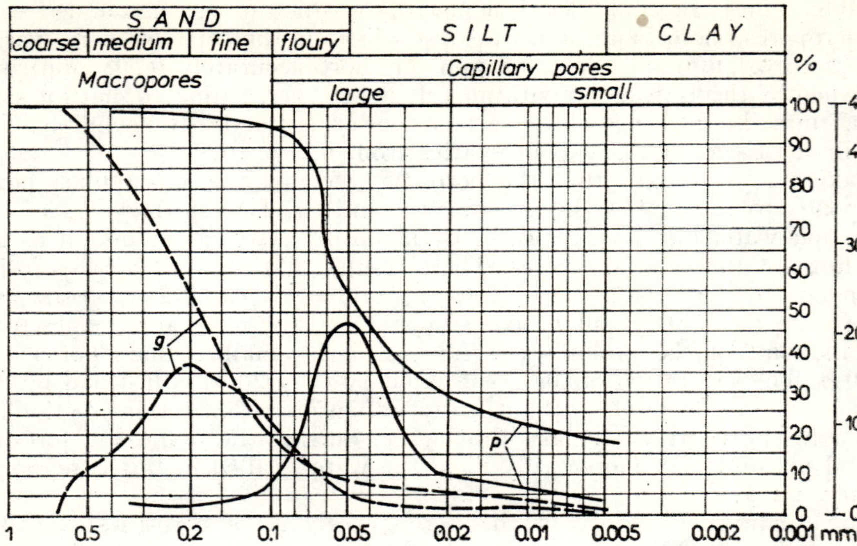

Figure 131. Distribution and frequency curves of grain-size (g) and pores (p) of sample C (disturbed)

With the help of the suction — degree of saturation curves (Figure 119), the porometry cumulative curves were drawn for the three sand samples C, N and Z in a disturbed state and the highest packing degree (Figures 131, 132 and 133).

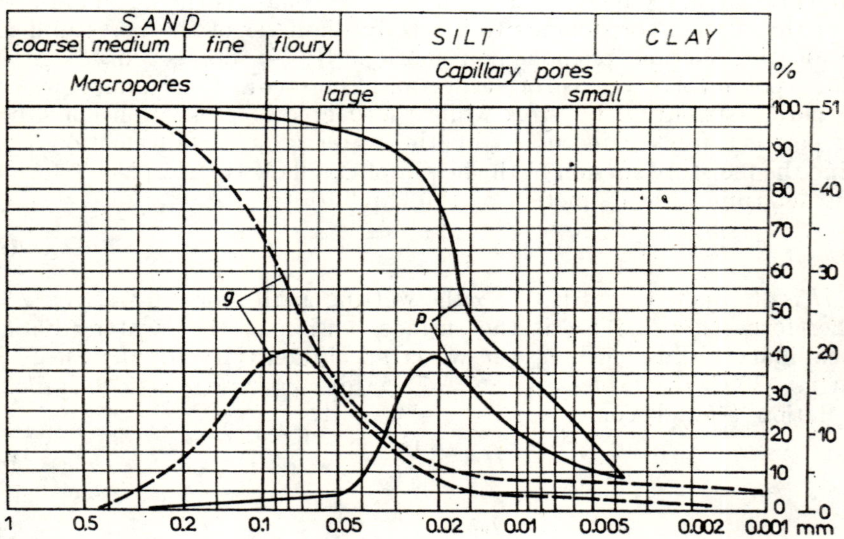

Figure 132. Distribution and frequency curves of grain-size (g) and pores (p), of sample N (undisturbed):

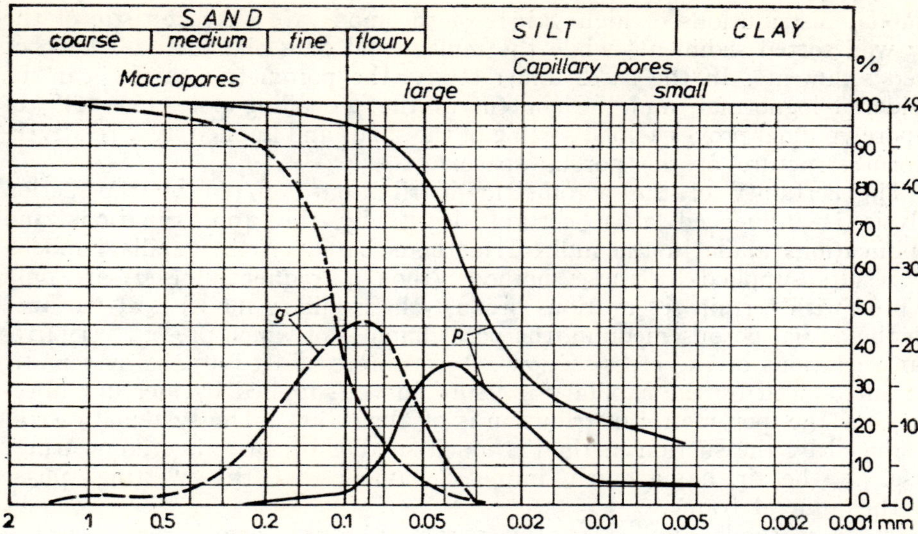

Figure 133. Distribution and frequency curves of grain-size (*g*) and pores (*p*) of sample *Z* (disturbed).

They only could be drawn for $d > 4.6$ μ owing to the higher limit of the suction determined in the laboratory (650 cm H_2O). The relative distribution of the pore types is given in Table 79. This table shows the concordance between the grain-size composition and the pore distribution parameters. Although

Table 79

Pores type	Sample porosity					
	Sample *C* (fine-medium sand)		Sample *Z*(fine-floury sand)		Sample *N* (floury-silty sand)	
Small capillaries	29	12.5	31	15.0	74	37.3
Big capillaries	66	29.3	64	31.9	23	11.9
Macropores	5	2.0	5	2.2	3	1.8
Total	100	43.8	100	49.1	100	51.0
Prevailing pores:						
Content	33	14.5	33	15.5	44	22.6
Diametre [μ]	65		47		28	

the macroporosity variation is insignificant, the capillary porosity distribution between the two classes is characteristic of each sample. The ultra-microporosity could not be estimated but its value is negligible as shown by the trend of the cumulative curves.

In performing of the porometric analysis, a particular attention must be given to the size and content of the prevailing pores, the values of which strongly influence the hydraulic properties. The literature mentions such cases of unimodal and multimodal porometric distributions. Obviously the unimodal distributions of high values of the mode are characteristic of the very well sorted sediments while the multimodal ones characterize the poor sorted sediments. In the cases under study, the porometric curves can be considered lognormal, and the content of the prevailing pores (33—44% of the pore volume) proves a good sorting of the sands and implicitely a relatively uniform structure of the porous space.

The certainty of the porometric distributions offered by the suction analysis is influenced at a certain degree by the approximations and simplifications made in the indirect methods based on the capillary model. The results supplied by these methods (though frequently used) are only accepted with certain reservations because the capillary model is at variance with the actual porous medium, the pore channels of which present very irregular variations of the cross-sections. According to Fatt [10] the systematic error introduced by the capillary pressure curve (supplied by mercury injections) on the porometric distribution is of about 30%. The systematic error introduced by the suction method is much smaller because the "nonwetting" phase, i.e. the air, has a much simpler behaviour than the "wetting" phase and the mineral frame.

Other methods applied to the porometric study are those based on a more or less direct visual examination. Since the pore sizes vary widely and at random and the number of measurements is very high, the description of the porous space is better made by the methods of the mathematical statistics. Their superiority consists in the fact that the notion of pore size has a strict geometric meaning unrelated to any other imposed model or to the observation of any physical phenomena.

On the thin section under the microscope, we can determine *the hydraulic diameter* of a pore channel, defined as the ratio of the cross-section to the wetted perimeter of each cutted pore; this is the nearest to the actual model but often the lack of a clear-cut contour of the grains checks this determination. But even if the two geometric elements of each pore could be measured, the vast working volume involved would obstruct a large application of this method. That is why the random measurement is recommended of quantity l, which is the measured distance along a straight line segment between any two points on the surface of two particles (Figure 134).

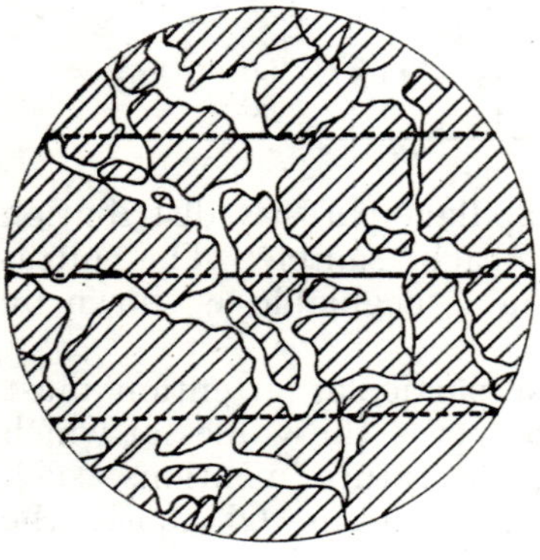

Figure 134. Measurement of the thin section of quantity l.

Since the sand of sample N has a certain adherence cohesion, this sample could be taken even in the undisturbed state. Then, the sample with undisturbed structure was injected with bakelite (fluidified by means of certain organic components) at a vacuum pressure of 200 mm Hg, after which it was polymerized during 48 hours at $t = 48°C$. By so doing, the slightly cemented sand was transformed into a strongly cemented rock, from which it was possible to detach thin sections. The colour photographs of these thin sections were enlarged $20\times$. Later, by projecting them onto a screen, enlargements of $526\times$ were obtained, 1 mm on the projection representing 1.9 μ from the natural scale. The tests made on black and white films gave practically no results because owing to the argillaceous cement, the pore contours were too indistinct. Even the red colour of the bakelite was sometimes with difficulty identified, owing to the dark hue due to the argillaceous cement. Trebin mentions that the bakelite, while clogging up the pores responsible for the effective porosity, does not indicate the corresponding limit diameter. From the measurement made on the section projection there resulted that the smallest sizes were 2 mm on the projection, i.e. 3.8 μ on the natural scale, which means that the pore size distribution includes equally part of the small capillary pores. Probably, by increasing the vacuum, the results would come nearer to the total porosity and by a $1000\times$ total enlargement of the projected section, pores of 1 μ could be brought into relief.

The quantity l was given by random measurements made on equidistant perpendicular alignements on one and the same thin section. In such a case the accuracy of the distribution law only depends on the total number of measurements. In paper [2], it is recommended not to exceed the number of 1000 measurements; this conclusion is based on an example where the difference is hardly perceptible between $N_1 = 425$ and $N_2 = 846$ measurements made upon the same sample. Taking all this into account, 740 measurements were effected upon six thin film sections of the bakelited sample N and porometrically studied. The large volume population of quantity l was subjected to the statistical computation for the estimation of the structure parameters of the porous space, by applying the method of the moments of unequal intervals (Table 80).

The conventional moments are:

$$V_1 = +18.99; \qquad V_2 = 1723.52$$

The sample mean value is:

$$\bar{x} = x_0 + V_1 = 28.50 + 18.99 = 47.49 \ \mu \tag{500}$$

and the sample variance

$$S^2 = V_2 - V_1^2 = 1723.52 - (18.99)^2 = 1362.90 \ \mu^2$$

$$S = \sqrt{1362.90} = 36.92 \ \mu \tag{501}$$

The verifications proved that no deviations appeared with regard to the bulk of the observed values.

Table 80

a — b	$x_i (\mu)$	n_i	f_i	$\zeta_i = x_i - x_0$	$f_i\zeta_i$	$f_i\zeta_i^2$
3.8	3.80	10	1.35	— 24.70	— 33.34	823.50
3.8 — 9.5	6.65	35	4.73	— 21.85	— 103.35	2 258.20
9.5 — 19.0	14.25	111	15.00	— 14.25	—213.75	3 045.94
19.0 — 38.0	28.50	236	31.91	0	0	0
38.0 — 57.0	47.50	133	18.00	+ 19.00	+342.00	6 498.00
57.0 — 76.0	66.50	77	10.40	+ 38.00	+395.20	15 017.76
76.0 — 95.0	85.50	63	8.51	+ 57.00	+485.07	27 648.99
95.0 —114.0	104.50	35	4.72	+ 76.00	+358.72	27 262.72
114.0 —133.0	123.50	14	1.89	+ 95.00	+179.55	17 057.25
133.0 —152.0	142.50	12	1.62	+114.00	+184.68	21 053.52
152.0 —171.0	161.50	5	0.67	+133.00	+ 89.11	11 851.63
171.0 —190.0	180.50	4	0.54	+152.00	+ 82.08	12 476.16
190.0 —209.0	199.50	1	0.13	+171.00	+ 22.23	3 801.33
209.0 —228.0	218.50	2	0.27	+190.00	+ 51.30	9 747.00
228.0 —247.0	237.50	1	0.13	+209.00	+ 27.17	5 678.53
247.0 —266.0	256.50	0	0	0	0	0
266.0 —285.0	275.50	1	0.13	+247.00	+ 32.11	7 931.17
		740	100.00		+1898.78	172,351.70

The value of the coefficient of variation is:

$$V = \frac{36.92}{47.49} \cdot 100 = 77.7\%$$

From the performed measurements and computations we get for the structure parameters of the porous space the following values of selection:

— mean value $l_m = 47.5\mu$
— mode value $l_M = 285.5\mu$
— maximum value $L = 285\mu$
— standard deviation $S = 36.8\mu$

The small value of the coefficient of variation makes conspicuous a relatively homogeneous structure of the pores. Both the dispersion graph (Figure 135) and the polygon of the relative frequencies (Figure 136) display an asymmetrical distribution where the high values of parameter l prevail, which presumes the existence of a lognormal distribution. So as to verify this assumption, the probabilistic diagram was used where the points are determined by the normalized values z_1 while the centres of the intervals are expressed by a new variable $x_i = \log x_i = \log l$. The computation of the normalized values (Table 81) is made with the help of the distribution function and of Laplace's function

$$F_n(x_i) = F(z_i) = 0.5 + \Phi(z_i) \tag{502}$$

$$\Phi(z)_i = F_n(x_i) — 0.5 \tag{503}$$

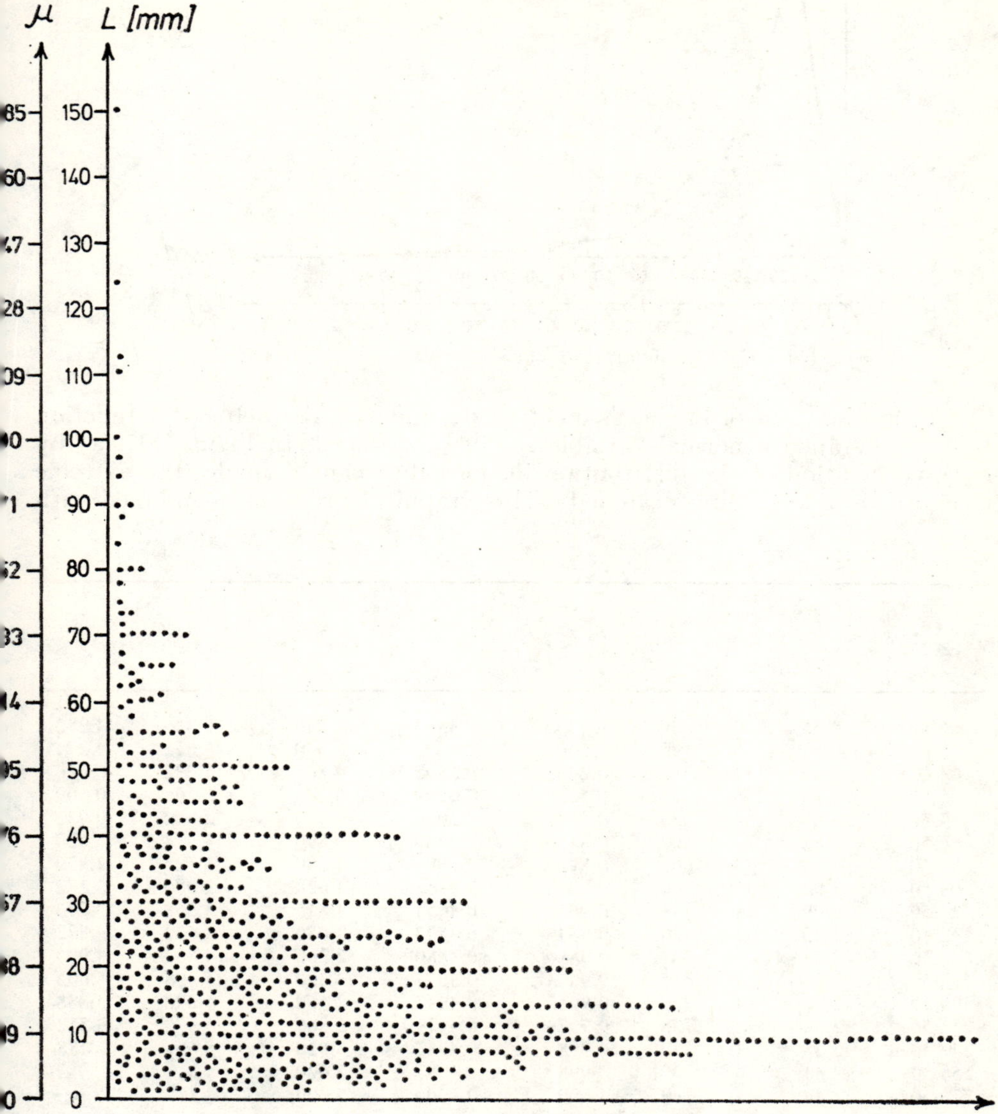

Figure 135. Dispersion graph of quantity l (sample N undisturbed);

L – values on the enlarged scale.

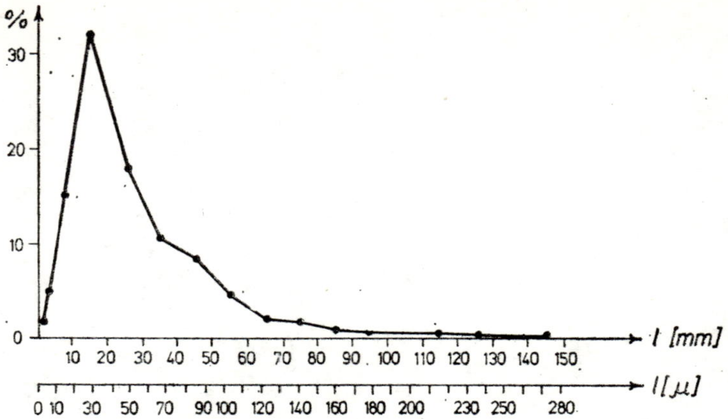

Figure 136. Relative frequency polygon of quantity l.

From the value of Laplace's function Φz_i and with the help of the function table, the standard normal variable z_i will be deduced. In Figure 137, a "by eye" verification of the distribution lognormality can be made (the extreme interval $x'_i = 2.44$ being excluded). This graphical procedure combined with

Table 81

x_i	$x'_i =$ $= \log x_i =$ $= \log l$	n_i	Σn_i	$F_n(x'_i)$	$\Phi(z_i)$	z_i	x'^2_i	$x'_i z_i$
3.80	0.58	10	10	0.0135	−0.4865	−2.210	0.336	−1.282
6.65	0.82	35	45	0.0608	−0.4392	−1.550	0.672	−1.271
14.25	1.15	111	156	0.2108	−0.2892	−0.805	1.322	−0.926
28.50	1.45	236	392	0.5299	0.0299	0.075	2.102	0.109
47.50	1.68	133	525	0.7099	0.2099	0.555	2.822	0.932
66.50	1.82	77	602	0.8139	0.3139	0.890	3.312	1.620
85.50	1.93	63	665	0.8990	0.3990	1.280	3.725	2.470
104.50	2.02	35	700	0.9462	0.4462	1.620	4.080	3.272
123.50	2.09	14	714	0.9651	0.4651	1.815	4.368	3.793
142.50	2.15	12	726	0.9813	0.4813	2.080	4.622	4.472
1.50	2.21	5	731	0.9880	0.4880	2.260	4.884	4.995
180.50	2.26	4	735	0.9934	0.4934	2.480	5.108	5.605
199.50	2.30	1	736	0.9947	0.4947	2.560	5.290	5.888
218.50	2.34	2	738	0.9974	0.4974	2.800	5.476	6.552
237.50	2.38	1	739	0.9987	0.4987	3.020	5.664	7.188
256.50	(2.41)	0	739	0.9987	0.4987	(3.020)*		
275.50	(2.44)	(1)	740	1.000	0.5000	(5.000)		
Σ^1	27.18	739				+16.870	53.783	43.417

* The values in brackets were neglected owing to their large deviations from the straight line.

the method of least squares is further recommended for estimating parameters $\log \gamma$ (sample mean) as well as the logarithmic variance $\sigma^2_{\log}$. These two parameters represent the mean and the variance of random variable $x'_i = \log l$. The straight line equation in the case of a normal distribution is deduced from equations:

$$z_i = \frac{x'_i - \log \gamma}{\sigma_{\log}} \tag{504}$$

$$z_i = \frac{1}{\sigma_{\log}} x'_i + \left(-\frac{\log \gamma}{\sigma_{\log}}\right) \tag{505}$$

$$z_i = a + bx'_i$$

where coefficients a and b are determined. From the comparison of these equations:

$$\sigma_{\log} = \frac{1}{b}; \quad \log \gamma = -\frac{a}{b} \tag{506}$$

To obtain an efficient estimation of the distribution parameters the method of least squares is applied. The two normal equations are:

$$na + b\sum x'_i = \sum z_i \tag{507}$$

$$a\sum x'_i + b\sum x'_i = \sum x'_i z_i$$

For $n = 15$ intervals, there results that $a = -4.02$ and $b = 2.84$, hence the adjusted straight line equation (Figure 137) is:

$$z = -4.02 + 2.84x'_i \tag{508}$$

The distribution parameters of variable $\log l$, according to relations (506), are:

$$\log \gamma = -\frac{-4.92}{2.84} = 1.41$$

$$\sigma_{\log} = \frac{1}{2.84} = 0.35$$

The geometric mean and the median of variable l is $\gamma = 25.70\ \mu$. These values permit us to compute the lognormal mean $\bar{x}_l$ and the variance σ^2_x of lognormal variable l, by means of relations:

$$\bar{x}_l = \exp(2.3 \log \gamma + 2.65\ \sigma^2_{\log}) \tag{509}$$

$$\bar{x}_l = \exp(2.3 \times 1.41 + 2.65 \times 0.12) = 35.2$$

$$\sigma^2_x = \bar{x}^2_l\ (e^{5.3\ {}^{2}_{\log}} - 1)$$

$$\sigma^2_x = 35.2^2\ (e^{5.3 \times 0.35^2} - 1) = 1115\ \mu^2 \tag{510}$$

$$\sigma_x = 33.4\ \mu$$

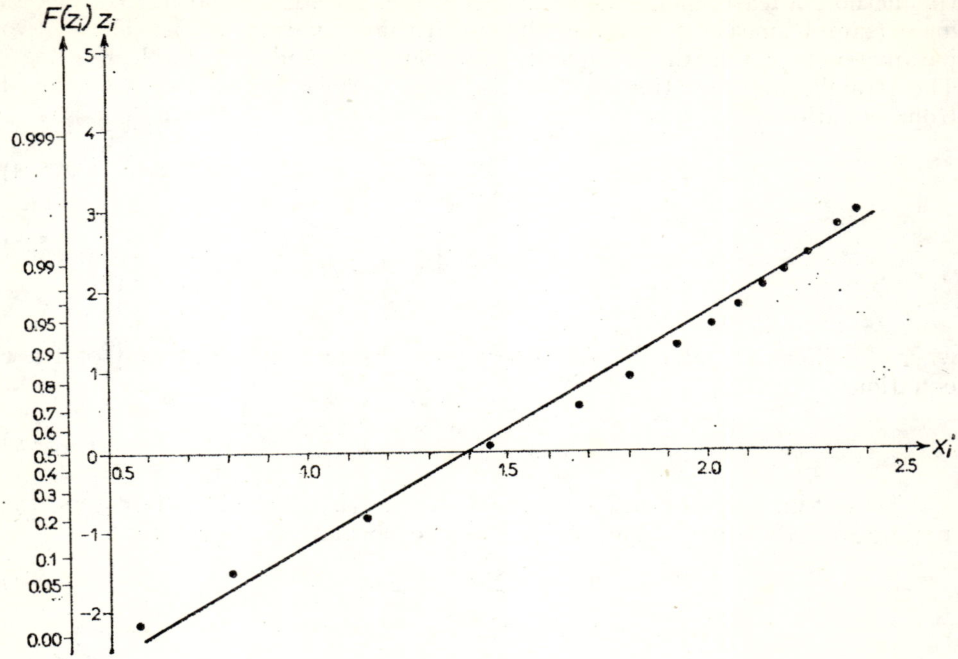

Figure 137. Verification of the porosity lognormal character.

Let us remark that although the sample volume is very large, noticeable differences appear between the values of the characteristics computed with the help of the empirical moments and those computed by the relations specific of the lognormal distribution. We should particularly note the overestimation of the mean value resulting from the sampled computation $l_m = 4.75\ \mu$. On the other hand the coefficient of variation is greater than that supplied by the sample selection

$$V_x = \frac{\sigma_x}{\bar{x}_l} = \frac{33.4}{35.2}\, 100 = 94.9\%$$

Theoretical frequencies n_i' were computed by relations (502—504) where the parameters of the adjusted distribution $\log \gamma = 1.41$ and $\sigma_{\log} = 0.35$ (Table 82) were included. With the help of values n_i' the frequency curve of quantity $\sigma_{\log}$ was drawn (Figure 138).

Taking into account that the pore-size distribution supplied by the direct method does not correspond to the total porosity, both the distribution and frequency curves were corrected by fraction $l < 3.8\ \mu$ which according to the suction method was estimated at 10%. Obviously, the latter is an approximation which takes into account the disturbed state of the sand subjected to the suction; the porosity values in both these cases — disturbed

Table 82

x_i	$x'_i = \log x_i$	Empirical frequencies		z'_i	$\Phi(z'_i)$	$F(z'_i)$	$F(z'_{i-1}) - F(z'_i)$	Theoretical frequencies	
		n_i	$f_i(\%)$					n'_i	$f'_i(\%)$
3.80	0.58	10	1.4	−2.35	−0.4906	0.0094	0.0094	7	1.0
6.65	0.82	35	4.7	−1.70	−0.4554	0.0446	0.0352	26	3.5
14.25	1.15	111	15.0	−0.73	−0.2673	0.2327	0.1881	140	19.0
28.50	1.45	236	32.0	+0.15	+0.0596	0.5596	0.3269	242	32.8
47.50	1.68	133	18.0	+0.75	+0.2734	0.7734	0.2138	158	21.4
66.50	1.82	77	10.4	+1.15	+0.3749	0.8749	0.1015	75	10.1
85.50	1.93	63	8.6	+1.50	+0.4332	0.9332	0.0583	43	5.8
104.50	2.02	35	4.7	+1.75	+0.4599	0.9599	0.0267	20	2.7
123.50	2.09	14	1.9	+1.95	+0.4744	0.9744	0.0145	11	1.5
142.50	2.15	12	1.6	+2.15	+0.4842	0.9842	0.0098	7	1.0
161.50	2.21	5	0.7	+2.30	+0.4893	0.9893	0.0051	4	0.5
180.50	2.26	4	0.5	+2.50	+0.4938	0.9938	0.0045	3	0.4
199.50	2.30	1	0.1	+2.60	+0.4953	0.9953	0.0015	1	0.1
218.50	2.34	2	0.3	+2.70	+0.4965	0.9965	0.0012	1	0.1
237.50	2.38	1	0.1	+2.82	+0.4976	0.9976	0.0011	1	0.1
		739	100.0					739	100.0

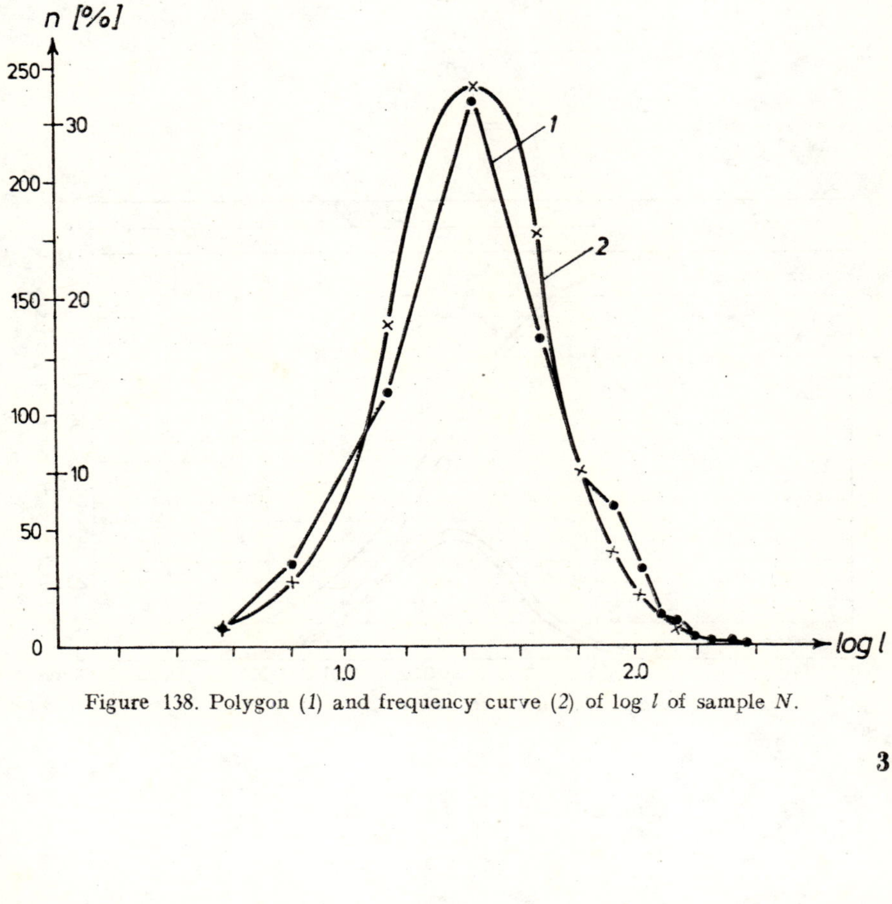

Figure 138. Polygon (*1*) and frequency curve (*2*) of log *l* of sample *N*.

and undisturbed state — were however very close to each other. The pore size distribution so corrected (Figure 139) has the following composition:

— small capillary pores	50%
— large capillary pores	46%
— macropores	4%

The comparison between the two pore-size distributions of sample *N* supplied by the suction method (on a disturbed sample) and by the direct method (on an undisturbed sample) permits us to draw certain conclusions. Thus, both these distributions are alike by their lognormal character. The distribution curve supplied by the direct method is shifted towards the high values of l which leads to like contents of both the large and small capillary pores, in contrast with distribution supplied by the suction method where the small capillary pores prevail. The characteristics of the prevailing pores show also different values: diameters of 28.5μ and 21μ and contents of 30% and 35% respectively, by the direct and by the suction methods.

So as to be able to appreciate the error due to the suction method, the ratio can be used of pore size l to grain size d. In the case of sample *N* the ratio was considered between the quartiles of the two distributions — granulometry and porometry — the results being as follows:

a) for the distribution supplied by suction (Figure 132):

$$\frac{d_{25}}{l_{25}} = \frac{0.040}{0.0065} = 6.2$$

$$\frac{d_{50}}{l_{50}} = \frac{0.073}{0.015} = 4.9$$

$$\frac{d_{75}}{l_{75}} = \frac{0.120}{0.020} = 6.0$$

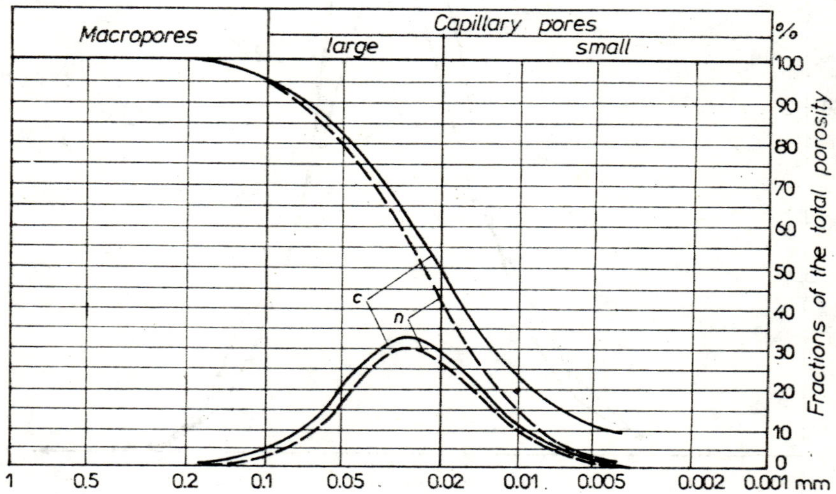

Figure 139. Distribution and frequency curves of sample *N* porosity.

b) for the distribution supplied by direct measurements (Figure 139)

$$\frac{d_{25}}{l_{25}} = \frac{0.040}{0.011} = 3.6$$

$$\frac{d_{50}}{l_{50}} = \frac{0.073}{0020.} = 3.6$$

$$\frac{d_{75}}{l_{75}} = \frac{0.120}{0.039} = 3.1$$

The difference between the two groups of values can be attributed to the imperfection of the suction method as well as to the great influence of the structure upon the pore-size distribution.

In the case of sample N it is estimated that the contribution of the second factor is larger owing to the high content of the pelitic material (30%), which determines certain structure peculiarities of the porous space in its natural state; these peculiarities cannot be established in the disturbed state even if the porosities values are equal. Consequently, we appreciate that the suction method, applied to porometry could only give satisfactory results in the case of pure sands ($d > 0.05$ mm), a domain where the structures of the porous space (for the same value of porosity) corresponding to the two states — disturbed and undisturbed — can be considered as having similar characteristics.

The knowledge of the porometry characteristics of the sands is quite important for solving practical problems of hydrogeology. Of late, besides the hydrodynamic methods, frequently, some analytical and empirical solutions are used, based on the structure factors of the porous space. Thus, concerning the importance of the prevailing pores, A. A. Hanin [28] established experimentally the dependence of the permeability upon the proportion and diameters of the pores type collectors (Figure 140). It follows from the graph that in the case of pore-size distributions with small contents (up to 10%) of the prevailing pores, the coefficient of permeability does not depend on their

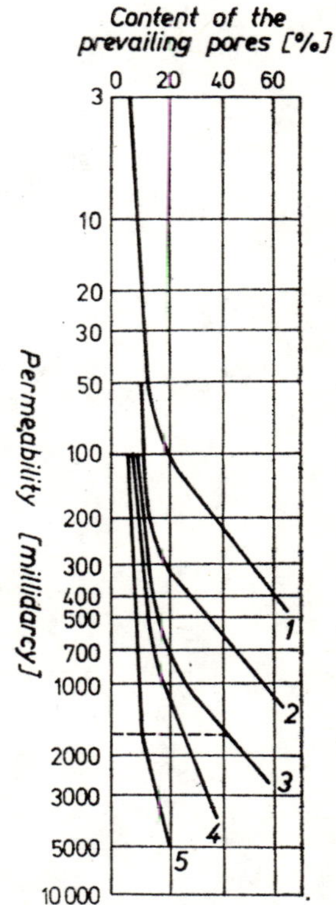

Figure 140. Permeability dependence on the prevailing pores content (after Hanin, 1963):

1 — 3—12·5 μ poros; 2 — 12·5—20 μ pores; 3 — 20—30μ pores; 4 — 30—40 μ pores; 5 — —40—100 μ pores.

size since their maximum values are 50—100 millidarcy. At higher contents of prevailing pores, the permeability depends moreover on their size so that the same value of permeability can be obtained for several sand samples of different granulometry and porometry. By applying this graph to the three samples, the following results were obtained e.g. the 1800 mD permeability, sample *N* (floury sand) should have a proportion of about 42% of prevailing pores; sample *Z* (floury sand) 20% and sample *C* (fine-medium sand) 12%; as a matter of fact, these proportions are larger so that the permeability is equally greater.

4

CORRELATIONS BETWEEN NATURAL FACTORS AND AQUIFEROUS POTENTIAL OF HYDROGEOLOGICAL STRUCTURES

4.1 Generalities

To solve any problem of applied hydrogeology, it is necessary to know the contribution of all the natural factors to the formation of the aquiferous potential of the structure, of the aquifer horizon and of the proposed area for capture or drainage. From the genetic standpoint, the natural factors can be divided into: climatic, morphological, hydrological, geological and pedological ones. Among them, the geological factors always play the principal part in defining the aquiferous potential. In certain cases, their role could be prevailing, as e.g. in the case of the deep aquifers with underground recharge and discharge, when the contribution of the other factors is insignificant. The phreatic aquifers are an example of an approximately equal weight of all the above-mentioned natural factors, owing to the fact that these aquifers are hydraulically linked with the drainage network.

Only after a good appreciation of the weight of the natural factors in the formation of the water yielding potential is it possible to look on the formation and accumulation process of the ground waters in the respective structure as a complex and natural process, allowing to schematize the aquifers and to define the hydrodynamic schemes of the capture or drainage systems of the ground waters. The non-concordance between the designed and the achieved parameters is chiefly due to the one-sided broaching of the hydrogeological problem, to wit from the hydrodynamic standpoint only.

In the problems of mining hydrogeology, the water-yielding potential is more comprehensive. That is why the notion of the mineral deposit *inundation degree* is proposed. The inundation degree of a mineral deposit (or of a working layer) can be defined as the totality of the factors favouring the penetration of the water inflow into the mining works as well as of all consequences arising from this situation, such as: the pressure increase on the mining works, the yielding of the protective shield of the hard and impermeable rocks, the formation of the havy sands, the increase of the discharge up to the eruptions, etc. Naturally, any mineral deposit occured under the groundwater level possesses a certain inundation degree but it is not directly proportional to its depth; there are cases when a shallow mineral deposit displays a higher inundation degree than a more deeply situated one. Obviously, beside the natural factors, the artificial ones intervene i.e. the technical and mining conditions, that owing to the alteration of the rock formation structure are liable to induce an increase of the inundation degree. Therefore, in order to evaluate the inundation degree the hydrogeological conditions should be well known such as: the occurance conditions of the ground waters, the lithological composition and the water yielding properties of the water bearing formations, the flow, the quality and quantity of the ground waters, as well as the régime peculiarities both under natural conditions and under the influence of the artificial factors.

The role of the climatic factors in the formation of the ground waters and of their régime in a certain structure diminishes at the same time as their depth and their remoteness from the surface recharge areas. Let us note that the climatic factors influence the grandwater flow both directly and through the intermediate of other natural factors such as the soil, the vegetation, the run-off and the surface geological structure.

As a function of the morphological conditions of the ground, the drainage network can have a higher or lower density. The great density (expressed in km/km^2) of the terrains with a low permeability leads to a high value of the surface run-off at the expense of the ground water run-off The latter can have a high value in the case of a reduced density of the natural drainage network in terrains with a high permeability.

The hydrological factors influence the process of groundwater formation both by the balance elements of the respective catchment basin and by the hydraulic link between the aquifer and the surface water. For instance, the existence of such a hydraulic link in the neighbourhood of a mineral deposit liable to be dewatered can lead to an increase of the inundation degree and implicitely to great difficulties in achieving the process of mining drainage.

The climatic, morphologic and hydrological factors possess a potential character as to the recharge of the groundwaters and to the formation of the subsurface drainage. The decisive role is played by the geological factors; the quantity of infiltrated water depends on the characteristics of the geological outcrop. On the other hand, the processes of formation, migration, accumulation and discharge of the groundwaters are determined by the characteristics of the collecting rocks, by the local and regional tectonics as well as by several thermal and physico-chemical deep processes.

An important natural factor that conditions the inundation degree of the coal deposits is the presence of impermeable formations in the direct top and bottom of the working layers. The existence of these protecting shields with great constant thicknesses can notably diminish the inundation degree of a deposit.

There are many cases when the protecting impermeable layer is of small and very variable thickness, or when the coal bed is in direct contact with the water-bearing sands. Such protecting shields however do not lead to important reductions of the inundation degree.

In the case of bedded deposits in underground mining, the inundation degree should be estimated and treated for each working layer, since certain geological factors (lithology and tectonics chiefly) are frequently variable along the vertical and determine various values of the inundation degree of each layer and area.

In this chapter, an analysis will be made of those natural factors the contribution of which to the aquiferous potential or to the inundation degree of the mineral deposits raises particular problems of interpretation and estimation. So as to explain more factually the processes, notions and terms used in this book, the problems dealt with will be mainly illustrated by examples of data concerning the hydrogeological structure of Șotînga-Mărgineanca lignite deposit (Romania) which I have studied thoroughly.

4.2 Influence of the tectonic structure on the formation of the hydrogeological conditions

The notion of hydrogeological conditions is very comprehensive. It includes the mineralogical composition and grain sizes, the transmissivity, the hydraulic diffusivity and the storage capacity of the aquifer, the value and the distribution of the ground water pressure, the nature of the contact with the surrounding formations on the horizontal as well as on the vertical, the permeability of these formations, the possible recharge source and the volume of the natural resources, the nature of the interaction between the ground waters and the drainage network.

The local or regional tectonic structure is likely to have a decisive role in the formation of the hydrogeological conditions i.e. in the behaviour of an aquifer under natural or influenced conditions. There are cases, in the strongly folded areas, where important dislocations can divide a regional aquifer into several hydrogeological structures with several régimes more or less independent from one another.

Hereafter, the principal tectonic elements of the Șotînga-Mărgineanca area are presented which in the following paragraphs will be used to define the local and regional hydrogeological conditions.

The major structure of the Miocen-Pliocene formations in the Șotînga-Mărgineanca area is represented by an anticlinorium within the limits of which several structures can be distinct that are emphasized by the surface mapping and the exploration wells. From north to south these structures are (Figure 2): anticline Mărgineanca, syncline Valea Popii, anticline Cărbunaru, syncline Valea Ilfovului and anticline Gheboeni-Șotînga. Anticline Cărbunaru and syncline Valea Ilfovului are locally developed and only brought into relief between Dîmbovița and Ialomița rivers. Between anticline Mărgineanca and syncline Izvoarele Brănești towards Valea Ialomiței, the pericline shut off of two folds — syncline Valea Roșca and anticline Vulcana-Pandele-Lăculețe — appear, within the limits of which the late Pontian outcrops. A geometric characteristic of these folds in the horizontal plane is the large ratio between length and width; they belong to the group of elongated folds. The surface mapping supplied enough elements for delimiting the important structures and dislocations. The reduced spacing of wells (at an average 400 m) permitted moreover to establish several minute elements related to the geometry of the dislocations and deep structures.

Anticline Mărgineanca is the major structural element of the anticlinorium since it has the highest elevation. It is an asymmetrical fold, the northern flank of which is more developed in the horizontal plane than the sourthern one. Its great asymmetry appears also in the fold height: on the northern flank its height is about 700 m, in relation to the bottom of syncline Izvoarele-Brănești, while on the sourthern flank it is about 100 m in relation to syncline Valea Popii. After the axial culmination in Plaiul Moga area anticline Mărgineanca sinks towards Valea Dîmboviței at a mean inclination of 20°.

From the correlation between the coal layers it follows that on the northern flank the inclinations vary between 18 and 25°. In the eastern sector, the strike of the layers changes from W—E to SW—NE and this determines the appearance of two folds; syncline Valea Reșca and anticline Vulcan-Pandele-Lăculețe (Figure 141).

Syncline Valea Popii is outlined by the surface mapping. At coal layer II level, the fold is slightly outlined in the area of that valley (Valea Popii). Westward as the distance from fault F_1 increases, the fold is clearer and it is sinking toward Valea Dîmboviței. The syncline is affected by the reverse, longitudinal fault (F_1) emphasized by the exploration work and ascertained by the mining works. This fault with a great dipping — about 80° northward — offers a maximum throw of 150 m in the eastern sector; this throw decreases gradually westward and the fault disappears in the axial area of anticline Mărgineanca. Since it separates the two major anticlines Mărgineanca and Gheboeni-Șotînga, syncline Valea Popii has a much higher structural level than syncline Brănești-Izvoarele and structure Aninoasa.

Anticline Gheboeni-Șotînga, situated at a lower structural level than anticline Mărgineanca, is well studied both by exploration wells and by mining works; this indicates a rather intricate minute structure. This anticline consists of two branches — the principal one Cărbunaru, northward, and the secondary one Șotînga, southward; both these branches reunite in sector Gheboeni and continue by a plunge towards Valea Dîmboviței. In its axial part, anticline Cărbunaru is affected by longitudinal fault F_2 along which the northern flank goes down into syncline Valea Popii by throws comprised between 30 m in the western section and 100 m in the eastern section. The fault plane tilts 55—60° northwards. This normal longitudinal fault links up with cross fault F_{IV} which becomes in this case a reverse fault. The vertical throw of this fault increases from 15 m in the linking up area to 100 m at the cross-bearing with fault F_3. Between the two above-mentioned syncline branches Valea Ilfovului develops by very small slopes which, as it approaches the anticlinal dome Gheboeni, vanishes.

Secondary anticline Șotînga with a lower structural elevation than the principal branch Cărbunaru represents together with syncline Valea Ilfovului undulations of the principal anticline. The southern flank of anticline Șotînga

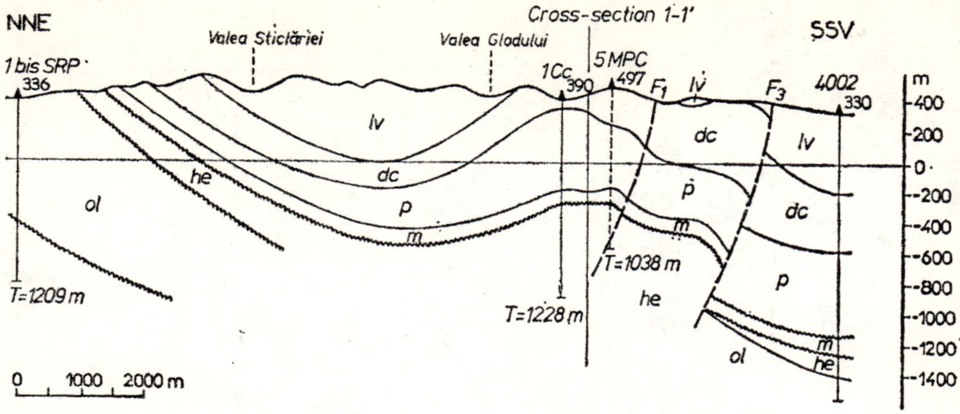

Figure 142. Geological cross-section $I-I'$.

is affected by an intricate tectonics. Thus, as it approaches fault F_3, the slope of this flank increases, up to 40°. This fault can be followed up to the overthrust line Cîmpina-Ocnița-Godeni where it reaches a throw of about 1000 m. Westward, the throw diminishes: 700 m in geological section II—II' (Figure 143) and about 200 m in section *IV—IV'* (Figure 145). Fault F_3 is affected from three transverse faults, emphasized by means of exploration wells F-I, F-II (from which yet another fault branches determined by mining works F-III) and F-IV. These faults present a minimum throw in the axial area of anticline Șotînga and a maximum one (about 100 m) at the intersection with fault F_3.

In the geological sections in Figures 142—144 appears also the global lithology of the Dacian deposits. As a whole, the tectonics of the Șotînga-Mărgineanca area can be considered as complicated, which bestows on it a principal role in the formation, accumulation and régime of the groundwaters occuring in the Dacian and Levantine formations as it will be shown in the following paragraphs.

4.3 Definition of the hydrogeological stratification

The hydrogeological exploration supplies a considerable volume of geological and hydrogeological information, the systematization of which asserts itself in the interpretation frame. The distinction of the groundwater bodies must be made on the basis of a complex of geological and hydrogeological features. The definition of the hydrogeological stratification should rely on the wateryielding properties of the permeable formations determined by: their lithological and facial composition, their physical and hydraulic properties, their conditions of bed and position in the section, their age and geographic location. According to the analogy or the diversity of above mentioned parameters as well as according to the aquiferous

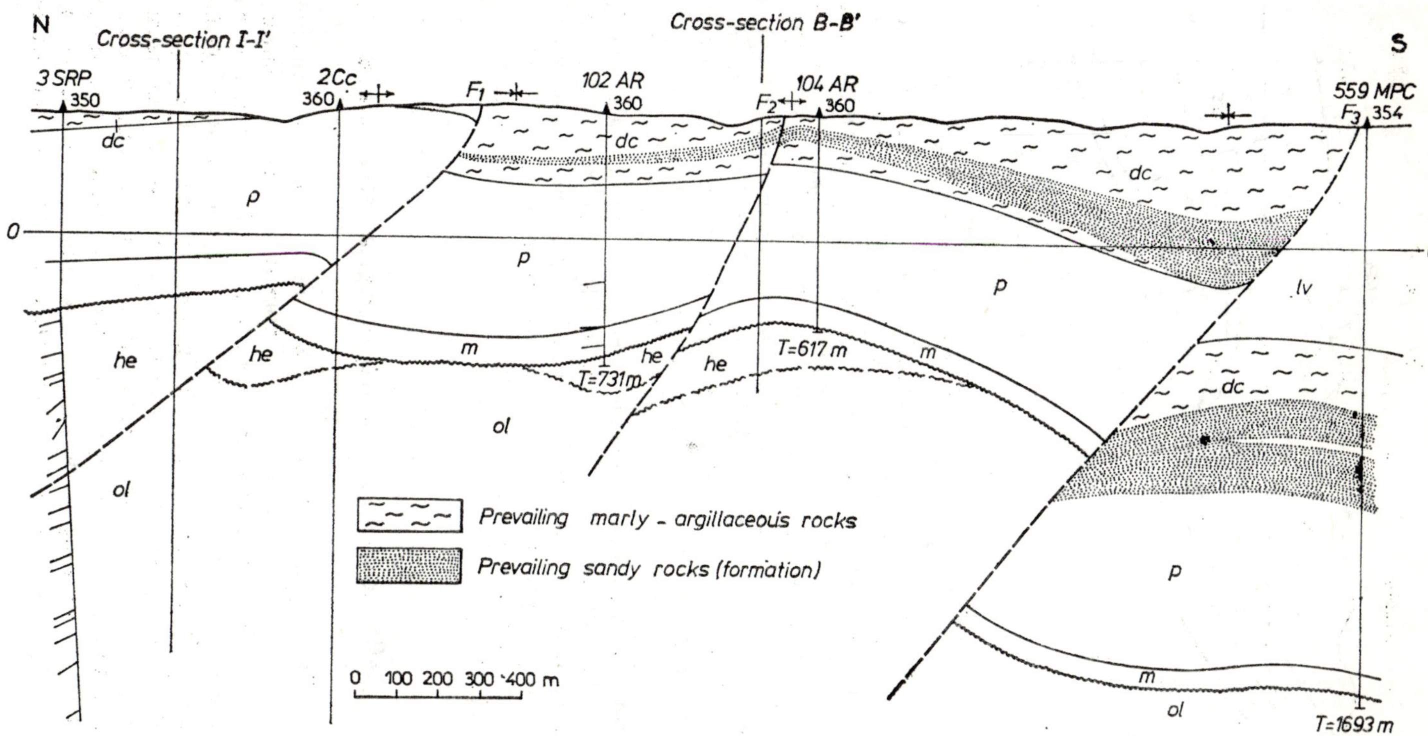

Figure 143. Geological cross-section $II-II'$.

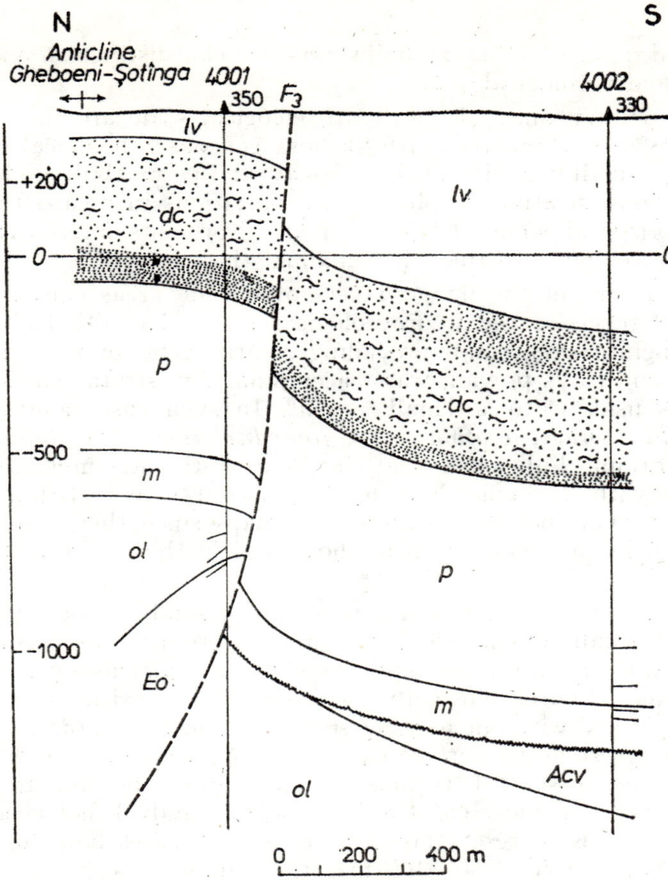

Figure. 144. Geological cross-section $III-III'$.

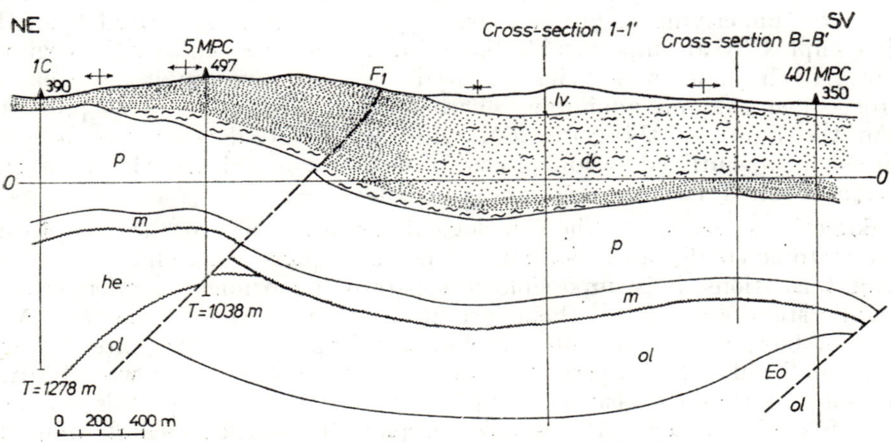

Figure 145. Geological cross-section $IV-IV'$.

character and quality of the groundwaters, we can distinguish various units within the hydrogeological scale.

The elementary units of the hydrogeological scale are the *stratum* and the *aquiferous lens*, characterized by a homogeneous lithological composition as well as by a small variation of the porosity and permeability. The thickness of the aquiferous stratum displays small variations in regard to its extent along the horizontal while the lens thickness presents large variations, its length not exceeding 1000 times its maximal thickness.

The continuity of the lignite layers over wide areas can constitute an important reference of the hydrogeological scale. The lithological interval between two lignite layers can be prevailingly marly-clayey or prevailingly sandy. When marls and clays prevail, water accumulates in strata, lenses or isolated intercalations usually to a small extent. In such cases a *hydrogeological structure with a sporadic location of the groundwaters* may be considered, since no direct hydraulic link exists and the hydrostatic pressures are variable. These areas with sporadic location of ground-waters within a certain lithological interval should be marked on maps since they raise particular problems both in the exploration methods and in the designing of drainage systems.

When the lithological interval is prevailingly sandy and the impermeable bodies are of small extent, the accumulated groundwaters constitute an *aquifer*. It represents the total aquiferous strata and lenses with analogous lithofacies and hydrogeologically characteristics, with large extents, hydraulically linked with one another so that it can be in practice considered that a unitary piezometric surface exists. The aquifer can include lithofacial types belonging to several rythms provided they pertain to the same sedimentary cycle. In most cases the prevailing sandy lithological intervals between lignite beds represent aquifers. In certain cases however, when the intervals are very thick (60—100 m), the sandy deposits are separated by impermeable layers into two or more levels that form as many aquifers with distinct piezometric surfaces. Such an example is offered by the aquifer underlying of Șotînga coal layer I, which is separated from the basic complex by an impermeable bed of 15 m thickness and with a regional extent. Each aquifer is characterized by its own hydrogeodynamic, hydrogeochemical and aquiferous characteristics. Owing to these premises, within the limits of the Șotînga-Mărgineanca lignite deposits the follwing aquifers can be outlined: below of coal layer I; between I and II coal layers; between II and III coal layers; above coal layer III.

Sometimes, owing to the lithological compositions, to the complexity of the tectonic conditions or simply to the inadequate research of the water-bearing formations, it is impossible to separate independent aquifers from a hydraulic standpoint, all of them forming a so called *aquifer complex*. As a matter of fact, in the case of a preliminary exploration of lignite deposits with a small number of layers, the hydrodynamic study of all waterbearing formations in the coal-bearing complex are usually made in this manner. In the following stage — the detailed exploration — the research must be resumed separately for each aquifer. On the north side of Mărgineanca anticline, the sequence of sands and marls in the top of coal layer III (Dacian + Levantine) reaches its maximum thicknesses of 40 m (see Figure 3). Owing to the wide variations of the lithofacies, it is very difficult to separate aquifers so that the research should be made within the whole aquiferous complex.

A sequence of aquifers semipervious and impermeable, homogeneous or with rythmical sedimentation, formed in the same phase of a tectonic and sedimentary cycle, constitutes an *aquiferous series*. In the Pre-Carpathian Neogene area a Pliocene aquiferous sequence and a Pleistocene aquiferous series (Cîndeşti strata) can be separated. The separation of these two units (characterized by their sedimentary continuity) is especially justified by their distinct facies peculiarities. Both these series constitute the great aquifer of the upper Molasse, separated from another great aquifer of the lower Molasse by a stratigraphic gap.

The major hydrogeological structure for the deep groundwaters is the *artesian aquifer basin* also sometimes called *artesian basin*

The separation of the above-mentioned hydrogeological structures — complex, series and basin — presents at first sight a rather theoretical importance but regarding the estimation of the regional hydrogeological conditions these notions are quite valuable.

Therefore in defining the hydrogeological stratification (actually an aquifer system) we must also take into account the hydrodynamic elements. By so doing, the aquiferous system can be defined as an area where all the natural or artificial actions interfere more or less with one another as a function of the scale of the discontinuities and heterogeneities that affect the water-bearing formations (facies changes, tectonic elements, etc.). [43].

4.4 Regional hydrogeological conditions

To estimate the water balance elements and to define accurately a hydrodynamic scheme of capture or drainage, the regional hydrogeological conditions must be known. The hydrogeological research within the limits only of the area proposed for capture or drainage could however lead to gross errors in designing the considered works. The extent of the hydrogeological frame (in depth and area) depends both on the complexity of the geological and hydrogeological conditions and on the ratio between the aquifer potential of the structure and the yield to be captured or also on the size of the required drainage. For instance, if a small yield is wanted with regard to the dynamic resource of the aquifer, a mere sketch would be sufficient of the regional hydrogeological conditions on a relatively small area.

The regional hydrogeological conditions are often studied indirectly, through the hydrogeological interpretation of the geological and geophysical documentations concerning the considered areas. In Romania, this is facilitated by the existence of synthesis reports worked out on the information supplied by full cored and geophysical logged boreholes. By selecting and interpreting the material and by stressing the lithological and structural conditions, the hydrogeological forecasts become possible.

In solving computation schemes for determining the water inflow towards the mining and dewatering works, one has to keep in mind: the distance from the mining field to the recharge and discharge domains, the extent of the drained aquifer system, as well as the relations between it and other aquifer systems (surface or underground ones), etc. All these characteristics determine on the whole the quantity of the natural resources of the groundwaters and implicitly the dewatering possibilities. But their knowledge can only be acquired from geological and hydrogeological material concerning the explored area. It follows that an accurate way of dealing with the dewatering problem can only be given by the regional hydrogeological conditions as a whole. There are indeed working coal deposits (Șotînga, Filipeștii de Pădure) quite close to the surface recharge domain (outcrop zones of the Dacian sands), of relatively small areas. Because of these conditions, the dewatering works are subjected to an unsteady hydrodynamic régime: the rate of discharges and hydraulic head diminish steadily, so that after a certain time period (6 months to 2 years) the dewatering is practically finished. In the lignite basin Rovinari, the rather great distance from the surface recharge domain and the large extent of the Dacian (where sands are prevailing) determine the existence of vast natural resources of groundwaters (Figure 146). Under these conditions, the hydrodynamic régime of the dewatering works is steady-state and the hydraulic head recover soon after the well drainage is finished. These examples show the necessity of knowing the hydrogeological conditions on a regional scale.

The hydrogeological division by zones of a territory should take into account the geostructural characteristics, the actual relief and the hydrogeological conditions as a function of which many confined aquifer basins can be defined.

According to M. E. Altovski a confined aquifer basin consists in a sequence of aquifers located within the limits of a large division of the stratigraphic scale, or having a common genesis of the reservoir rocks. A. M. Ovchinikov established a correspondence between the basin types and their areas. Thus, the widest confined aquifer basins with more than 100 000 km^2 area are to be found in platform areas. The confined aquifer basins in the lateral depression and in the great intermontane depressions, that are considered as middle-sized, have areas of 10 000—100 000 km^2. Among these great and middle-size basins, small basins of less than 10 000 km^2 can be separated.

On the basis of the structural and lithological data as presented on a regional scale in § 4.3, a tentative hydrogeological division can be made. The whole Mio-Pliocene subzone between Ialomița and Danube rivers can be considered as a vast confined aquifer basin owing to its geostructural and geomorphological characteristics that impart a tendency of uniformity to the regional hydrogeological conditions. On its internal border, the basin contacts the mountainous region by a quite normal stratification or is affected by simple faults. This structural and geomorphological occurrence determines the existence of a continuous outcrop band of the Mio-Pliocene formation which from the hydrogeological standpoint represents the external recharge domain of the basin. The extent of this recharge domain up to 1000 m heights (Depression Cîmpulung-Muscel) which includes areas with a high quantity of annual precipitations creates favourable conditions for continuous regeneration of the groundwater resources. The eastern limit of the basin goes along Ialomiţa river. Thus, from the internal zone of the Flysch up to and including syncline Brăneşti-Izvoarele the formations point out an axial raising in this valley with eastward and westward dipping axials. Consequently, the above-mentioned syncline which presents a particular hydrogeological importance between Valea Prahovei and Valea Dîmboviţei is divided into two hydrogeological units by a structural threshold that acts as hydrogeological watershed for the waters in the Dacian and Levantine formations. At lower structural levels, it is possible that a hydraulic relation could exist between the two basins. South of syncline Brăneşti-Izvoarele the height of the structural threshold is smaller and this determines a continuity of the Dacian aquifers on both sides of Valea Ialomiţei. East of Valea Ialomiţei there is a tectonic outline formed by the overthrust line Ocniţa-Glodeni and fault Şotînga-Doiceşti (F_3). This outline which goes up to Valea Reşca presents an opening portion along Valea Ialomiţei. Taking into account that the throw of these tectonic lines is 700—1000 m their outline can be considered as the limit of the confined aquifer basin. Fault F_3 diminishes its westwards throw so that below Valea Dîmboviţei it is reduced to about 300 m. As a matter of fact, westward of Valea Dîmboviţei the tectonics of the Mio-Pliocene subzone is more quiet and the depression is generally wider.

The southern limit of the Mio-Pliocene confined aquifer basin can be estimated by the structural relations between the wide peripheral syncline of the Pre-Carpathian Depression and the Moesic Platform. I. Gavăt [21] gives this great syncline the name Peri-Carpathian Neogene Depression (Sarmatian-Pliocene). On the geological map (Figure 141) the stratigraphic and tectonic elements are represented of this depression between Valea Vîlsan and Valea Prahovei. It is considered that the limit between the two large units is the parallel from which the quicker dip begin northward of the pre-tertiary formation in the platform zone. Along this limit (following approximately the Craiova—Slatina—Bucureşti— etc. line) the subsurface discharge occurs of the waters from the Mio-Pliocene formations into the prevailingly calcareous formations of the platform. The hydrogeological relations between the depression and the platform are even more striking in Oltenia region. For instance, in a N—S section along Valea Oltului (Olt river), the Cretaceous formation of the platform zone touches successively all the Neogene (stratigraphic) terms and near the Danube river it is directly covered by the Levantine grounds. The lignite explorations made between rivers Jiu and Motru show that, in the hill zone, the Dacian is prevailingly sandy; over 400 m thick it constitutes a secondary artezian basin containing important groundwater resources. In the natural régime period (before the dewatering works started in basin Rovinari) the flow direction was north-south, with a hydraulic gradient of about 1%. Under these conditions, it is obvious that a subsurface discharge occurs of the waters in the Dacian depression, namely from the platform into the Cretaceous (or Sarmatian). It is to be remarked that the Sarmato-Cretaceous aquifer complex possesses an external recharge domain, probably in the hilly zone of the Balkan Mountains.

In the process of hydrogeological structure formations, under more or less independent conditions, the impermeable layers with a regional extent play an important part. Among them, we mention the gray gypsiferous marls of the Helvetian, the Pontian marls and the argillaceous-sandy complex of the early Levantine. In this basin, reservoir rocks appear in the whole Neogene series: conglomerates and sandstones in the Helvetian, sands and sandstones in the Meotian, thick complexes of sands in the Dacian and "the Cîndești strata" in the Pleistocene. One more remark should be made concerning the early Pontian which, towards the recharge domain between Ialomița and Dîmbovița, changes into a permeable facies comprising sands, sandstones, limestones and even conglomerates.

In the geological sections in Figures 143—145 the lithological complexes of the Dacian appear as a global image of the spatial variation of the aquiferous potential.

The presence of oil fields with active water within the Helvetian Sarmatian, Meotian and Dacian formations confirm the above-mentioned hydrogeological characteristics of the Mio-Pliocene aquifer basin.

These regional hydrogeological elements permit us to conclude that the Șotînga-Doicești lignit field is situated at the eastern border of a great artesian aquifer basin. At least at the Levantine and Dacian level, this limit does not represent however a recharge boundary, but is particularly important for the reduction of the inundation degree of the lignite field.

4.5 Hydrogeological behaviour of the faults

In the case of pre-quaternary hydrogeological structures, the faults can have a great influence upon the aquiferous potential. Broadly speaking, from the hydrogeological standpoint, the faults are either *open*, i.e. non-water-tight, or *closed*, viz. water-tight. The open faults are as a rule gravitational; at their formation tensile forces occurred. The closed faults (barrier) are of overlappings characterized by compression forces. In the geometrical classification, the closed faults belong to the longitudinal type and the open ones to the cross type. These characteristics have been verified in the Pre-Carpathian Depression by the oil explorations and exploitation wells. In mining hydrogeology problems (especially in the case of underground working) the hydrogeological behaviour of the faults should be dealt with at a more thorough level, within the frame of the detailed exploration. In the case of closed faults, each tectonic compartment displays a hydrogeological structure the hydrodynamic and hydrochemical characteristics of which differ from those of neighbouring compartments. The open fault achieves a hydraulic link between its two compartments owing chiefly to the hydraulic conductivity of the fault plane itself. Evidently, in the natural state many situations occur intermediate between the two fundamental types. From the hydrogeological standpoint, the notion of *closing degree* of the fault is the more proper one. It can be quantitatively determined by steady state pumpings in experimental groups with a high number of observation wells along at least two alignments, one of which must cross the fault. The analysis of the piezometric curve at the end of the pumping allows us to determine the faults throw corresponding to the fault which from the hydraulic standpoint represents a local charge loss. Positive conclusions can equally be drawn from compared analysis of the piezometric curves corresponding to the two alignments.

It is possible that in the case of observation wells in the neighbouring compartment (with the one in which the experimental well is placed) no drawdown could be recorded; in these cases it is considered that the fault is practically closed. Of course, such a method is expensive because of the number and depth of the wells.

The principal geological factors influencing generally the hydraulic conductivity of the dislocations are the time and their formation mechanism as well as the physico-mechanical properties of the affected rocks. High values of the hydraulic conductivities are usually to be found in rocks with considerable mechanical strength. It follows therefore from this standpoint that generally the faults affecting the Dacian sandy-argillaceous deposits with small mechanical strengths possess low hydraulic conductivities. A fault, beside its own plane, possesses a crushing area, very fissured — and therefore permeable in the case of high resistent rocks — and very broken, possessed of many smooth surfaces and friction planes in the case of sandy-argillaceous rocks.

Beside the geological factors, the hydraulic conductivity of the faults is influenced by other parameters such as length, throw, dip, etc. As a function of the fault length and throw, big dislocations (with throw exceeding 100 m along tens of kilometers) as well as medium and small ones can be defined.

From observations made in lignite mining operations where the workings met faults with small throw there results that the area of faulting does not present an open fissuration and is practically impermeable. The width of the crushed area can reach tens of centimeters; this area is strongly scarred by argillaceous rocks.

When the lithological sequence is formed by an alternation of sandy and argillaceous layers of small thicknesses, the fault becomes water-tight owing principally to the continuity of the crushing area. When the sandy and argillaceous layers are very thick, two cases can occur: 1) the throw is greater than the sandy layer thickness and then the fault is water-tight, 2) the throw is smaller than the layer thickness and then the hydraulic link is made through the intermediate of the so-called "hydraulic windows". The artesian aquifer in coal interval III—IV in Rovinari basin is such an example. The whole coal complex is affected by a system of normal longitudinal faults (I—II—III) with throw varying between 40 m towards the eastern limit and zero at the western limit of the perimeter (Figure 146).

The thickness of the artesian aquifer being about 80 m allows for the existence of a few hydraulic windows of at least 40 m. The stress relieving

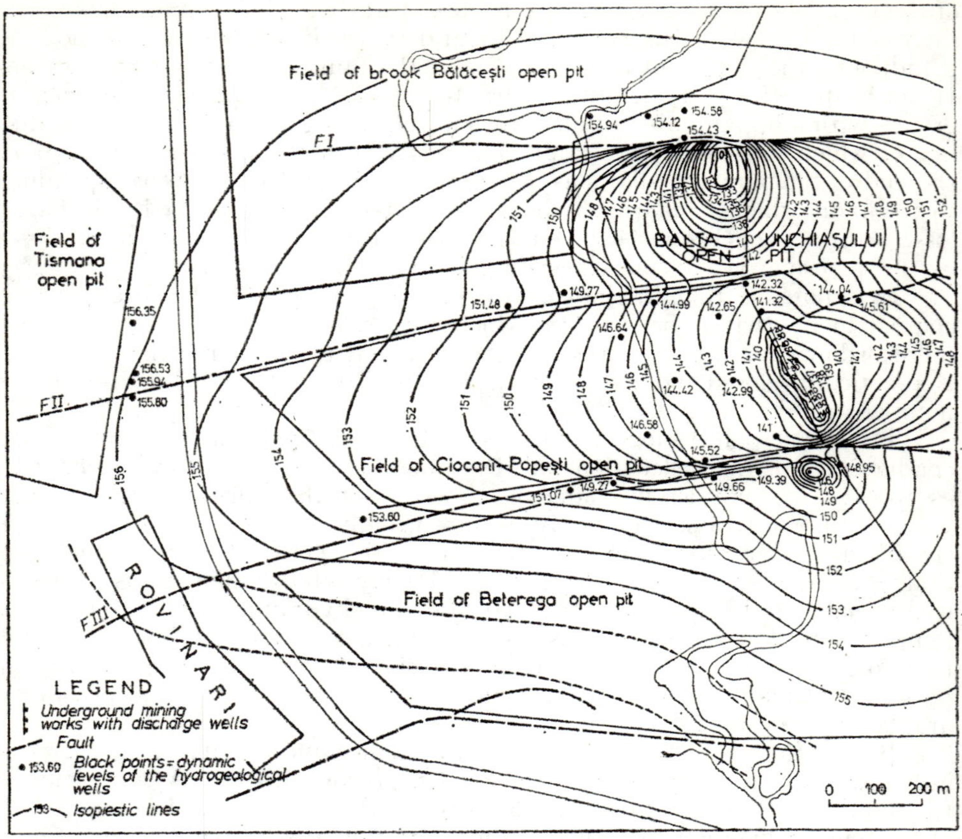

Figure 146. Piezometric surface of the artesian aquifer in the Rovinari area on October 30, 1963 (reduction from the ICEMIN map at 1: 500).

of this aquifer was achieved in the 1958—1965 period by means of the screenless wells made on the working cut-off trench Balta Unchiaşului and of the subterranean wells in Cicani area. As a consequence of this strong drainage a great depression zone appeared, characterized by a marked asymmetry as well as by isopiestic line discontinuities. The effects are made conspicuous by the isopiestic map made by ICEMIN at the end of October 1965 when the quarrying of Balta Unchiaşului came to an end. The asymmetry of the depression area is made obvious by the extreme values of the radius of influence: maximum along the east-west direction of the geological structure and of the faults, and minimum on the north-south direction. In the eastern area, discontinuities or changes of direction of the isopiestic lines appear which lessen westward concomitantly with the vanishing tendency of the faults. The configuration of the isopiestic lines of the Balta Unchiaşului quarry shows on fault F_I a charge-loss of about 8 m. This is due to the throw between the two principal compartments and to the presence of a lid fallen between the fault planes, which increases the degree of hydrogeological shutting off. In conclusion, the influence of the faults on the inundation degree in Rovinari basin at the level of the working coal complex is small. It was experienced to a certain extent on the north-south direction in the eastern area of the perimeter.

In the Șotînga-Mărgineanca perimeter, the faults influence much more intricately and more conclusively the hydrogeological conditions of the coal deposit. Thus, fault F_3 (Șotînga-Doicești) represents a major dislocation with a 200—700 m throw within the explored perimeter and a length of over 20 km. This regional dislocation has a water-tight character which determines, in the explored perimeter, the separation of an independent hydrogeological structure. The other two faults F_1 and F_2 can be considered as medium dislocations, their throw being generally less than 100 m. Since the maximum jumps occur in Valea Ialomiței it is probable that they continue equally eastward. The degree of hydrogeological closing of both these faults can be estimated on the basis of natural piezometric levels. The analysis period of this régime corresponds to the 1962, 7 IX—24 XII for the aquifer between II—III coal layers; on the basis of level variation measurements graphs were made for each well (Figure 147) and the maps of piezometric surface

Figure 147. Natural régime of the piezometric levels in aquifer between II—III coal layers (Sotînga mining field).

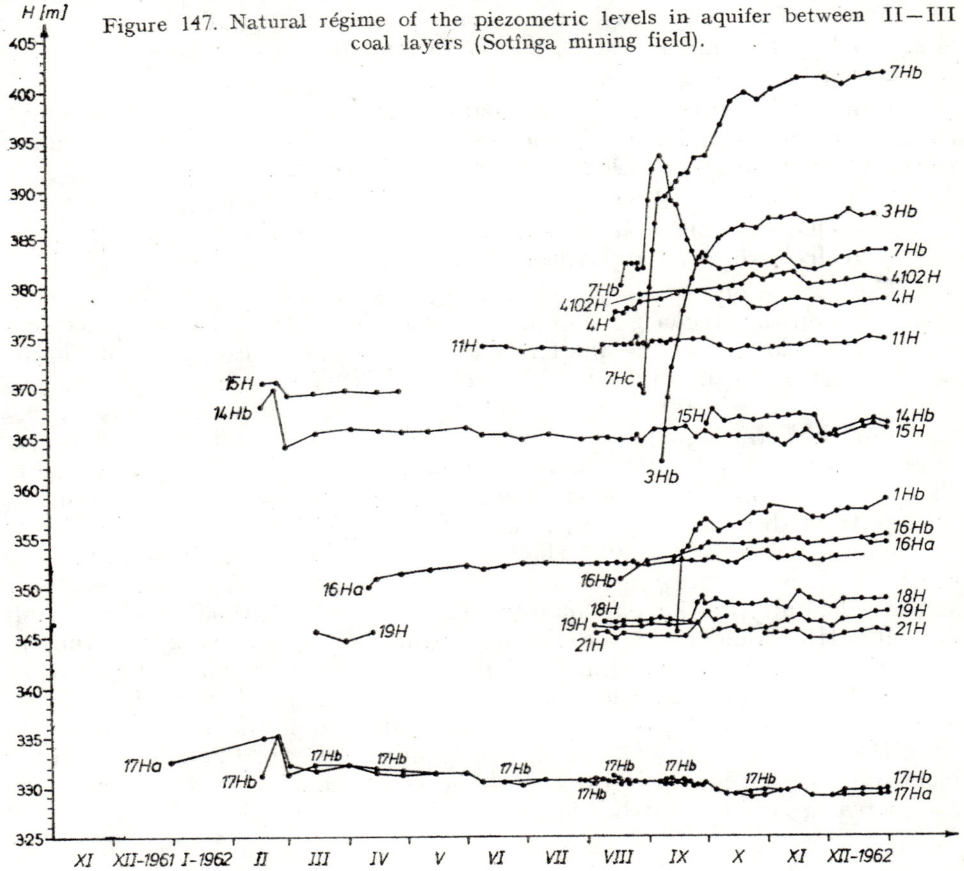

were made at the beginning and end of the interval (Figure 148 and 149). Thus, the isopiestic lines cross fault F_2 at a small angle which means that a recharge exists coming from the south compartment direction; this configuration is maintained on both the maps and is therefore valid during the whole period. Along fault F_1, the situation is more intricate and less clear owing to the low density of the observation points. On the piezometric map of 1962 7 IX, in the western sector (with regard to the fault crossing with the Valea Popii syncline axis) the south-north flow direction is obvious as also is the hydraulic conductivity in this sector. In the eastern sector no appreciation can be made in the absence of sufficient data. Nevertheless, a certain tendency of a hydrogeological closing, accompanied by a reversal of the flow direction — that becomes north-south — is conspicuous. Owing to the sinking of the block situated between faults F_1 and F_2, a contact occurs in the east area (with maximum vertical throws), the aquifer between coal layer I—II come into contact with the very strong aquifer in the bottom of coal layer I. From the previously presented data it follows that it is not possible to count upon a permeability of the fault plane likely to achieve a hydraulic link between aquifers situated at different levels. The hydraulic link of the two flanks is only achieved when the positions on the vertical of two aquifers coincide on the fault plane. This proves the absence of the very argillaceous crushing area nearby the aquifers. These considerations are equally valid in the case of transverse faults $F_I — F_{IV}$.

Figure 147 represents the fluctuations of the piezometric levels under natural régime in wells. As a function of the fluctuation amplitude, during period of July-December 1962 the wells can be grouped as follows:

1. Wells with small amplitudes (4 102 H, 15 H, 11 H, 17 Ha)
2. Wells with moderate amplitudes (16 Hb, 18 H, 19 H, 21 H)
3. Wells with large amplitude (7 Hb, 7 Hc, 3 Hb, 1 Hb).

This variation of the level fluctuations under the conditions of a relatively constant recharge can only be explained by the existence of tectonic or lithologic closings of the aquifer. The distinction of these types of closing is sometimes more difficult when the wells are situated in the neighbourhood of the faults. For instance, wells 4102 H and 15 H confirm the existence of a hydraulic link through fault F_2. The case of wells 3 Hb and 16 Hb is however more intricate since fault F_1 could be open while the closing could be due to the die out of the sandy layer. But the exploration data pointed out the continuity of the sandy aquifer, thus excluding the assumption of the closing by die out. On the other hand, the conclusion was previously reached that in the western sector fault F_1 achieves an obvious hydraulic link with the northern sector. The only possible explanation could be that the lithologic shutting-off of well 3 Hb should be due to the diminishing of the permeabilities. Indeed, aquifer transmissivity determined in this well is the lowest ($kM = 0.8$ m²/day). The fault influence should however be equally considered, subsidiarily. This is the only way to explain the increasing of piezometric level, ΔH up to 22 m H_2O, during a period of about two months. Well 16 Hb is situated in the vicinity of a local line of dying out of the sandy aquifer, which accounts for the increasing of the piezometric level ΔH by 4 m H_2O.

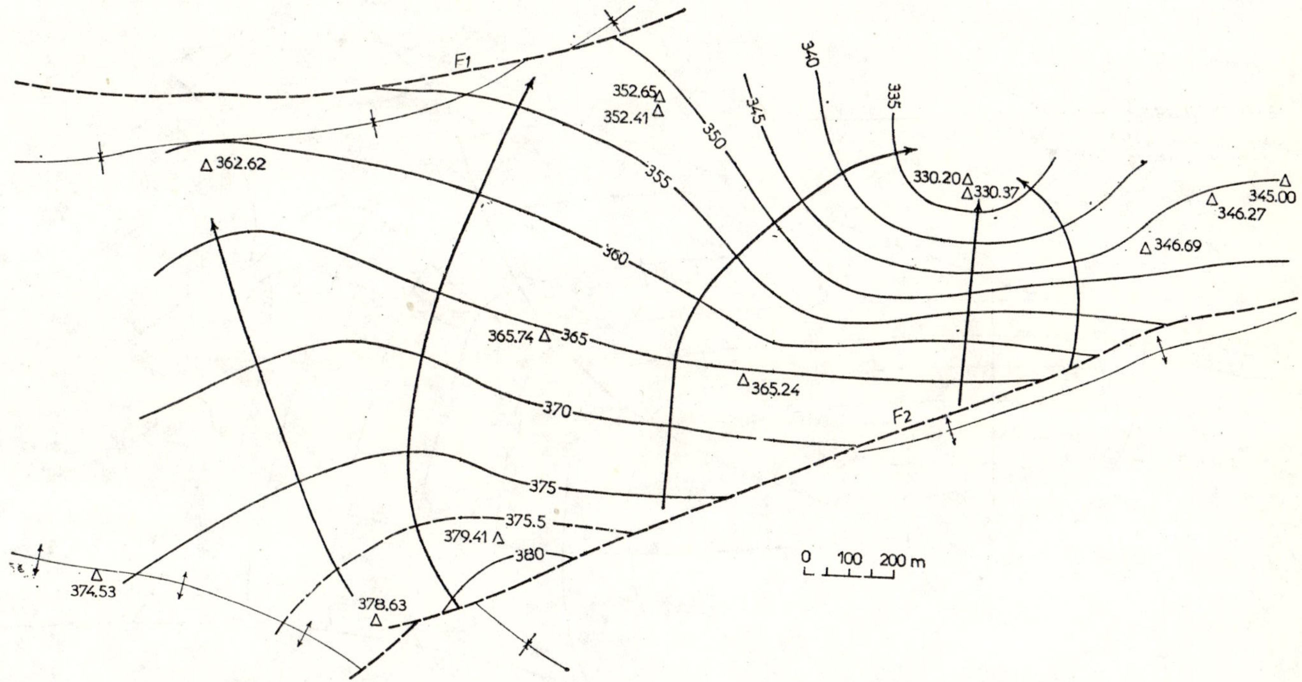

Figure 148. Piezometric surface aquifer between II and III coal layers (Şotinga) on September 7, 1962.

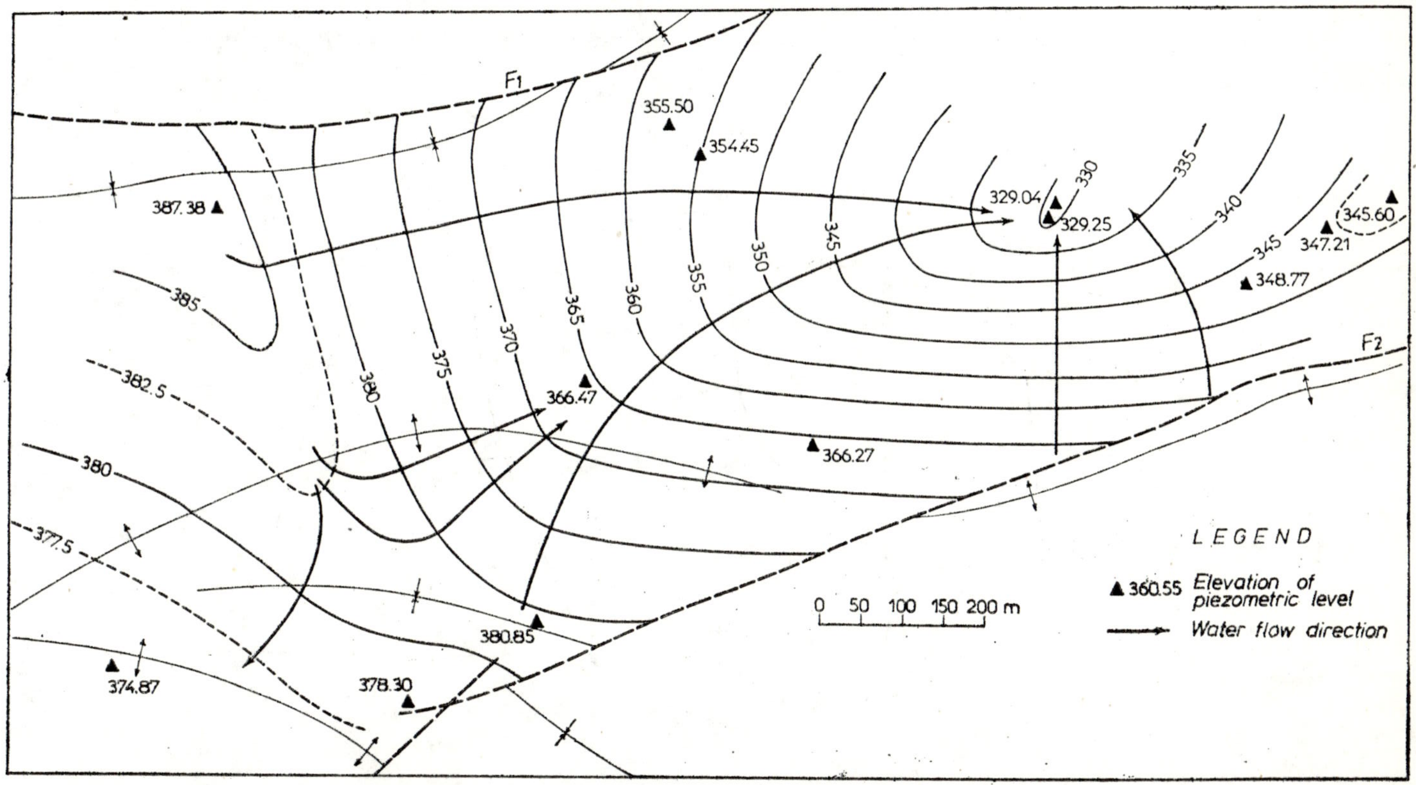

Figure 149. Piezometric surface of aquifer between coal layers II and III (Șotînga) on December 24, 1962.

In conclusion, the faults affecting the prequaternary aquifers present hydrogeological behaviour very different from one another and could only be highlighted by detailed exploration. The additional expense of the operations required to this end are justified since they permit us to define the aquiferous potential as well as the flooding degree of the mineral deposit.

4.6 Estimation of the water balance elements

According to the hydrodynamic conditions, the natural resources of ground waters can be static, dynamic and elastic. The importance of each type of resource is different as a function of purpose. Thus, captures in the case of potable or industrial water as well as of irrigation and thermo-mineral water, the dynamic resources are the most important ones, since they must exceed the capture rate of discharge; the more the dynamic resource of an aquifer surpasses the necessary yield, the easier the capture conditions are.

The static and elastic resources (which expressed in units of volume can also be called *reserves*) are less important in the case of captures, since they can only be relied upon for short periods, to wit: on the one hand when hydrostatic levels are high of the unconfined aquifers — in the case of static resources — and on the other hand in the initial period of exploitation of the confined aquifer in the case of elastic resources.

In the drainage (dewatering) process high values of the dynamic resource cause great difficulties due to the inflow of the groundwater provided by the natural recharge. A drainage work is only efficient if its yield exceeds that of the dynamic resource; if this is the case we can proceed either to exhausting the elastic resource by stress relieving, or the static resource in the case of a proper dewatering.

The computation formula of the elastic and static resources (reserves) consists in multiplying the drained or stress relieved volume by coefficient n_e of elastic storage or respectively by coefficient n_a of gravitaty storage. The method of determining these two parameters was dealt with in chapters 2 and 3.

The aquiferous potential of a hydrogeological structure is closely related to the hydrological (global) balance of the surface stream system, especially when water-bearing formations are recharged by precipitations in the outcrop domains.

In the hydrological balance equation the term *groundwater runoff* appears as synonymous with the dynamic resource; it is expressed quantitatively by the water column contributing to the aquifer recharge during a certain period (a hydrological year as a rule). The methods for estimating the groundwater runoff (the dynamic resource) are [4]:

1. The hydrogeological method that determines the aquifer flow rate by means of the wells.

2. The estimation of the aquifer recharge on the basis of data supplied by lysimeters.

3. The separation on the river hydrograph of the two components, i.e the surface and the groundwater runoff.

4. The use of the water balance equation.

The hydrogeological method is the most accurate one, because it can be applied to small units (layer, aquifer) by taking into account all the surface and subsurface sources of recharge and discharge; it permits us finally to draw a map with the isolines of the unitary flow rate.

The estimation of the groundwater runoff by the infiltration of the precipitation into the recharge domain can be achieved by lysimeters. The method has certain shortcomings, because it does not take into account the indirect subsurface recharge of the aquifer. The measuring of the groundwater runoff by means of the hydrograph is only satisfactory in areas with a permanent stream system which, during a certain period of the year, enjoy only the subsurface recharge. In the other cases, the hydrograph is more intricate and its interpretation requires the corroboration of the other methods. The application of this method to the estimation of the groundwater runoff from the Dacian formations within the Șotînga-Mărgineanca perimeter offers certain peculiarities because the Ialomița river, which sits across the geological structure, goes through these formations only between faults F_1—F_2 along about 1500 m, while Dîmbovița river goes entirely through the Levantine formations. The rest of the stream system in the Dacian outcrop zone has a torrential character, i.e. small discharges for the most part of the year and is partially supplied by the waters resulting from dewatering works.

In the exploration, the hydrogeological method can be used on a small area within the limits of which the groundwater runoff can be computed quite accuratelly. The running of capture and drainage works with large drawdowns and great radii of influence likely to exceed by far the limits of the explored perimeter requires, however, the knowledge on a regional scale of the groundwater runoff; this parameter can be estimated with the help of the water balance equation. This method widely applied today was chiefly improved by G. Castany [8] and B. I. Kudelin [39].

A hydrological division into zones now exists of the Romanian territory, the principal elements of the water balance being estimated for each zone [69]. According to this division, the Șotînga-Mărgineanca perimeter is in the area with a variable humidity of the Sub-Carpathian Hills; the balance elements possess the following mean annual values:

— Precipitations	X_0	= 650 — 750 mm/year
— Total runoff	Y_0	= 150 — 200 mm/year
— Surface runoff	S_0	= 110 — 165 mm/year
— Groundwater runoff	U_0	= 40 — 55 mm/year
— Evapotranspiration	E_0	= 500 — 550 mm/year.

The estimate of the balance elements was made with relation

$$X_0 = S_0 + U_0 + E_0 \tag{511}$$

In this relation, S_0 and U_0 are determined by the hydrograph method and evapotranspiration E_0 is the difference between the precipitations and the total runoff (Y_0). This method is only correct when the groundwater basin practically coincides with the catchment area and cannot be used in hydrogeological cases similar to the area between rivers Dîmbovița and Ialomița, north of Șotînga. As was pointed out in § 4.4, this area constitutes a surface recharge domain of a large confined aquifer basin. In the outcropping areas of the Pontian, Dacian and Levantine, a large part of the precipitations infiltrate, while the natural discharge of the aquifers in the Pontian and the Dacian occurs in the subsurface at hundreds of kilometers distance. The omission of these losses leads to an overestimate of the evapotranspiration, which was however determined by difference.

Equation (511) is valid in the flat country areas, as well as in all the phreatic aquifers devoid of any hydraulic connection with the deep groundwaters. If the outcrop areas correspond to surface discharge domains, the error is due to the underestimation of the evapotranspiration. For an accurate definition of the problem the groundwater runoff should be divided into several components:

— the phreatic runoff (F_0) which contributes entirely to the supply of the drainage network;

— the open deep groundwater runoff (D_0) concerning the confined aquifers recharged and discharged within the same catchment basin;

— the closed deep groundwater runoff (I_0) concerning the confined aquifers with discharge in the subsurface or in other catchment basins.

The division into these types of groundwater runoff corresponds to the hydrogeological conditions of the area between the Dîmbovița and Ialomița rivers. Therefore certain corrections should be applied to the value of the water balance in the Șotînga-Mărgineanca perimeter. For instance groundwater runoff U_0 is in fact only a phreatic runoff. The closed groundwater runoff (I_0) (that will be estimated later on) should also be introduced as a balance element. The evapotranspiration should be reduced by its value I_0. Taking into account the hydrogeological conditions, B. I. Kudelin proposes a few particular expressions for relation (511):

1. For catchment basins (or parts of them) situated within the recharge domain of the confined aquifer

$$X_0 = Y_0 + E_0 + I_0 \tag{512}$$

where I_0 is the annual mean of the infiltration in the deep aquifer.

2. For catchment basins (or parts of them) situated in the discharge domain

$$X_0 = Y_0 + E_0 - Z_0 \tag{513}$$

where Z_0 is the annual mean of the aquifer discharge into the catchment basin.

3. For catchment basins including both recharge and discharge domains:

$$X_0 = Y_0 + E_0 + I_0 - Z_0 \tag{514}$$

The method for determining the balance elements is not too intricate and does not require special researches; it is based upon geological, hydrogeological, hydrological and climatic data, generally to be found in the literature.

A first problem is the division of the recharge and discharge domains of the confined aquifer by balance zones. To this end the multiannual means of the hydrometeorological and hydrological characteristics can be used.

Hereafter, the estimation will be made of the recharge domain of the Dacian in perimeter Șotînga-Mărgineanca. Within this perimeter, the watershed line between the two catchment basins is close to the river Dîmbovița so that all the longitudinal valleys are tributary to Ialomița river. This is due to the fact that the Ialomița thalweg is lower than the Dîmbovița one (along the same parallel). Approximately on the Gheboeni-Șotînga line, the watershed line suddenly changes its position and draws nearer to the basin of Ialomița river so that all the transverse rivers (Valea Trestiei, Valea cu Plopi, Valea Porcarului) are tributary to Dîmbovița river. Consequently, most part of the Dacian recharge domain is included in the basin of Ialomița river (24.5 ha) and much less in the basin of Dîmbovița river (6.7 ha).

When analysing the recharge domain, two cases have been distinguished:

1. The permeable formation outcrops so that the infiltration is determined by the characteristics of the aeration zone.

2. Part of the outcropping area (the case of Șotînga-Mărgineanca perimeter) is concealed by the alluvia of the river bottom and of the three terraces. The permanent presence of a phreatic aquifer in the lower terrace and in the water-meadow determines a recharge under constant head of the aquifers in the coal complex. This is the situation on the right side of Ialomița river between faults F_1 and F_3 and the plunge from Dîmbovița of the anticline Mărgineanca The land between Dîmbovița and Ialomița rivers is situated in a region of raised epeirogenetic terraces above their normal relative altitudes, thence the great altitude differences, 40—50 m, between them. The hanging position of the middle and higher terraces determine the formation of seasonal phreatic aquifers gradually drained by the early Dacian formations; as a consequence, the contribution of these terraces to the recharge of the Dacian aquifers is smaller; during the precipitations period, they store a larger volume of water than they yield afterwards gradually to the bottom aquifers the permeability of which is smaller.

The outcrop lines of the three lignite layers are traced on the geological map (Figure 2). The area between them corresponds to the surface recharge domain of the respective aquifer. For the aquifer in the bottom of coal layer I, the recharge domain is comprised between coal layer and the Dacian-Pontian limit, while for the aquifer in the top of coal layer III the higher limit is the Dacian-Levantine limit. Table 83 presents the structure of the recharge domains by basins and lithologic intervals. It follows the small area assignable to river basin Dîmbovița, the large (with regard to the other ones) recharge domains for the top of coal layer III, as well as the small contribution of the areas situated below the alluvial deposits of the terraces.

For the perimeters corresponding to the surface recharge domains, a characterization is moreover necessary of the hydrometeorological factors. For the perimeter under study, these factors were determined in posts Voinești and Mănești in Dîmbovița basin and in posts Pucioasa and Doicești and station Tîrgoviște in Ialomița basin. Among all these, only station Tîrgoviște and post Voinești operated permanently during the last 20 years.

Table 83

Recharge pattern of the aquifer system in the carbonaceous complex

Lithological interval	Dîmbovița catchment basin			Ialomița catchment basin			The whole perimeter		
	In outcrops	Below aluvial dep.	Total [ha]	In outcrops	Below aluvial dep.	Total [ha]	In outcrops	Below aluvial dep.	Total [ha]
In top of coal layer III	3.49	1.93	5.42	9.05	1.42	10.47	12.54	3.35	15.89
Between coal layers II and III	1.05	0.18	1.23	3.04	0.38	3.42	4.09	0.56	4.65
Between coal layers I and II	0.08	—	0.08	1.61	1.53	3.14	1.69	1.53	3.22
Below coal layer I	—	—	—	5.00	2.49	7.49	5.00	2.49	7.49
Total	4.62	2.11	6.73	18.70	5.82	24.52	23.32	7.93	31.25
%	68.6	31.4	100.0	76.2	23.8	100.0	74.5	25.5	100.0

In their totality, the distribution of the precipitations is characteristic both in time and in space. Thus, during the period of 1945—1955, the distributions of the monthly mean values (Table 84, Figure 150) permit us to find out a period of maximum precipitations in months V—VIII. This period is less clearly outlined for station Tîrgoviște which is practically in the plain area.

The sketch with isohyets (Figure 151) shows an obvious increase trend of the precipitations in the Ialomița basin. In the period of 1954—1956, a meteorological post has worked in Doicești; the total values of the precipitations are comprised between those determined in Voinești and in

Table 84

Station	Monthly records							
	I	II	III	IV	V	VI	VII	VIII
Pucioasa	38.44	37.91	28.55	50.53	72.34	78.51	58.50	79.73
Voinești	55.67	58.47	47.20	69.84	106.30	74.40	91.15	90.01
Tîrgoviște	39.36	38.03	27.86	39.48	61.35	95.62	83.25	66.69
Mănești	45.60	55.40	36.75	34.85	94.89	94.14	87.73	68.57

Station	IX	X	XI	XII	Yearly records [mm]	Period
Pucioasa	35.77	55.36	51.03	43.14	629.81	1945—1955
Voinești	41.25	46.68	66.47	50.22	702.66	1949—1955
Tîrgoviște	32.53	50.30	43.69	38.97	617.13	1945—1955
Mănești	26.68	59.10	53.30	63.60	730.61	1945—1955

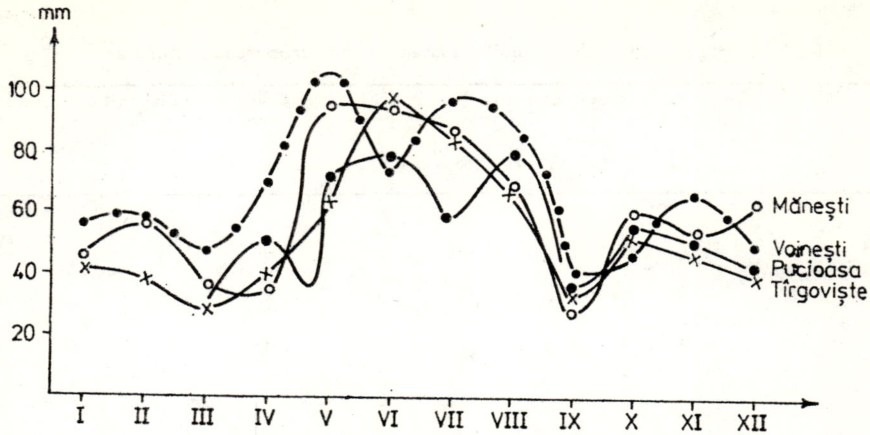

Figure 150. Distribution of the average monthly precipitation.

Tîrgoviște with a tendency to be nearer to the former ones (Table 85). Since the piezometric régime was analysed for year 1962, we also took into consideration the precipitation in Tîrgoviște and Voinești during 1961—1963. Figure 151 gives, besides the values of the monthly precipitations, the mean curve established with the help of the mobile means method (a single levelling). The evident parallelism of the mean curves underlines the constancy in time of the distributions of the precipitations in both the basins. Another remark is the increase of the period of the maxima precipitations in the interval 1962—1963 chiefly in Voinești.

The *effective infiltration* represent the infiltrated precipitations contributing to the recharge of the aquifers. It can be determined either directly by using lysimeters, infiltrometers, moisture profiles, etc. or indirectly

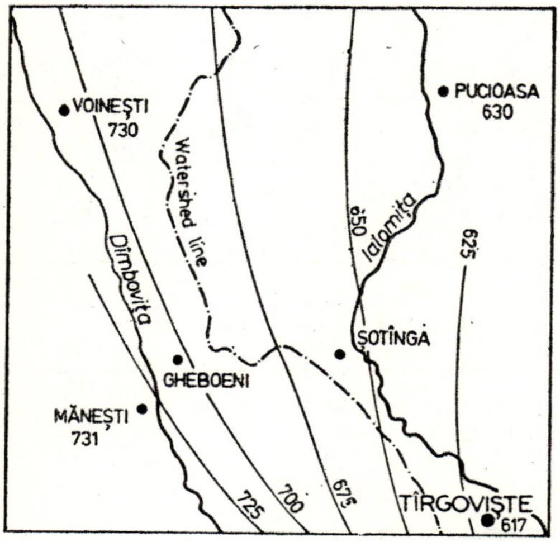

Figure 151. Sketch of the isohyets — Scale 1: 200 000.

Table 85

Climatic factor	Year	Station Tîrgoviște (elev. 293 m)	Station Doicești	Station Mănești	Station Voinești (elev. 380 m)
Precipitation [mm/year]	1954	8C8.6	1 000.6	1 130.0	1 022.7
	1955	737.4	925.7	964.5	1 021.1
	1956	683.2	673.2	930.8	788.2
	1961	675.5	—	—	663.2
	1962	541.3	—	—	766.7
	1963	571.1	—	—	827.8
Yearly mean temperature	1954	+ 9.2	—	—	—
	1955	+ 9.7	—	—	+9.2
	1956	+ 8.6	—	—	—
	1961	+10.2	—	—	+9.6
	1962	+ 9.8	—	—	+9.2
	1963	+ 9.8	—	—	+8.4

by computing the difference with regard to the other water balance elements. The period of effective infiltration of the precipitations can be estimated by means of the distribution in time of the potential evapotranspiration and of the piezometric level fluctuations. The practice showed that the precipitations and the effective annual infiltration (I_c) are connected by a correlation such as

$$I_c = (X - b)\, a \tag{515}$$

It follows that beginning from a certain value of the precipitations (b is the pluviometric threshold) the effective infiltration appears, which is a linear function of pluviometry. Constants a (angular coefficient of the straight line) and b are the infiltration factors. Their values depend on the hydrogeological and hydrometeorological factors as well as on the characteristics of the ground surface (topographic slope, vegetable cover). The pluviometric threshold (b) computed in certain areas in France varies from 111—583 mm, which means a noticeable dependence of the effective infiltration upon the above-mentioned factors.

When analysing the hydrogeological factors, an important problem is the dynamic of moisture content in the aeration zone and its correlation with the effective infiltration. By applying radiometric methods of measuring the moisture within the wells to the droughty regions of southern Ucraina, this problem was solved and the annual quantity of effective precipitations (X_{ef}) was finally established which contribute to the supply of the ground waters; it was found that $X_{ef} = 113$ mm, i.e. about 30% for a total precipitation $X = 381$ mm, at $T = 9.5$°C.

The *mean runoff* is an accurately enough determined parameter with the help of the hydrograph. The *runoff coefficient,* defined as the ratio of the mean

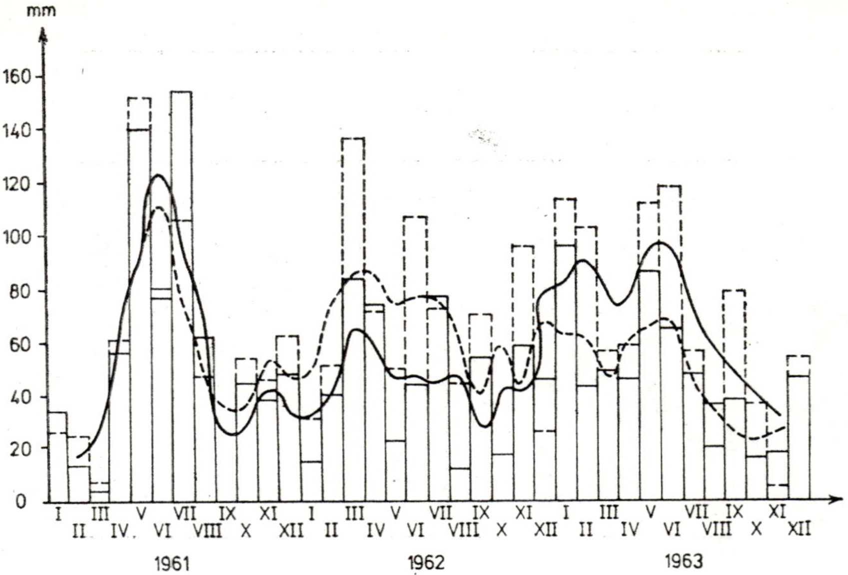

Figure 152. Curve of the mean monthly precipitation at stations Tîrgoviște (full line) and Voinești (interrupted line).

runoff of the surface streams to the mean quantity of the precipitations was estimated to $\eta_0 = 0.20$ in perimeter Șotînga-Mărgineanca.

The *potential evapotranspiration* represents the evapotranspiration that would be produced if the required moisture should be permanently ensured. One of the empirical formulae which is easily used is M. N. Ivanov's:

$$E_p = 18.4 \ d \ [\text{mm/month}] \qquad (516)$$

where d is the monthly mean value of the water vapour saturation of the air.

With the data in Table 86, the mean curves were traced of the monthly precipitations and evapotranspirations in years 1954—1956 at station Tîrgoviște. The maxima of the two curves are likely to be the same only in 1954; in 1955 a one-month lag between the maximum values of these two curves appears and in 1956 a three-months lag (Figure 153). By superimposing the two mean curves we find the months with effective precipitations by considering that they correspond to $X > E_p$. The periods and values of the effective precipitations are recorded in Table 87.

The fact that half of the total precipitations are contributing to the groundwater recharge shows that, from the climatic standpoint, this situation can be appreciated as good in certain years. So as to be able to appreciate the recharge conditions of the groundwaters resulting from precipitations it is moreover necessary to know the *coefficient of aridity* (k_a), defined as the ratio of evaporability to precipitations (Table 86). As a matter of fact, the months having $k \geqslant 1$ correspond to periods of effective precipitations.

Potential evapotranspiration and aridity coefficient at Tîrgoviște station *Table 86*

Month	Precipitation X [mm]			Saturation deficiency d [mm]			Potential evapotranspiration E_p [mm]			Aridity coefficient k_a		
	1954	1955	1956	1954	1955	1956	1954	1955	1956	1954	1955	1956
I	44.7	56.8	22.2	0.3	0.2	1.2	5.5	3.7	22.1	0.12	0.07	1.00
II	104.2	26.9	80.7	0.2	0.4	0.3	2.7	7.4	5.5	0.03	0.28	0.07
III	8.1	67.2	23.1	0.8	0.8	1.1	14.3	14.7	20.2	1.77	0.22	0.87
IV	41.4	76.4	16.7	1.6	2.1	3.9	29.4	38.6	71.6	0.71	0.51	4.28
V	77.1	85.6	145.0	3.8	5.9	3.6	70.0	109.0	66.2	0.91	1.27	0.50
VI	56.0	102.5	179.1	5.9	6.1	6.2	108.0	113.0	114.0	1.94	1.10	0.64
VII	160.7	103.6	11.1	5.2	5.4	9.9	96.0	99.5	182.0	0.60	0.96	16.45
VIII	103.6	58.1	57.1	6.8	6.4	10.6	125.0	118.0	195.0	1.20	2.04	3.40
IX	46.3	86.2	2.4	3.7	5.3	7.4	68.0	98.0	136.0	1.47	1.14	56.60
X	50.8	15.6	43.1	1.4	2.7	4.1	25.8	49.7	75.3	0.51	3.20	1.78
XI	62.7	49.7	31.8	0.3	1.0	0.9	5.5	18.4	16.6	0.09	0.37	0.52
XII	53.0	8.8	72.8	0.3	1.0	1.1	5.5	18.4	20.2	0.10	2.10	0.22
	* 708.6	* 737.4	* 683.2	** 1.8	** 3.1	** 4.2	* 555.7	* 688.4	* 924.7	—	—	—

* Total
** Arithmetical means

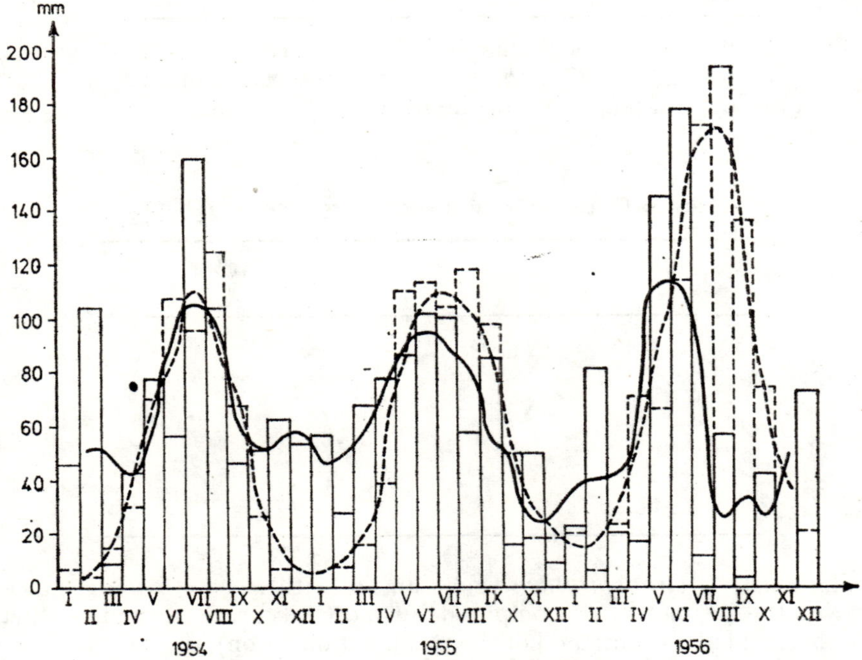

Figure 153. Average curve of the mean monthly precipitations (full line) and potential evapotranspiration (interrupted line) at station Tîrgoviște.

Table 87

Effective precipitations (X_{ef}) recorded at station Tîrgoviște

Year	Months	X_{ef} mm	%
1954	I–IV X–XII	198 166	
		364	51
1955	I–IV XII	227 9	
		236	32
1956	I–V XI–XII	289 104	
		393	57

Table 88 shows that more than half of the three year period corresponds to sufficient and rich humidity ($k \geqslant 1$) though the period of humidity shortage is also high enough (about a 1/3 of the number of months).

Table 88

Aridity coefficient recorded at station Tîrgoviște in 1954–1956

k_a	Humidity degree	Number of months	%
0.8	Abundent	17	47.3
0.8–1.0	Sufficient	4	11.1
1.0–1.2	Moderate	3	8.3
1.2	Scarce	12	33.3
		36	100,0

The *real evapotranspiration* (E_r) can be either directly determined by means of the lysimeters or computed by more accurate formulae (method of the thermal balance, method of the turbulent diffusion), that require however quite laborious researches. In default of these data empirical relations will be used, based on: the precipitation quantity, the mean temperature and sometimes the insolation. A formula satisfactorily used in France is L. Turc's one [6]:

$$E_r = \frac{X}{\sqrt{0.9 + \frac{X^2}{L^2}}} \tag{517}$$

where E_r and X are expressed in mm/year and $L = 300 + 25\,t + 0.05\,t^3$ (t being the annual mean temperature of the air).

This relation allowed us to estimate the actual evapotranspiration during periods 1954—1956 and 1961—1963 at stations Tîrgoviște and Voinești, to wit 58.5—74.0% at Tîrgoviște and 49.0—67.5% at Voinești of the total precipitation.

Since during the period 1961—1963 the climatic data are only available for stations Tîrgoviște and Voinești we shall consider the arithmetic mean of these data (precipitations and evaporation). The water balance of this perimeter is recorded in Table 89.

Table 89

Water balance in perimeter Șotînga-Mărgineanca

	X		S		E_r		I		U		$I + U = I_s$	
	mm	%	mm	%	mm	%	mm	%	mm	%	mm	%
1961	670	100	98	14.6	455	68.0	81	12.0	36	5.4	117	17.4
1962	654	100	96	14.6	439	67.0	84	13.0	35	5.4	119	18.4
1963	700	100	102	14.6	442	63.0	118	17.0	38	5.4	156	22.4

The surface and phreatic runoff (S and U) were considered as constants and we operated on the multi-annual values; the deep groundwater runoff (I) was derived from their difference.

The territorial distribution map of the evapotranspiration [69] shows that in the investigated perimeter the normal value is 550 mm/year, therefore higher than the estimated one. This difference is chiefly due to the various manner of considering the water balance. Even by admitting that L. Turc's formula offers an accuracy of about 50 mm [7], the deep groundwater runoff (I), which is greater than that, still remains an actual problem. Evidently, the accuracy of the water balance can increase by the determination of I and the difference would permit us to estimate the evapotranspiration. But, for determining I we should know the dynamic resources of all the aquifers and this condition is not fulfilled in the case of the explored perimeter.

According to Table 89, the total groundwater runoff (phreatic and deep) vary from 17.4 to 22.4%; if expressed in mm/year this parameter can also be called the *infiltration rate*. From the examples mentioned in the literature, it follows that in the case of sandy-clayey formations, the most frequent values (experimentally determined) are comprised within the limits 0.15—0.20%.

Alongside the estimation of the water balance the pluviometry-effective infiltration correlation should be established. However, this correlation is not recommended when the balance is estimated on the basis of the climatic data and of the empirical relations, i.e. at a lower accuracy. The correlation is nevertheless indispensable when using lysimeters as well as when the groundwater runoff is determined by the hydrogeological methods.

We know that in 1962, we had $I = 84$ mm and that the total area of the recharge domain (Dacian formations) is $\Omega = 312\,500$ m^2; the direct recharge from the atmospheric precipitations will therefore be

$$Q_a = 312\,500 \times 0.084 = 26\,250 \text{ m}^3/\text{year} = 0.9 \text{ l/s}$$

From the isoline map of the unitary flow rate (Figure 154), it follows that the dynamic resource of the aquifer between coal layers II and III along fault F_2 is $Q = 2$ l/s; but this is only one part of the aquifer dynamic resource. If to these values we add those corresponding to the other aquifers, we see that the height of the direct recharge from the surface to the formation of the ground water runoff of the aquifer system in the carbonaceous complex is small. Therefore, we also have to admit the indirect recharge from the Levantine aquifers (the outcropping area of which is much larger) and from the Dacian and Pontian hydrogeological structures east of Ialomiţa river. In fact, from the hydrogeological standpoint no accurate limit could be drawn between Dacian and Levantine owing to the similar characteristics of the lithofacies This could facilitate an important downward recharge of the Dacian aquifers; this problem will be dealt with in § 4.8.

In the hill and mountain zones, the non-concordance in the plane between the hydrographical basin and hydrogeological one arise difficulties about the estimation of the water balance. So as to obtain the correct solution of the problem, we must know the surface geological conditions and define the recharge and discharge domains of the various aquifer systems. Naturally, the water balance of a hydrographical basin has on overall character since it includes the groundwater runoff of all the aquifers in the basin. Thence, the necessity of detailing the balance for each aquifer of some importance (layer, horizon, complex, etc.). In this manner, the accuracy of the water balance increases and the hydrogeological balance becomes a very useful instrument for establishing a capture or drainage scheme. These problems are thoroughly dealt with in paper [9].

4.7 Hydrogeological confinement of structure

The notion of *hydrogeological confinement* synthesizes the dynamics or the intensity of the exchange in natural régime of the water in a geological structure (a collector). M. K. Ignatovitch distinguishes, along the vertical, three geohydrodynamic zones: 1) with active exchange, 2) with slow exchange and 3) with a practically stagnant aquifer régime. In connection

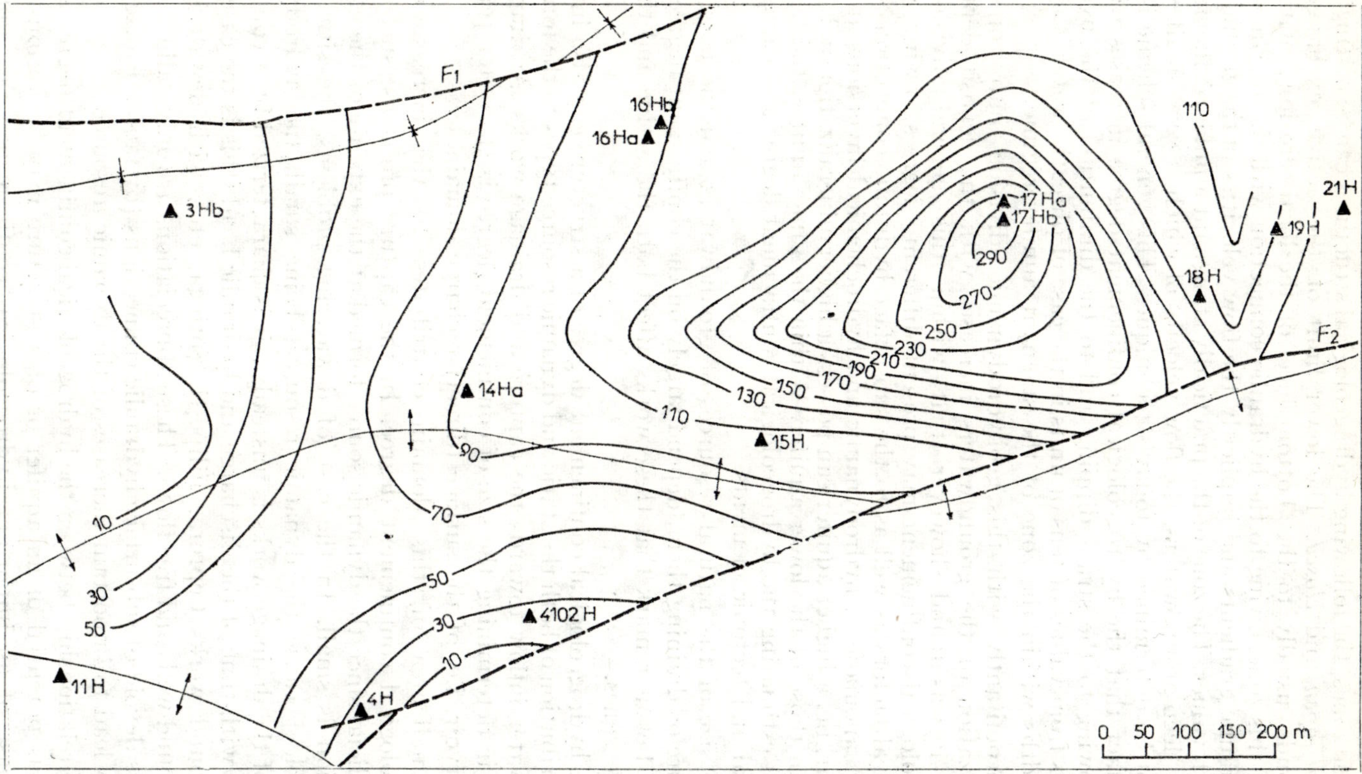

Figure 154. Unitary flow rate q (l/day) isolines of aquifer between II and III coal layers (Şotînga) on December 24, 1962.

with these zones, the following hydrogeological structures can be distinguished: *open*, *half-open* and *closed*. The lower limit of the active exchange zone corresponds usually to the bottom level of the local erosion; within the limits of this zone, owing to the hydraulic connection with the stream system, the hydraulic gradients and implicitly the flow velocities of the ground waters have high values. The zone of the practically stangnant aquifer régime includes the very deep groundwaters, the flow velocity of which is very small, so that it cannot be determined by today's techniques. This general scheme is based on the fact that the hydrogeological confinement degree increases with the depth, owing to the strata dipping and to the diminution of the hydraulic gradients from values of tens of units in the case of phreatic waters to tens of thousandths within the zone of the stagnant aquifer régime.

Concomitantly with this geohydrodynamic division into zones, the mineralization of the groundwaters increases with the deposit depth. In practice it was found however that within this general division the hydrogeological confinement may present differenτ degrees, owing to the lithological factors as well as to the subsurface hydraulic connections. Owing to these "anomalies", hydrodynamic and hydrochemical inversions are likely to appear, i.e. a higher aquifer can possess a higher mineralization and a slower flow than those of the lower aquifer. Within the same aquifer, the necessity occurs of estimating the hydrogeological confinement degree according to the structural and tectonic elements.

The size of the natural resources of groundwaters and the possibilities of capture or draining them, depend largely on the confinement degree of the aquifer. Let us remark that these works affect both the active and the slow exchange zones.

The hydrogeological confinement degree of a structure can also influence the contribution of both the static and dynamic resource types to the formation of the water inflow towards the capture and drainage works. An important element in determining the size of confinement degree is the outcrop area of the reservoir, i.e. of the surface (direct) recharge domain.

A small area of this domain contributes to the increase of the hydrogeological confinement degree. In the capture problems, such a case is prejudicial since the dynamic source formed at the expense of the effective infiltration is small. In the case of dewatering problems, the situation is favourable because the dynamic resources being small, the hydrodynamic régime of the drainage works is unsteady; it is characterized by a continuous decrease with time of the discharge and hydraulic head (this is the case of the underground works Ceptura, Filipești, Șotînga, etc., with dewatering by underground wells and free flow). These appreciations are only valid when the aquifer is devoid of subsurface hydraulic connections (by faults, leakage, etc.) likely to lead to important increases of the dynamic resource.

The relationship between the hydrogeological confinement degree and the aquiferous potential of an aquifer or of a structure may be various. Thus, in the case of an important hydrogeological confinement, the aquiferous potential is always small. When the structure is hydrogeologically open owing to a wide outcrop area, the aquiferous potential will only be high if the hydrometeorological conditions are favourable (large precipitations, small evapotranspiration, etc.). A great denudation area is therefore only a potential factor of the formation of the natural resources and of the deposit inundation degree.

Various other formulations were proposed to express the hydrogeological confinement. N. A. Gatalski proposed for instance, for the depression structures to use the mineralization gradient which is the ratio of the total mineralization (mg/1) to the depth of the respective aquifer.

Attempts were made to estimate the hydrogeological confinement of aquifer between II—III coal layers (Șotînga-Doicești deposit) by making use of the chemical composition data (Table 90) of the groundwaters. The total mineralization being considered equal to the fixed residuum, it was possible to trace the isomineralization lines (Figure 155). We can remark at the sight of the great mineralization variation (194-1683 mg/1) on a small area, owing to the great heterogeneity of the hydrogeological confinement degree and of the hydrodynamic factors. Although the studied perimeter is small, an abvious tendency appears in the horizontal plane of the mineralization decrease from south towards the north, coinciding with the principal direction of flow. This presupposes an important confinement degree near fault F_2 and its diminution towards the centre of this tectonic compartment.

So as to obtain more values, we also took into consideration the mineralization gradients of the aquifer between coal layers I and II. From the

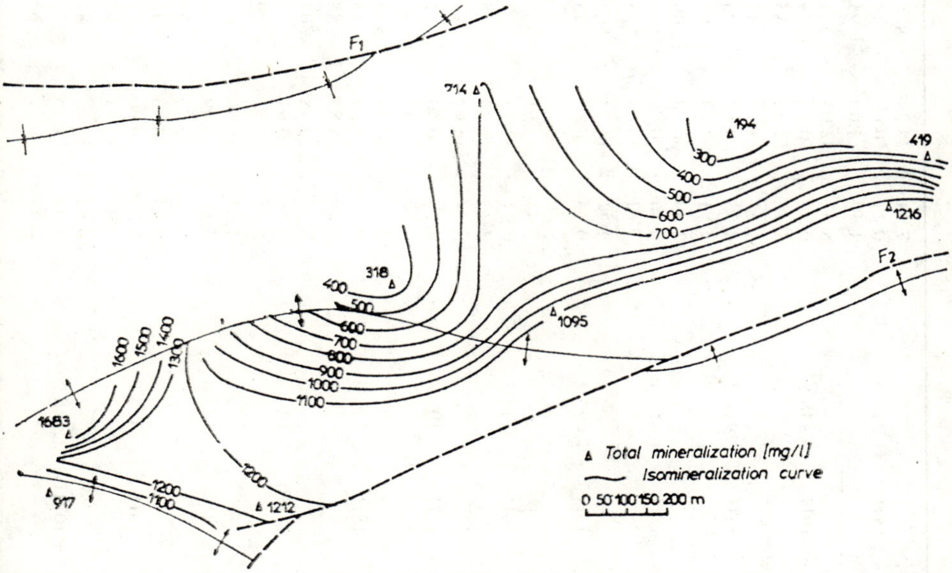

Figure 155. Total mineralization (mg/l) map of the groundwater in aquifer between II and III coal layers (Șotînga).

Table 90

Chemical composition of the groundwaters in mining field Șotînga-Doicești *

	Well symbol	Total mineralisation [mg/l]	pH	Cl^{+}	SO_4^{--}	HCO_3^{--}	CO_3^{--}	Total anions	Na^{+} K^{+}	Ca^{++}	Mg^{++}	Total cations	Total hardeness [degrees]
Aquiferous horizon between coal layers II and III	4H	1 212.0	7.5	7.54 0.21 1.40	33.29 0.69 4.60	875.05 14.05 94.00	—	897.88 14.95 100.00	268.00 11.67 78.00	15.70 0.78 5.22	30.40 2.50 16.78	314.10 14.95 100.00	9.18
	19H	418.8	7.2	8.37 0.24 4.47	32.05 0.67 12.48	271.45 4.46 83.05	—	311.87 5.37 100.00	46.85 2.04 37.98	50.02 2.50 46.56	10.04 0.80 15.46	106.91 5.37 100.00	9.32
	11H	917.2	7.5	11.51 0.33 2.84	28.77 0.60 5.25	640.50 10.50 91.91	—	680.78 11.43 100.00	155.01 6.75 59.05	69.96 3.08 26.95	19.43 1.60 14.00	236.40 11.53 100.00	13.10
	12H	1,683.1	7.3	12.89 0.36 1.72	36.40 0.76 3.63	1 209.38 19.82 94.65	—	1 258.67 20.94 100.00	194.80 8.47 40.45	195.82 9.77 46.65	33.82 2.70 12.90	424.44 20.94 100.00	34.91
	15H	1,095.4	8.5	4.52 0.13 0.96	11.51 0.24 1.77	805.20 13.18 97.27	—	821.23 13.55 100.00	167.79 7.30 53.87	77.18 3.85 28.41	29.25 2.40 17.72	274.22 13.55 100.00	17.50
	16Ha	714.4	7.5	7.31 0.21 2.23	101.51 2.11 22.40	433.10 7.10 75.37	—	821.23 9.42 100.00	63.47 2.76 29.30	71.46 3.57 37.90	37.55 3.09 32.80	172.48 9.42 100.00	18.65
	17Ha	193.8	9.0	3.72 0.10 2.46	23.83 0.50 12.34	88.45 1.45 35.80	60.00 2.00 49.40	176.00 4.05 100.00	42.31 1.84 45.43	22.15 1.11 27.40	13.32 1.10 27.17	77.78 4.05 100.00	6.19
	18H	1 216.0	7.2	12.73 0.36 2.29	66.58 1.39 8.85	850.95 13.95 88.86	—	930.26 15.70 100.00	63.93 2.78 17.71	164.38 8.20 52.23	57.42 4.72 30.06	285.73 15.70 100.00	36.17
	14Hb	317.7	7.5	8.37 0.24 5.71	29.18 0.61 14.52	204.35 3.35 79.77	—	241.90 4.20 100.00	21.39 0.93 22.14	37.16 1.85 44.05	17.24 1.42 33.91	75.79 4.20 100.00	9.16
Aquiferous horizon between coal layers I and II	2H	515.6	7.4	11.31 0.32 4.62	28.36 0.59 8.53	367.00 6.01 86.85	—	406.67 6.92 100.00	6.12 0.37 5.35	58.89 2.94 42.49	43.90 3.61 52.16	108.91 6.92 100.00	18.34
	5H	573.0	7.2	9.43 0.27 3.94	43.15 0.90 13.14	46.60 5.68 82 92	—	399.18 6.85 100.00	12.10 0.52 7.59	74.40 3.71 54.16	31.80 2.62 38.25	118.30 6.85 100.00	17.72
	4 102H	1.128.5	7.4	36.84 1.04 72.5	55.07 1.15 8.02	741.15 12.15 84.73	—	833.06 14.34 100.00	202.00 8.77 61.16	65.03 3.24 22.59	28.38 2.33 16.25	295.41 14.34 100.00	15.60

* Anions and cations are expressed in mg/l, millivalents and percent-equivalents

graph (Figure 156) it follows that down to the depth of 90 m no well-defined reduction tendency appears of the mineralization gradient. This anomaly can be explained by the effect either of lithological confinement (due to the transmissivity reduction or even by wedging-out) or of the faults. The hydrochemical anomalies (corresponding to the strong mineralizations) are accompanied moreover by the water type change of the cations. The presence of the three water types (sodic, calcareous, magnesian) and the prevalence of the hydrocarbonated type (75—94% equivalent) point out the importance of the cations exhange between waters and rock as concerns the chemical composition of the groundwaters.

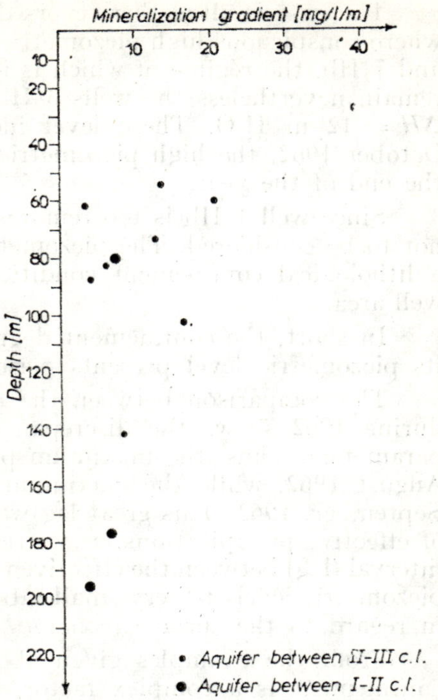

Figure 156. Variation with the depth of the mineralization gradient.

The hydrogeological confinement degree can also be estimated with the help of the piezometric level variations measured during a long enough period. This method is based on the fact that the small piezometric variations correspond generally to good hydrogeological characteristics: transmissivity, permeability, porosity, storage coefficient, etc., or inversely. In the case of a confined aquifer, a head increase in the recharge domain due to increased precipitations would be transmitted in the form of a wave and would thus increase the field pressure. The velocity and amplitude of this pressure wave depend on the hydraulic and elastic properties of the aquifer. A decrease of the piezometric level would be due to the decrease of the elastic resource, as a consequence of the fact that groundwater runoff exceeds the recharge rate. Reciprocally the increase of the piezometric level means implicitly an increase of the elastic resource due to the fact that recharge rate grows larger than the subterranean runoff.

The fluctuations of the piezometric levels of the aquifer between II—III coal layers (Figure 147) point out the variation of the hydrogeological confinement degree. Thus, in well 17 Ha (with constant measurements during 1962) $\Delta H = 332.4 - 329.0 = +3.4$ m. This shows that the aquifer recharge exceeds the groundwater runoff in the well area. The greatest increase occurred in wells 17 Ha and 14 Hb where in 1962, January-February, were recorded 5.5 and 4.0 m H_2O, respectively. During the rest of the year, the highest amplitudes did not exceed 1 m.

The case is altogether reversed for wells 7 Ha, 7 Hb, 3 Hb and 1 Hb, where conspicuous high piezometric levels appeared. Setting aside wells 7 Ha and 7 Hb, the régime of which is influenced by the presence of gases, there remain nevertheless the wells 3 Hb with $\Delta H = 22$ m H_2O and 1 Hb with $\Delta H = 12$ m H_2O. These level increases were recorded in September—October 1962, the high piezometric level remaining practically constant till the end of the year.

Since well 1 Hb is too remote from faults F_1 and F_2, their influence is not to be considered. The piezometric level increase in these wells is due to a lithological confinement conditioned by the aquifer wedging out in the well area.

In short, the confinement degree of an aquifer is very important when its piezometric level presents periods of large variation amplitudes.

The comparison between the precipitations and the piezometric levels during 1962 shows the discrepancy between the maximum of the two parameters. Thus, the maximum precipitations period occurred in March—August 1962, while the maximum piezometric levels began to appear in September, 1962. This great lag was explained by the much overdue period of effective precipitations, as we remarked in § 4.7. Therefore, the real time interval (lag) between the effective precipitation and appearance of the highest piezometric levels is very small (about a month) owing to the small distance in regard to the surface recharge domains.

From the examples given above, it follows that the hydrogeological confinement is a complex factor, specific of the confined aquifers in the quaternary formations (e.g. alluvial cones) and chiefly in the pre-quaternary ones with important lithofacies variations and tectonic disturbances liable to determine confinements, either lithological or due to the transmissivity or to the faults, etc. or to a combination of these factors.

In the case of aquifers placed at great depths and implicitly at great distances from the surface recharge domains, the hydrogeological confinement cannot be appreciated anymore by the piezometric level variations since, owing to the wave damping, their amplitudes are very small. In such cases, we must take into consideration the flow velocity and the total mineralization or else a principal ion. On the special maps, we can separate zones with various hydrogeological confinement degrees.

The hydrogeological confinement degree must be established for each structural-tectonic element. In the case of detailed investigations, characteristical zones must be separated even within each structural element. Obviously, it is practically indicated that the capture works should be placed in areas with a small hydrogeological confinement degree, which are characterized by a high aquiferous potential. But, generally, the works designed on the dewatering or drainage cannot always be placed in areas with a small degree of confinement. Therefore the density of the works will depend on the confinement character; if it is chiefly due to the transmissivity, the density will be higher than in the case of a lithological confinement or of one due to faults and the aquifer will be considered as a limited one in the hydrodynamic schemes.

Within the general frame of the three zones established by M. I. Ignatovitch, the hydrogeological confinement degree can be specified as a function of a certain parameter.

4.8 Natural hydrodynamic conditions

The dynamics of an aquifer are determined, among other things, by the conditions of recharge, pressure formation and discharge of the hydrogeological set to which it belongs.

The recharge of the groundwaters can be made directly by infiltration of the precipitations or of the surface waters in the outcrop areas of the considered water bearing formation, or indirectly by contribution of other hydrogeological structures or of adjacent aquifers. The domains of surface recharge can be, according to their position in the plane, external, i.e. outside the spreading area of the aquifer, or internal, i.e. within its limits. On a regional scale, all the area between Dîmbovița and Ialomița rivers (north of Șotînga) constitutes an external recharge area both for the Dacian and the Pontian formations. On a local scale however, owing to the outcropping in the form of "isles" of the sandy deposits, the recharge areas of the groundwaters in the carbonaceous complex have an internal character.

In the hydrogeological practice, the terms *hydrostatic surface* and *hydrostatic level* are sometimes improperly used. According to H. Schoeller [60], the hydrostatic surface is the horizontal plane passing through the recharge area of the aquifer and should not be mistaken for the initial piezometric surface corresponding to the levels reached by the water in the wells at any point of the aquifer. Therefore, the notion of hydrostatic level is only proper for a shallow recharge domain where the free surface of the aquifer is practically horizontal.

Let us remark that, within the same hydrogeological structure, the hydrostatic level of an aquifer can have various elevations along the structure, owing to the diverse conditions of recharge and discharge. In the rest of the structure, where the aquifer is confined, the level is the initial piezometric one.

Characteristic of the aquifers in the explored perimeter Șotînga-Mărgineanca is the changing positions of the piezometric levels, the heights of which decrease continuously as the depth increases. Only informatively, we give here the piezometric levels variations in the period 1958—1961:

— aquifer above coal layer III	340—430 m
— aquifer between coal layers II—III	320—410 m
— aquifer between coal layers I—II	290—400 m
— aquifer below coal layer I	260—350 m.

The variation of the piezometric levels of the various aquifers within the same confined basin can be due to several causes: various positions of the surface recharge and discharge domains, the existence of a local zone of admission or discharge, the variation along the horizontal plane of the aquifer lithofacies and its thickness. The weight of each of these factors could only be appreciated on the basis of regional hydrogeological researches. Presumably, in the case of aquifers in the top of coal layer I the chief cause is the variation in the plane of the lithofacies and of the thickness, i.e. the transmissivity decrease as the depth grows. This interpretation is no more valid for the aquifer in the bottom of coal layer I. This aquifer is compact and has a continuous development in the plane, with the result of smaller hydraulic resistances; it is probable that a subsurface discharge domain creates in this aquifer a field of lower piezometric levels.

The lithofacies variation, the tectonic elements and the piezometric levels, as well as their connections with the adjacent structures are illustrated by hydrogeological section *A—A'* (Figure 157) of the map in Figure 2. Under such hydrogeological conditions (superposed aquifers with very different piezometric levels seperated from one another by semipervious layer of small thickness) a hydraulic connection can be established between the aquifers (by leakage): the aquifer with the higher level undergoes a subsurface discharge, while the other with lower level gets an indirect (underground) recharge. Therefore the interaction on the vertical of the aquifers should be estimated in the natural régime since during the capture or dewatering process it has a greater rate, contributing to the increase of the dynamic resource. This in turn contributes to the increase of the innundation degree of the deposit. So as to solve the hydrodynamics of the interaction along the vertical, an alignment is required of at least three wells (in the case of a one-dimensional current) where piezometric measurements can be made during a long time period. On the other hand, we must know the positions of the aquifers in the top and in the bottom of the aquifer under study. These requirements are fulfilled by alignment 4102 H — 14 Hb — 16 Hb (Figure 158) for the principal aquifer between the coal layers II—III. The great difference (about 43 m H_2O) between the piezometric level of this aquifer and that of the lower aquifer (between I and II coal layers) gives rise to a downward percolation. The relatively simple computation scheme starts from equality

$$W = Q_v \tag{518}$$

where W is the aquifer recharge or discharge during time $\sum(\Delta t)$ by unit area and Q_v the difference between the inflow and the outflow rates within the same time period.

Particularly, in the case of three wells and a one-dimensional flow, we shall have [4]

$$Q_v = \frac{(q_a - q_d)_1 \,\Delta t_1 + (q_a - q_d)_2 \,\Delta t_2 + \ldots + (q_a - q_d)_n \,\Delta t_n}{0.5(l_1 + l_2)} \tag{519}$$

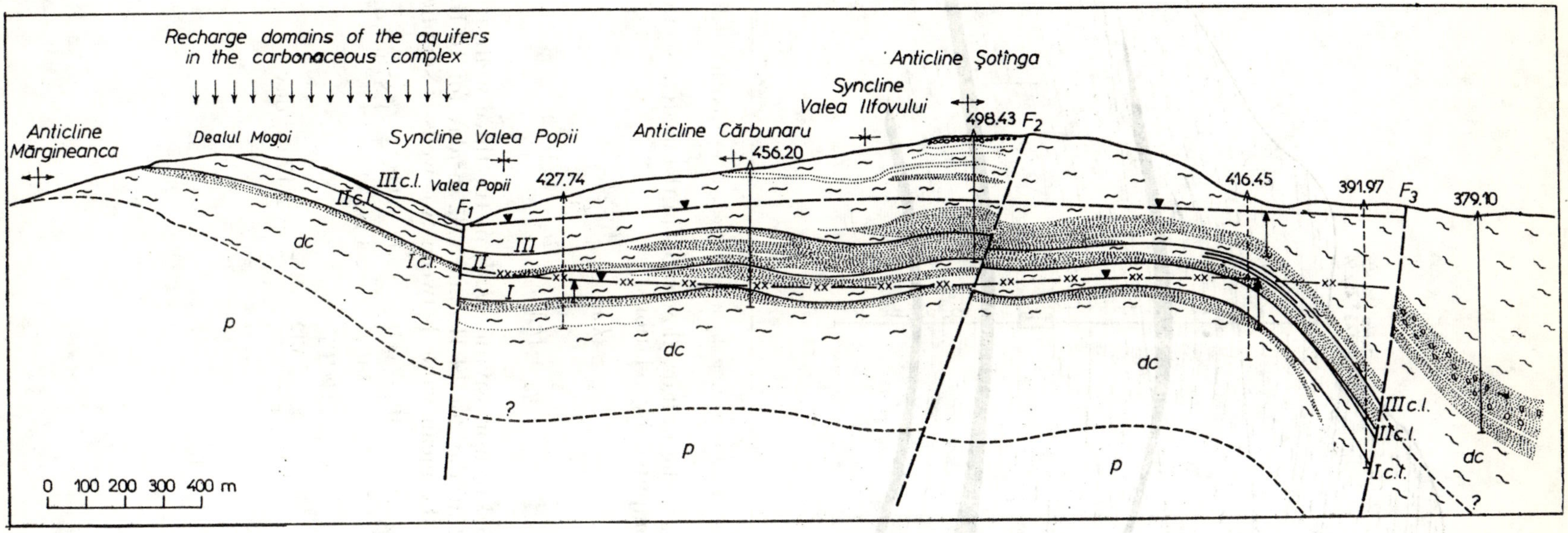

Figure 157. $A-A'$ general hydrogeological cross-section.

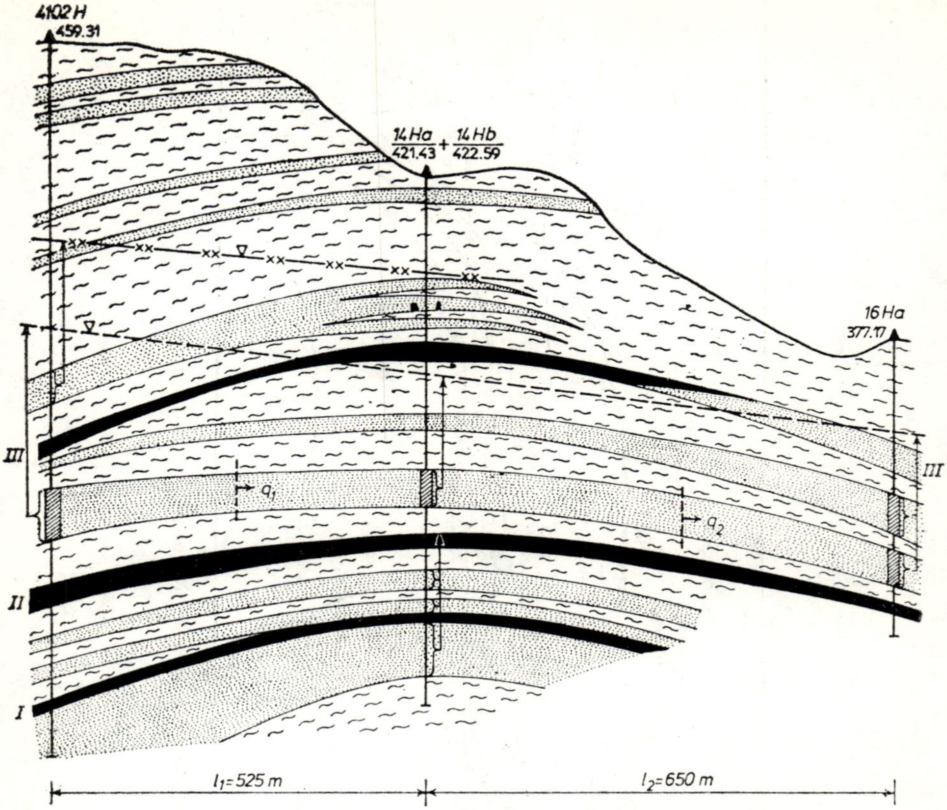

Figure 158. Computing hydrogeological cross-section (Șotînga).

where q_a and q_d are the unitary flow rates in the median section between the wells and are computed by relations

$$q_a = (kM)_{1-2} \frac{H_1 - H_2}{l_1} \tag{520}$$

$$q_d = (kM)_{2-3} \frac{H_2 - H_3}{l_2} \tag{521}$$

The computed transmissivities (kM) of the three wells are: 1.00 m²/day for 4102 H; 3.12 m²/day for 14 Hb and 1.42 m²/day for 16 Ha. Assuming a linear variation, the mean value between two wells is supplied by relation:

$$(kM)_{1-2} = \frac{(kM)_1 - (kM)_2}{\ln \frac{(kM)_1}{(kM)_2}} = \frac{3.12 - 1.00}{2.3 \log 3.12} = 1.87 \text{ m}^2/\text{day} \tag{522}$$

$$(kM)_{2-3} = \frac{3.12 - 1.42}{2.3 \log \frac{3.12}{1.42}} = 2.17 \text{ m}^2/\text{day}$$

The results are recorded in Table 91. They show that the discharge across 1 km² during the entire computation period ($t = 59$ days) is:

$$V_F = Q_v F = -WF = 0.0017903 \times 100\,000 = 1790\ \text{m}^3/\text{km}^2.$$

The discharge (respectively the recharge) modulus of the aquifer between coal layers I—II is $M = 1790 \times 100/(59 \times 24 \times 60 \times 60) = 0.35$ l/s · km².

Table **91**

Time	Δt [days]	Weter-table elevation [m]			q_a m³/day	q_d m³/day	$Q_v = \sum_{i=1}^{n} \frac{(q_a - q_d \Delta t)}{0,5(l_1+l_2)}$	
		H_1 (4102H)	H_2 (14 HB)	H_3 (16 Ha)			m	mm
1962								
28 IX	—	379.69	265.39	352.68				
30 IX	2	379.71	365.51	352.72	0.051	0.042	0.0000307	0.0307
18 X	18	380.31	365.01	352.76	0.055	0.040	0.0004907	0.4907
24 X	6	381.15	365.05	353.39	0.058	0.038	0.0006951	0.6951
30 X	6	380.88	365.11	353.54	0.057	0.038	0.0008893	0.8893
4 XI	5	381.06	364.58	353.05	0.060	0.038	0.0010766	1.0766
9 XI	5	381.25	354.19	352.92	0.062	0.037	0.0012895	1.2895
15 XI	6	381.50	364.97	353.07	0.060	0.039	0.0015041	1.5041
20 XI	4	380.52	365.36	352.79	0.041	0.055	0.0015995	1.5995
27 XI	7	380.37	364.76	352.73	0.056	0.040	0.0017903	1.7903

Once the direct and indirect recharges and discharges of an aquifer in a given perimeter are known, the balance equation can be established from which the mean hydraulic coefficient is deduced of the semi-pervious layer.

The semipervious formations are represented by the lignite layers and by the various kinds of marly-clays. Sometimes, in certain areas the marls and clays are sandy; in these areas, also called hydrogeological windows, the hydraulic link on the vertical is naturally more important.

The succession on the vertical of the piezometric levels in Șotînga-Mărgineanca perimeter is favourable to the mining exploitation, because a successive downward dewatering of the aquifers diminishes the hydraulic conection between them. In a contrary case (i.e. if the piezometric levels would increase towards the lower aquifers) the dewatering conditions would grow more difficult since the dewatering of a higher horizon would lead to its recharge from the bottom.

To conclude, the leakage parameters of the confined aquifers may have a high weight as concerns the innundation degree of certain important lignite layers.

The geometrization of the hydrogeological features (forms, properties, processes) constitutes an effective method of synthesizing the hydrogeological informations and of estimating the aquiferous potential provided it is applied together with the mathematical statistics (as already told earlier) and with the methods of the underground geometry.

Hereafter, we shall deal with the geometrization problem of certain hydrodynamic features (piezometric level, transmissivity, hydraulic gradients and unitary flow rate) of the aquifer between coal layers II—III in the compartment delimited by faults F_2—F_3, as in the case of this aquifer the density of the hydrogeological wells is satisfactory. On the basis of the systematic measurements of the piezometric level during the period September 7 — December 24, 1962 (when important growths were recorded in certain wells) parametric maps were drawn at the beginning and the end of the interval. That permits us to make a few appreciations concerning the variation of the natural piezometric régime. The piezometric levels of some wells in the arching zone of field Ghẹboeni (8H, 12H, 7Hb) were not taken into account, because they were influenced by the presence of gases (in the form of accumulations or dissolved in the ground water). Wells 4102 H and 11 H were taken into account because the gasiferous manifestations in these wells were very weak.

The analysis of the piezometric levels distribution and of their variation with time constitutes an inexpensive, while very accurate, method for the behaviour analysis of an aquifer. The interpretation of the piezometric surface, beside permitting the computation of some hydrodynamic parameters, leads to important conclusions concerning the hydrogeological characteristics of the aquifer and the ratio recharge-discharge. On the piezometric map drawn on September 7,1962 (Figure 148), two elementary structures can be identified corresponding to the western and eastern sectors. In the western sector the flow can be considered as one-dimensional since the isopiestic lines have a very great radius of curvature and their concavity is up-stream directed. The flow lines, though simple, present some deviations from the principal south-north direction of the flow. From the map with hydraulic isogradients (Figure 159), it follows that their values vary in the western sector from 0.9 to 2.5% with a growing tendency eastwards. In the eastern sector, the hydrodynamic conditions are more complex: the flow becomes radial-convergent and the values of hydraulic gradients are of 2.5 to 10%. This situation can be explained by the great dipping of the strata or by the existence of a local domain of underground discharge. Since in 17H well section in interval between coal layers I and II are not sands, the high values of hydraulic gradients are chiefly due to a considerable north-eastward dips of strata as shown by the map with isobaths at the level of coal layer II. As a matter of fact, in both these sectors a dependence exists of the piezometric area upon the geological structure: the principal flow directions coincide roughly with the layer dip direction, while the hydraulic gradients and the inclinations of the strata are connected by a direct correlation.

The piezometric surface, drawn on December 24, 1962 (Figure 149), shows an important alteration at the western limit of the perimeter: the flow direction becomes north-south, while the flow assumes a radial-divergent structure. The other alterations of the piezometric area morphology are insignificant.

From the map with isotransmissivities (Figure 160), it follows that this complex parameter has a wide variation interval ($kM = 0.8 — 7.6$ m^2/day)

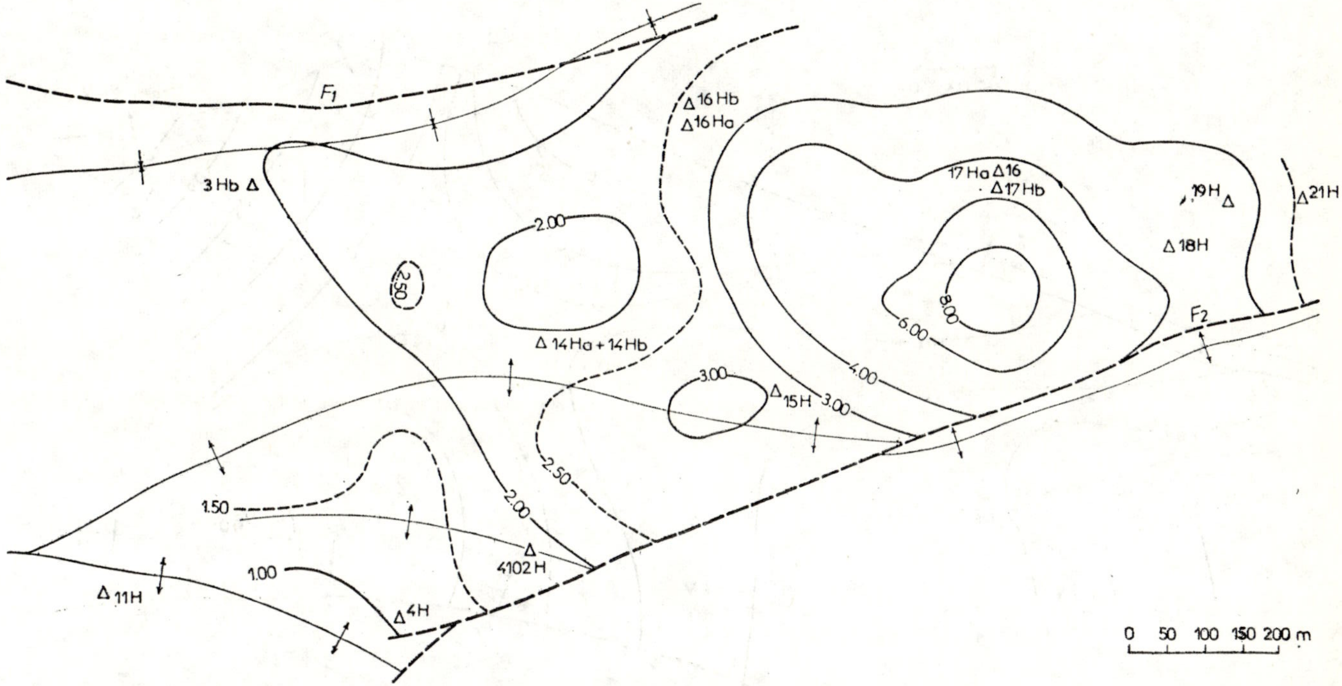

Figure 159. Hydraulic isogradients of aquifer between II and III coal layers (Șotînga) on September 7, 1962.

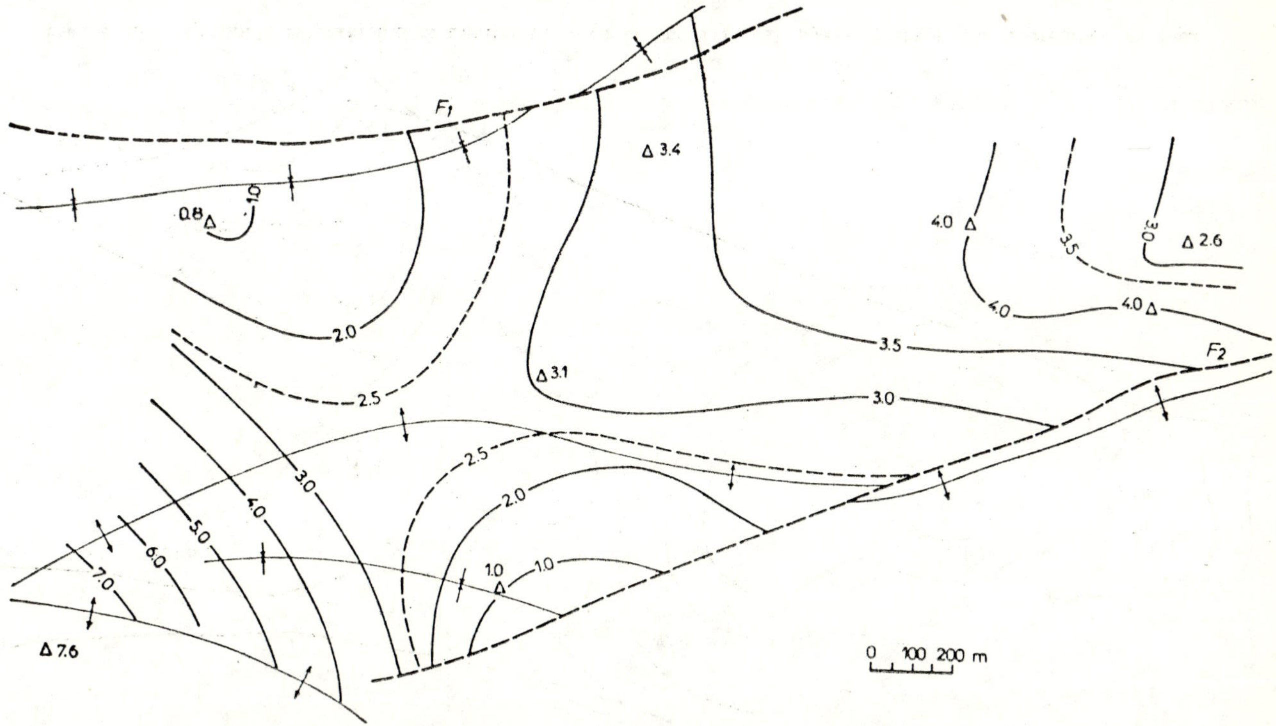

Figure 160. Map of isotransmissivities $T = kM$ (m^2/day).

and a preferential distribution along certain directions. The values magnitude shows a very weak to weak potential. Let us remark that in the transmissivity computation we could not make use of the schemes given in chapter II, because of the lack of the corresponding primary data. As a consequence, these values are generally underestimated, because they were computed with the schemes of the steady flow without taking into account the seepage resistance of the well.

The change of the flow direction on the western limit of the perimeter, during an interval of 108 days is due to the fact that the piezometric levels of wells 4H and 11H remained practically constant while that of 3Hb underwent a rise of about 25 m H_2O. A cause of this nonuniform rise of the
a rise of about 25 m H_2O. A cause of this nonuniform rise of the
piezometric level is the large variation of the transmissivity: 0.8 m^2/day (the lowest value in the perimeter) in 3 Hb and 7.6 m^2/day (the highest value) in 11H. Evidently, under these conditions, the ratio of the recharge to the groundwater runoff can be altered with time while involving at the same time the change in the flow direction. This change was moreover favoured to a certain extent by the undulating shape of the layers.

A synthetic parameter of the aquiferous potential and of the inundation degree is the *unitary flow rate,* which is expressed by the following relation

$$q = kMI \tag{523}$$

It is easily established by means of the transmissivities and of the hydraulic gradients maps. The unitary flow rate represents in fact the dynamic resource on 1 m width of the aquifer, which is a decisive factor of the water inflow to the capture and drainage works. In Figures 161 and 154, the isolines of the unitary flow rates are drawn, expressed in l/day, as computed on September 7 and December 24, 1962. The cartographic images offered by these maps warrant the separation of the two above-mentioned sectors: the western sector with a small inundation degree ($q < 80$ l/day) and the eastern one with a high inundation degree ($q = 80 - 340$ l/day).

With the isolines of the unitary flow rates and the piezometric map, we can estimate the indirect recharge or discharge (from the top or from the bottom) of an aquifer. To this end, we consider a flow tube and compute by summing up the unitary flow rates, the inflow (on the up-stream isopiestic line) and the outflow (on the down-stream isopiestic line); the positive difference between the two unitary flow rates points out the discharge, while the negative one shows the recharge.

For the aquifer in the coal layer III top, the isopiestic lines were traced for a restricted sector in the formation between faults F_2 and F_3 (Figure 162). The resulting elementary structure points out a radial-convergent current with a south-eastward axial flow and hydraulic gradients of 0.5—3.6%. The axial flow presuppose the existence of a discharge structure through the Şotînga-Doiceşti fault.

The above presented hydrodynamic analyses permit us to draw conclusions of a more general order. The particular role is emphasized by the longitudinal faults in the formation of independent hydrogeological structures delimited on the north-south direction by these faults. The hydraulic connection of the

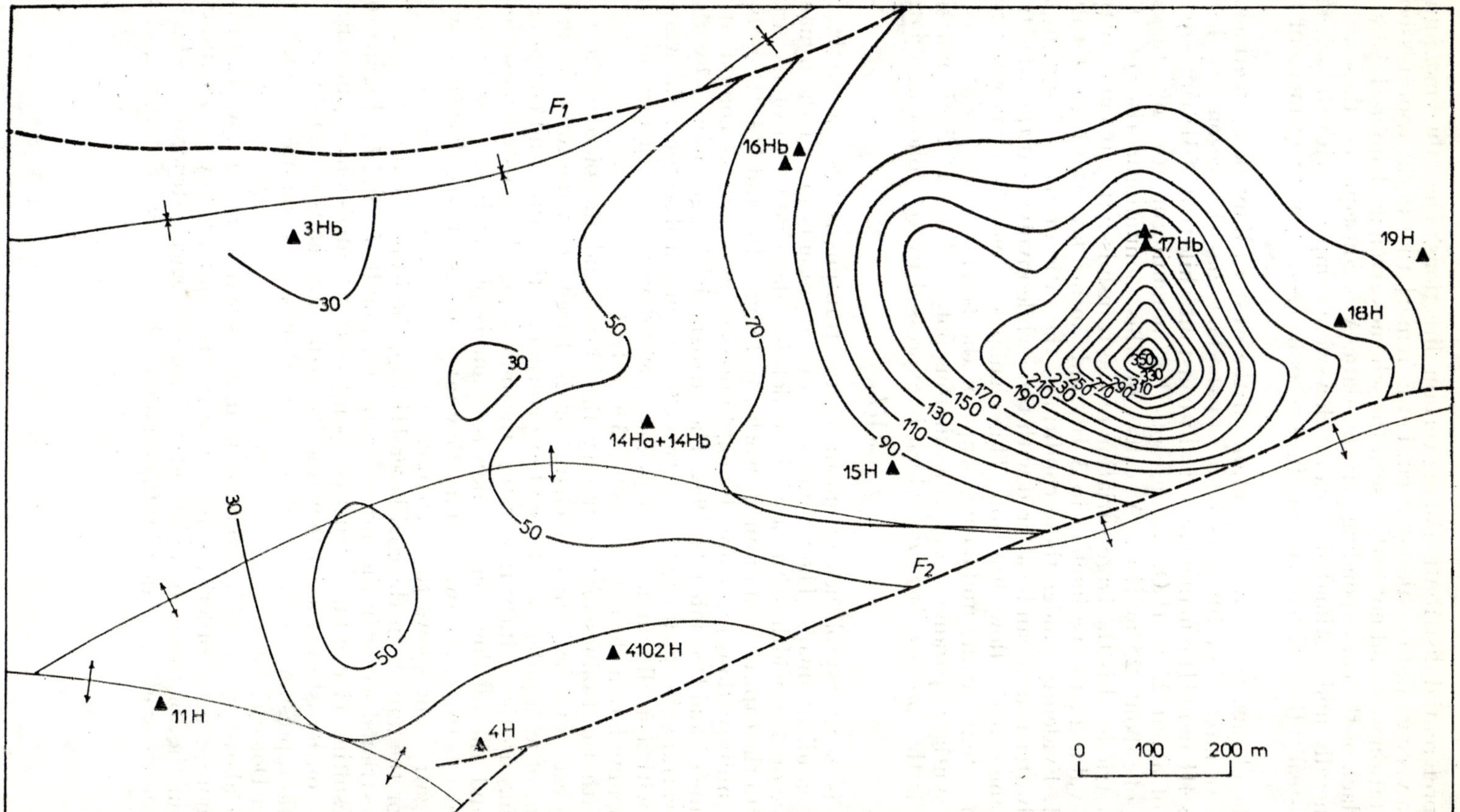

Figure 161. Map with unitary flow rate q (l/day) isolines on September 7, 1962.

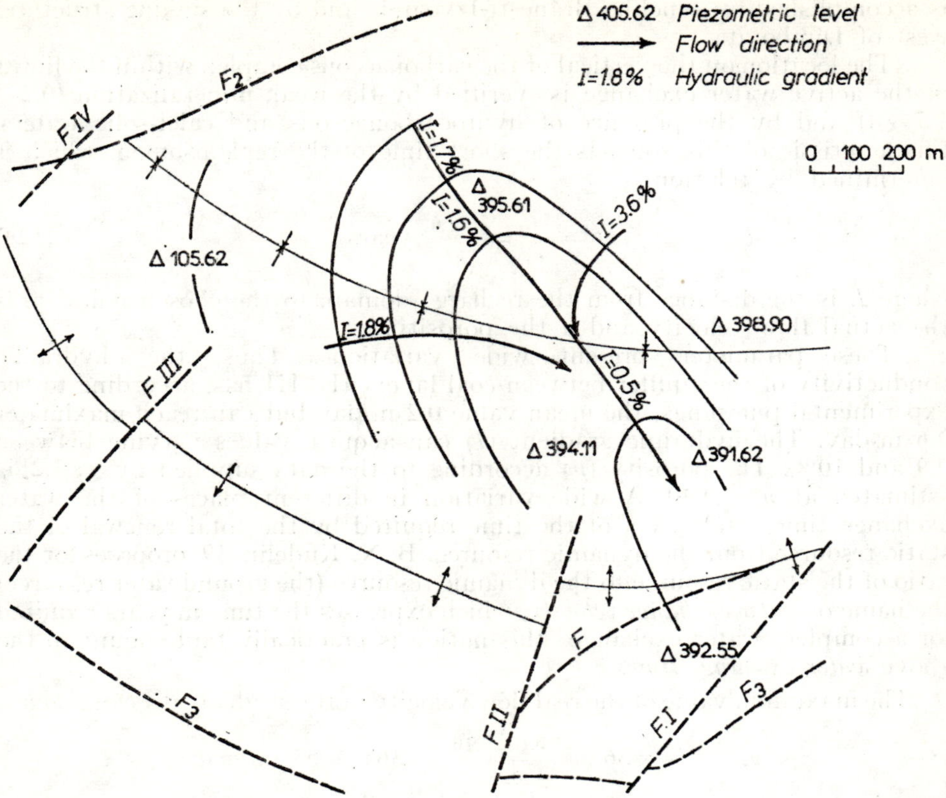

Figure 162. Piezometric surface of the aquifer between coal layers II and III (Șotinga).

tectonic compartment along the faults is obvious on some sections that, when this is the case, achieve the recharge or the discharge of the aquifers from one another.

The groundwater runoff can be either *local* or *of transit*. The separation of a transit groundwater runoff is quite important in the case of confined aquifer basins consisting of several hydrogeological structures. This is obvious at the south-eastern limit of the explored perimeter between faults F_1 and F_2. From the piezometric maps (Figures 148 and 149) it follows that the flow direction is south-north. That implies the existence of greater field pressures in sector F_2—F_3 which could only be due to an inflow under Ialomița river since fault F_3 is water-tight. This inflow constitutes a transit groundwater runoff the magnitude of which is appreciable with regard to the Dacian outcrop field east of Ialomița river. Taking into account the sandy facies of the late Pontian, the possibility is not excluded of a groundwater runoff from these formations eastward of river Ialomița towards the Dacian formations in the explored perimeter. The natural discharge of the groundwaters of deposit Șotînga-Mărgineanca (and of all Pliocene formations in this anticlinorium) is accomplished by syncline Brănești-Izvoarele and by the sinking structures west of Dîmbovița.

The location on the vertical of the carbonaceous complex within the limits of the active water exchange is verified by the weak mineralization (0.2—1.7 g/l) and by the presence of hydrocarbonaceous and calci-solic waters. Characteristic of this zone, is the short time of the replenishmen, which is determined by relation

$$t = \frac{L}{v_r} = \frac{L \cdot n}{kI} \text{[years]} \tag{524}$$

where L is the distance from the recharge domain to the chosen point, v_r is the actual flow velocity and n the porosity.

These parameters present wide variations. Thus, the hydraulic conductivity of the aquifer between coal layers II—III has, according to the experimental pumpings, the mean value 0.2 m/day but can reach maxima of 0.6 m/day. The hydraulic gradient (I) can acquire values varying between 0.9 and 10%. The porosity (n) according to the data supplied by § 3.2, is estimated at $n = 0.40$. A wide variation in different places of the water exchange time results, i.e. of the time required by the total renewal of the static resource from the dynamic resource. B. N. Kudelin [39] proposes for the ratio of the static resource to the dynamic resource (the groundwater resource) the name of *water exchange velocity*, which expresses the time in years required for a complete water exchange; this notion is practically tantamount to the above *water exchange time.*

The maximum value of the real flow velocity in the studied area can reach:

$$v_r = \frac{kI}{n} \times 365 = \frac{0.6 \times 0.10}{0.40} \times 365 = 55 \text{ m/year}$$

By combining the various values of k and I we can find minimum values as low as $v = 10$ m/year. This case is in concordance with the large variation of the total mineralization. The literature mentions similar values of the actual velocities of other aquifer systems in France and U.S.S.R.

Assuming as a conventional hydrogeological unit the tectonic compartment between faults F_1 and F_2 ($L = 1300$ m), the minimum water exchange time would be $t = 24$ years, which shows the actual negligible recharge of the studied aquifer system.

For the confined aquifers located in folded structures we to have establish the connections between the piezometric surface morphology and the geometric elements of the folds. The appearance of consequent currents (the flow direction of which is parallel to the strate diping) depends on the parallelism of the impermeable top and bottom strata. In many cases such a geometry of ato and bottom do not exist and hydraulic connections on the vertical (leakage) between the aquifers should occur (subsequent currents). This is the case of aquifer between coal layers II—III, where although the folds are generally inclined southward, the flow direction of the groundwaters is from south to north, because of hydraulic connections with the other aquifers along the faults or through the semi-pervious marly-clays.

Another hydrogeological peculiarity of anticlinorium Șotînga is the superposition of the partial discharge domain on the surface recharge domain of the same aquifer. The discharge is achieved by the appearance of small discharge springs in Popii river and Perilor river, within the recharge domains in the eastern field. This superposition can be explained by the quite heterogeneous lithological structure of the aquifer capable of an independent drainage of the upper part, while the lower part has only a recharge function. Such are the cases of aquifer between coal layers II and III and of that in the top of coal layer III. However, the small number of springs and the low intensity of the drainage, in general, do not justify the separation on the map of the respective domains.

BIBLIOGRAPHY

1. Andrei S. *Apa în pămînturile nesaturate*. Bucureşti, Editura tehnică, 1967
2. Ban A., ş.a. *Vlianie svoistv gornîh porod na dvijenie v nih jidkosti*. Moskva, Gostopthe dat, 1962.
3. Bindeman N. N. *Oţenka expluataţionîh zapasov podzemnîh vod*. Moskva, Gosgheoltehizdat, 1963.
4. Bocever F. M., Garmonov I. V., Lebedev A. V., Şestakov V. M. *Osnovî ghidrogheologhiceskih rascetov*, Moskva, Izdatelstvo Nedra, 1965.
5. Bocever F. M., Verighin N. N. *Metodiceskoe pasobie po rascetam expluataţionîh zapasov podzemnîh vod dlia vodosnabjenia*, Moskva, Gosstroiizdat, 1961.
6. Castany G. *Traité pratique des eaux souterraines*. Paris, Dunod, 1963.
7. Castany G. *Principe de l'établissement des bilans d'eaux*. Paris, Bulletin du Bureau de Recherches Géologiques et Minières, No. 3/1967.
8. Castany G. *Prospecţiunea şi exploatarea apelor subterane* (trad. din limba franceză), Bucureşti, Editura tehnică, 1972.
9. Cernov B. S., Bazlov M. N., Jukov A. I. *Ghidrodinamiceskie metodî isledovania skvajin i plastov*. Moskva, Gostoptehizdat, 1960.
10. Chauveteau G. et Thirriot Cl. *Régimes d'écoulement en milieu poreux et limite de la loi de Darcy*. La Houille blanche 2/1967.
11. Constantinescu L. şi alţii. *Prospecţiuni geofizice vol. I*. Metodele cîmpurilor naturale, Bucureşti, Editura tehnică, 1964.
12. Costache Gh., *Forajul şi exploatarea sondelor de apă*. Bucureşti, Editura tehnică, 1972.
13. Creţu I., Soare Al. *Cercetarea hidrodinamică a sondelor*. Bucureşti, Editura tehnică, 1967.
14. Davis N. Stanley, De Wiest J. M. Roger. *Hydrogeology*. John Wiley and Sons, Inc. New York—London—Sydney.
15. De Lury B. D. *Values and Integrals of the Orthogonal Polynomials up to n = 26*. Toronto, 1950.
16. Dementiev L. F. *Staticeskie metodî obrabotki i analiza promîslovogkeologhiceskih dannîh*. Moskva, Izdatelstovo Nedra, 1966.
17. Emsellem V., Prudhomme P. *Traitement des mesures hydrogéologiques à l'aide des modeles*. Annales des mines, 1/1967.
18. Engelhardt V. Wolf. *Der Porenraum der Sedimente*. Springer Verlag, Berlin—Gottingen—Heidelberg, 1960.
19. Ferris G. J., Knowles B. D., Brown H. R. and Stallman W. R. *Theory of aquifer tests*. Groundwater hydraulics. Geological Survey Water Supply Paper 1536-E Washington 1962.

20. Forkasiewicz J. *Interpretation des données des pompages d'essai pour l'évaluation des paramètres des aquiferes.* Aide-Mémoire BRGM = Paris, 1969.

21. Gavăt I. *Geologia petrolului și a gazelor naturale.* București, Editura didactică și pedagogică, 1964.

22. Gavrilko V. M., Alexeev V. S. — *O nekotorîh rezultatah vskrîtia filtrov skvajin gornîmi vîrabotkami.* Razvedka i ohrana nedr 6/1965.

23. Gheorghe Al. *Metode de asecare la exploatărilor de cărbuni.* Revista Minelor nr. 5—6/1964.

24. Gheorghe Al. și Bomboe P. *Hidrogeologie minieră.* București, Editura tehnică, 1963

25. Gheorghe Al. și Stănescu Dinu. *Influența condițiilor hidrogeologice regionale asupra proiectării sistemelor de asecare.* Revista Minelor nr. 2/1969, București.

26. Gheorghe Al. și Stănescu D. *Sisteme de asecare pentru Bazinul Rovinari,* Revista Minelor nr. 3/1969.

27. Hahn J. G., Shapiro S. S. *Statistical Models in Engineering.* John Wiley and Sons, Inc. New York, London, Sydney, 1967.

28. Hanin A. A., Koloskova M. I. *Zavisimost pronițaemosti gornîh porod ot razmera i soderjania dominiruiuscih por.* Razvedky i ohrana nedr nr. 2/1963.

29. Hantush M. S. *Analysis of data from pumping tests in leaky aquifers.* Trans. Am. Geophys. Union, 37, No. 6, 1956.

30. Hantush M. S. *Modification of the theory of leaky aquifers.* Journ. of Geophys. Res. 65, No. 11, 1960.

31. Hantush M. S. *Flow of ground water in sands of nonuniform thickness.* Journ. of Geophys. Res. 67, No. 2/1962.

32. Iosifescu M., Mihoc Gh., Theodorescu R. *Teoria probabilităților și statistica matematică.* București, Editura tehnică, 1966.

33. Kerkis E. E. *Opredelenie Koefițienta vodopronițaemosti i radiusa vliania po danim opîtnîh kustovîh otkacek metodom naimenșih kvadratov.* Zapiski gornovo instituta im. G. V. Plehanova, Tom. XXXIV, vipusk 2. 1958, Leningrad.

34. Kerkis E. E. *Voprosî metodiki probniîh i opitnîh otkacek.* Trudi VNIIG, vipusk XLVI, *Ghidrogheologhia solianîh mestorojdenii i mineralnie vodi,* Leningrad, 1964.

35. Kostiukovici P. N. *Sravnitelnaia oțenka formul opredelenia promejutka vîsacivania.* Zapiski gornovo instituta im. G. C. Plehanova Tom. XLVIII, vîpusk 2, Leningrad, 1965.

36. Krumbein W. C. *Regional and local components in facies maps.* Bul. An. Assoc. of Petroleum Geologists, vol. 40, nr. 9, 1956.

37. Krumbein W. C., Graybill A. Franklin. *An introduction to statistical models in geology.* New York McGraw-Hill Book Company.

38. Krumbein W. C. and Sloss L. L. *Application of moments to vertical variability of stratigrafic units.* Bul. Am. Assoc. of Petroleum Geologists, vol. 41, nr. 2/1957.

39. Kudelin B. I. *Prinţipi reghionalnoi oțenki estestvenîh resursov podzemnîh vod.* Izdatelstvo Moskovskovo Universiteta, 1970.

40. Leonards G. A. *Foundation engineering.* New York, McGraw-Hill Book Company, 1962.

41. Liteanu E., Constantinescu V., Croitoru M. *Cercetări hidrogeologice prin foraje manuale și încercări experimentale în zona Titu.* Studii Tehnice și economice seria E., nr. 8. Inst. Geologic, București, 1970.

42. Louis C. *Les drainages dans les zones fissurées.* BRGM, Hydrogéologie et géologie de l'ingénieur, section III, No. 1—1972.

43. Margot J. *Le role des eaux souterraines dans la gestion des ressources en eaux.* BRGM, Hydrogéologie et géologie de l'ingénieur, section III, No. 3—1972.

44. Margot J. *Progrès récentes des techniques de l'hydrogéologie*. Annales des mines, mars 1974.

45. Matheron G. *Traité de geostatistique appliqués*, t I et II. Mémoires du Bureau de Recherches géologiques et minières, 14, 1962, France.

46. Matuszak R. David. *Stratigraphic trend studies by electronic computer*, World-Oil, vol. 163, no. 2/1966.

47. Mihăilă N. *Introducere în teoria probabilităților și statistică matematică*. București, Editura didactică și pedagogică, 1965.

48. Miller L. Robert, Kan Steven James *Statistical analysis in the geological sciences*, John Willey and Sons, Inc. New York-London, 1962.

49. Murard R. *Probabilités et statistique*. Revue de l'industrie minérale. Numéro spécial 15 oc- tobre 1960.

50. Nosova O. N. *Rascet vodootdaci pescianih gruntov*. Gosenergoizdat, Moskva-Leningrad, 1962.

51. Oncescu N. *Geologia României*. Ediția III (apărută sub îngrijirea Acad. M. G. Filipescu) București, Editura tehnică, 1965.

52. Polubarinova-Kocina P. I. *Teoria dvijenia gruntovîh vod*. Gosudarstvenoe izdatelstvo teh- niko-teoreticeskoi literaturî. Moskva, 1952.

53. Polubarinova-Kocina P. L., Priajinskaia V. G., Emih V. N. *Matematiceskie metodî v va- prosah orosenia*. Moskva Izd. Nauka, 1969.

54. Rancu M. și Tövissi L. *Statistica matematică cu aplicații în industrie*. București, Editura Acad. 1963.

55. Rîjov P. A. *Gheometria nedr*. Moskva, Izdatelstvo Nedra, 1964.

56. Ruhin L. B. *Bazele litologiei. Știința rocilor sedimentare* (translation from Rusian). București, Editura tehnică, 1966.

57. Scelkacev V. N. *Razrabotka neftevodonosnîh plastov pri uprugom rejime*. Moskva. Gostop- tehizdat, 1959.

58. Scheidegger E. Adrian. *The physics of flow through porous media*. University of Toronto Press 1957.

59. Schmieder Antal. *Zuverlässigkeit und Vergleichbarkeit einiger Verfohren zur Bestimmung des Cickerfaktors*. Mitteilungen des Ungarischen Forschungsinstitutes für Bergbau. Nr. 7/1963—1964.

60. Schoeller H. *Les eaux souterraines. Hydrologie dynamique et chimique. Recherche, exploitation et evaluation des ressources*. Paris Masson et C-ie, 1962.

61. Șaghiev R. G., Portnov V. I. *Opredelenie parametrov plasta i nekotorîh osobenostei plastovoi vodonapornoi sistemi po krivîm vostanovlenia davlenia*. Inst. I. M. Gubkina, vipusk 33. Podzemnaia ghidrodinamika, Moskva, 1961.

62. Șarapov I. P. *Utilizarea statisticii matematice în geologie* (trad. din limba rusă). București- Editura tehnică, 1968.

63. Șestakov M. V. *Teoreticeskie osnovî ocenki podpora, vodoponijenia i drenaja*. Izd. Moskovs- kovo Universiteta, 1965.

64. Șestakov M. V., Kravcenko I. P., Poșkovskii I. S. *Prakticum po dinamike podzemnîh vod* Izd. Moskovskovo Universiteta, 1969.

65. Taylor W. Donald. *Fundamentals of soil mechanics*. New York—London, 1954.

66. Teodorescu A. și Popovici I. *Analiză numerică*. Centrul de documentare al industriei chimice și petroliere, București, 1968.

67. Todd David Keith. *Ground-water hydrology*, John Wiley & Sons New York, 1959.

68. Trebin F. A., Scerbakov, G. V., Iakovlev V. R. *Ghidromehaniceskie metodî issledovania skvajin i plastov*. Moskva, Nedra, 1965.

69. Ujvary I. *Geografia apelor României*, București, Editura științifică, 1972.

70. Verighin N. N. *Metodî opredelenia filtraționîh svoistov gornîh porod*. Moskva, Gostroiizdat, 1962.

71. Vernescu Al., *Mecanica zăcămintelor petrolifere*, București, Editura tehnică, 1966.

72. Walton W. C. *Ground-water Resource Evaluation*. McGraw-Hill Book Company, New York, 1970.

73. Yule Udny G., Kendell G. M. *Introducere în teoria statisticii* (traducere din limba engleză), București, Editura științifică, 1969.

74. * * * *Hydrogéologie des roches fissurées*. Actes du colloque de Dubrovnik. vol. I. 1965.

75. * * * *Manualul Inginerului*, vol. I, București Editura tehnică 1965.

76. * * * *Methodes d'études et de recherches des nappes aquifères*. Bureau de Recherches Geologiques et Minnères, Paris.

77. * * * *Methods of Determining Permeability, Transmissibility and Drawdown*. Ground-waterl Hydraulics Geological Survey Water-Supply. Paper 1536-1, Washington, 1963.

78. * * * *Razvitie issledovanii po teorii filtrații v S.S.S.R.* (1917—1967). Nauka, Moskva, 1969

79. * * * *Shortcuts and Special Problems in Aquifer Test. Methods of Aquifer Tests*. Geologica Survey Water-Supply. Paper 1545-C Washington, 1963.

80. * * * *Spravocinoe rukovodstvo ghidrogheologa*. Pod redakției V. M. Maksimova Tom 1, 2. Nedra, Leningrad 1967.

SUBJECT INDEX

NAME INDEX